Intermetallics 2016

Special Issue Editor
Ana Sofia Ramos

MDPI • Basel • Beijing • Wuhan • Barcelona • Belgrade

MDPI

Special Issue Editor
Ana Sofia Ramos
University of Coimbra
Portugal

Editorial Office
MDPI AG
St. Alban-Anlage 66
Basel, Switzerland

This edition is a reprint of the Special Issue published online in the open access journal *Metals* (ISSN 2075-4701) from 2015–2017 (available at: http://www.mdpi.com/journal/metals/special_issues/intermetallics2015).

For citation purposes, cite each article independently as indicated on the article page online and as indicated below:

Author 1; Author 2. Article title. *Journal Name* **Year**, *Article number*, page range.

First Edition 2017

ISBN 978-3-03842-630-1 (Pbk)
ISBN 978-3-03842-631-8 (PDF)

Table of Contents

About the Special Issue Editor

Ana Sofia Figueira Ramos is Post-Doc Researcher at the University of Coimbra. She belongs to the Centre for Mechanical Engineering, Materials and Processes (CEMMPRE). After completing her 5-year degree in Chemical Engineering she obtained a MSc in Materials Engineering. In 2002 she received a PhD in Mechanical Engineering, specialization in Materials Science. Her main research interests are the use of magnetron sputtering to produce nanostructured films for non-conventional applications such as joining/welding and self-healing, in particular Me1/Me2 (Me – metal) reactive nanomultilayers. Upon reaction, these energetic materials release heat with the formation of Me1xMe2y intermetallic phases.

Ana Sofia Ramos has participated in several projects funded by Portuguese Science Foundation and European Commission. She also conducted several experiments at European synchrotron facilities. She has 58 papers in international peer review journals, 52 in ISI Web of Knowledge, which give her an h-index of 17 (Scopus).

Preface to "Intermetallics 2016"

The combination of low density, high strength, and good corrosion resistance makes intermetallics promising for structural applications, especially at high temperatures and under severe environments. These materials have also potential for functional applications since some intermetallic phases have unique properties, such as shape memory or thermo electric effect. Intermetallic phases of interest include aluminides, silicides, Laves and Heusler phases, among others. Intermetallic compounds can be produced from metals that exothermically react with each other releasing energy useful for several applications, including joining. As a result of the increasing demand for novel/advanced materials with improved properties, recent years have been marked by the return of intermetallics. Proof of this is the ongoing and future conferences/symposia dedicated to intermetallics.

The quality and variety of the 17 papers published in Metals' "Intermetallics 2016" special issue reflects the interest in the intermetallics field. Processing and joining of intermetallics, their mechanical properties and oxidation/corrosion behavior, phase transformations involving intermetallics, modelling, and numerical simulation have been the topic of these papers. The joining topic stood out with five papers focusing both the joining of intermetallic materials and the formation of intermetallic compounds during soldering and diffusion bonding processes.

The following materials were studied, with particular attention on the aluminides:

- TiAl, FeAl, (Ni,Pt)Al, ZrCuAl and FeCrAl;
- Ta_5Si_3;
- Ti-6Al-4V;
- TiNi;
- Cu_6Sn_5 and Cu_3Sn;
- TiZr;
- Fe_2Mo;
- $Bi_2Te_{2.55}Se_{0.45}$, Sn–Te–Se–Bi intermetallics and Ni_3Sn_4.

Composites involving intermetallics were the subject of two papers, while two papers were dedicated to metallic glasses. For the materials investigated, the simulation and experimental works carried out allowed understanding the relation between microstructure and properties aiming at their use as structural materials at high and ultra-high temperatures, as well as their use in aerospace/aeronautic, automobile, electronic, tribological, and biomedical applications. Whatever the application, joining of intermetallics to other materials is of paramount importance which is reflected in this issue.

As guest editor of "Intermetallics 2016" special issue, I think it will contribute to the progress of the state of the art.

Ana Sofia Ramos
Special Issue Editor

metals

MDPI

Editorial

Intermetallics

Ana Sofia Ramos

CEMMPRE, Department of Mechanical Engineering, University of Coimbra, R. Luís Reis Santos, 3030-788 Coimbra, Portugal; sofia.ramos@dem.uc.pt; Tel.: +351-239-790700

Academic Editor: Hugo F. Lopez
Received: 10 October 2017; Accepted: 18 October 2017; Published: 20 October 2017

1. Introduction

The combination of low density, high strength, and good corrosion resistance makes intermetallics promising for structural applications, especially at high temperatures and under severe environments [1]. These materials have also potential for functional applications since some intermetallic phases have unique properties, such as shape memory or thermo electric effect. Intermetallic phases of interest include aluminides, silicides, Laves and Heusler phases, among others. Intermetallic compounds can be produced from metals that exothermically react with each other releasing energy useful for several applications, including joining [2–4]. As a result of the increasing demand for novel/advanced materials with improved properties, recent years have been marked by the return of intermetallics. Proof of this is the ongoing and future conferences/symposia dedicated to intermetallics.

2. Contributions

The quality and variety of the 17 papers published in *Metals*' "Intermetallics 2016" special issue reflects the interest in the intermetallics field. Processing and joining of intermetallics, their mechanical properties and oxidation/corrosion behavior, phase transformations involving intermetallics, modelling, and numerical simulation have been the topic of these papers. The joining topic stood out with five papers focusing both the joining of intermetallic materials and the formation of intermetallic compounds during soldering and diffusion bonding processes.

The following materials were studied, with particular attention on the aluminides:

- TiAl, FeAl, (Ni,Pt)Al, ZrCuAl and FeCrAl;
- Ta_5Si_3;
- Ti-6Al-4V;
- TiNi;
- Cu_6Sn_5 and Cu_3Sn;
- TiZr;
- Fe_2Mo;
- $Bi_2Te_{2.55}Se_{0.45}$, Sn–Te–Se–Bi intermetallics and Ni_3Sn_4

Composites involving intermetallics were the subject of two papers, while two papers were dedicated to metallic glasses. For the materials investigated, the simulation and experimental works carried out allowed understanding the relation between microstructure and properties aiming at their use as structural materials at high and ultra-high temperatures, as well as their use in aerospace/aeronautic, automobile, electronic, tribological, and biomedical applications. Whatever the application, joining of intermetallics to other materials is of paramount importance which is reflected in this issue.

As guest editor of "Intermetallics 2016" special issue, I think it will contribute to the progress of the state of the art.

3. Conclusions

A variety of papers have been published covering important aspects related to intermetallics. Nevertheless, there are still several challenges to overcome in order to optimize intermetallics' properties and enlarge their field of application. Therefore, further issues in this field would be welcome.

Acknowledgments: I would like to thank all the authors for their contributions, the reviewers for the comments, and the *Metals* editorial staff for the effort in putting up this special issue.

Conflicts of Interest: The author declares no conflict of interest.

References

1. Westbrook, J.H. Historical Sketch. In *Intermetallic Compounds—Principles and Practice*; Westbrook, J.H., Fleischer, R.L., Eds.; John Wiley & Sons: Chichester, UK, 1995; Volume 1, pp. 3–18. ISBN 0 471 94219 7.
2. Duckham, A.; Spey, S.J.; Wang, J.; Reiss, M.E.; Weihs, T.P.; Besnoin, E.; Knio, O.M. Reactive nanostructured foil used as a heat source for joining titanium. *J. Appl. Phys.* **2004**, *96*, 2336–2342. [CrossRef]
3. Ramos, A.S.; Vieira, M.T.; Duarte, L.; Vieira, M.F.; Viana, F.; Calinas, R. Nanometric multilayers: A new approach for joining TiAl. *Intermetallics* **2006**, *14*, 1157–1162. [CrossRef]
4. Cao, J.; Feng, J.C.; Li, Z.R. Microstructure and fracture properties of reaction-assisted diffusion bonding of TiAl intermetallic with Al/Ni multilayer foils. *J. Alloys Compd.* **2008**, *466*, 363–367. [CrossRef]

metals

MDPI

Article

Experimental and Numerical Studies on Self-Propagating High-Temperature Synthesis of Ta$_5$Si$_3$ Intermetallics

Chun-Liang Yeh *, Chi-Chian Chou and Po-Wen Hwang

Department of Aerospace and Systems Engineering, Feng Chia University, Taichung 40724, Taiwan; rita511235@gmail.com (C.-C.C.); pwhwang@fcu.edu.tw (P.-W.H.)

* Author to whom correspondence should be addressed; clyeh@fcu.edu.tw; Tel.: +886-4-2451-7250 (ext. 3963); Fax: +886-4-2451-0862.

Academic Editor: Ana Sofia Ramos

Received: 11 August 2015; Accepted: 28 August 2015; Published: 1 September 2015

Abstract: Formation of Ta$_5$Si$_3$ by self-propagating high-temperature synthesis (SHS) from elemental powder compacts of Ta:Si = 5:3 was experimentally and numerically studied. Experimental evidence showed that the increase of either sample density or preheating temperature led to the increase of combustion wave velocity and reaction temperature. The apparent activation energy, $E_a \approx 108$ kJ/mol, was determined for the synthesis reaction. Based upon numerical simulation, the Arrhenius factor of the rate function, $K_0 = 2.5 \times 10^7$ s^{-1}, was obtained for the 5Ta + 3Si combustion system. In addition, the influence of sample density on combustion wave kinetics was correlated with the effective thermal conductivity (k_{eff}) of the powder compact. By adopting $0.005 \leq k_{eff}/k_{bulk} \leq 0.016$ in the computation model, the calculated combustion velocity and temperature were in good agreement with experimental data of the samples with compaction densities between 35% and 45% theoretical maximum density (TMD).

Keywords: self-propagating high-temperature synthesis; Ta$_5$Si$_3$; numerical simulation; activation energy; Arrhenius factor; thermal conductivity

---

1. Introduction

Transition metal silicides are of great interest as high-temperature materials, due to their specific properties such as high melting point, good thermal stability, excellent oxidation resistance, good creep tolerance, and high mechanical strength at elevated temperatures [1–3]. Intermetallic silicides of the Mo–Si, Cr–Si, Ti–Si, V–Si, Nb–Si, and Ta–Si systems are potential candidates for high-temperature applications. In particular, the compounds of the Ta–Si system are among the most refractory silicides with melting points in the range of 2200 to 2500 °C [4]. Preparation of transition metal silicides has been conducted by a variety of fabrication routes, often combining two or more of them, like hot pressing, hot isostatic pressing, reactive sintering, combustion synthesis, mechanical alloying, thermal or plasma spraying, and vapor infiltration [5]. Combustion synthesis in the mode of self-propagating high-temperature synthesis (SHS) is one of the emerging cost-effective techniques and is merited by high energy effectiveness, fast reaction rates, and simple facilities. The SHS method has been applied for the preparation of many advanced materials, including borides, carbides, nitrides, carbonitrides, aluminides, silicides, and complex oxides [6–8].

By means of the classical SHS approach, many silicide phases in the Mo–Si, Ti–Si, Zr–Si, and Nb–Si systems were prepared from the reactant compacts made up of stoichiometric elemental powders [9–12]. In addition, formation of the silicides in the V–Si and Ta–Si systems was carried out by a modified SHS technique known as field-activated combustion synthesis (FACS) that applied an electric field perpendicularly to the propagation direction of the combustion wave [13,14]. Another modification called mechanically-activated SHS (MASHS), which pretreated the reactant powders by prolonged high-energy ball milling at a given frequency and impact energy, was utilized to produce $MoSi_2$, $TaSi_2$, and Ta_5Si_3 [15,16]. However, despite many studies devoted to the synthesis of intermetallic silicides, the fundamental aspects of reaction rate kinetics and heat transfer parameters on combustion synthesis of metal silicides are scarcely investigated. For the Arrhenius factor (K_0) of the rate function, Li [17,18] adopted $K_0 = 4 \times 10^{10}$ s^{-1} to calculate the flame propagation rate of the Ti + 2B combustion system. Yeh *et al.* [19] considered the particle size effect and suggested $K_0 = 3 \times 10^8 - 2 \times 10^9$ s^{-1} and $1.5 \times 10^8 - 4 \times 10^8$ s^{-1} for SHS formation of NiAl and CoAl, respectively, from elemental powder compacts. On account of the porous nature of the powder compact, Gennari *et al.* [20,21] introduced an effective thermal conductivity (k_{eff}) equal to $1/70$ of the bulk value (k_{bulk}) to their computation model for the study of the SHS process on the formation of transition metal aluminides. Based upon good agreement between measured and calculated combustion wave velocities and temperatures, Yeh *et al.* [19] further indicated that k_{eff} of the Ni–Al powder compacts with relative densities of 50%–65% falls within the extent of $k_{eff}/k_{bulk} = 0.016$–0.052.

The objective of this study is to perform a thorough investigation on combustion synthesis of Ta_5Si_3 from elemental powder compacts of Ta:Si = 5:3. In the experimental part, the effects of sample density and preheating temperature were studied on the combustion wave velocity, reaction temperature, and product composition. The apparent activation energy associated with SHS formation of Ta_5Si_3 was deduced. In the numerical part, the Arrhenius factor of the rate function for the 5Ta + 3Si combustion system was determined. The effective thermal conductivity of the powder compact was evaluated and correlated with sample density.

2. Experimental and Numerical Methods

For the experiments, tantalum (Aldrich Chemical, <45 μm, 99.9%, St. Louis, MO, USA) and silicon (Strem Chemicals, <45 μm, 99.5%, Newburyport, MA, USA) powders at an atomic ratio of Ta:Si = 5:3 were dry mixed in a ball mill. The reactant mixture was then cold-pressed into cylindrical compacts with a diameter of 7 mm and a height of 12 mm. The sample density (d_s) of the powder compact was a studied variable, which considered three values in terms of 35%, 40%, and 45% of the theoretical maximum density (TMD). In a stainless-steel windowed chamber under high-purity argon (99.99%), the experiment was performed with samples at different preheating temperatures (T_p) from room temperature (25 °C) to 300 °C. This study measured the combustion wave velocity and reaction temperature, and identified the phase constituents of SHS-derived products. Details of the experimental setup were reported elsewhere [22]. In addition, the apparent activation energy (E_a) of the synthesis reaction was determined from the dependence of combustion wave velocity on the reaction temperature [10,11].

In the simulation model, an energy equation expressed as Equation (1) was numerically solved to obtain the transient sample temperature profile during the propagation of a self-sustaining combustion wave [19].

$$\rho C_p \frac{\partial T}{\partial t} = k_{eff} \frac{\partial^2 T}{\partial x^2} + \rho Q \phi(T, \eta) - \frac{4h(T - T_0)}{d} - \frac{4\sigma\varepsilon(T^4 - T_0^4)}{d} \tag{1}$$

where C_p is the heat capacity of the product, k_{eff} the effective thermal conductivity, Q the exothermic heat of the reaction, ϕ (T,η) the reaction rate function, h the convection heat transfer coefficient, σ the

Stefan-Boltzmann constant, ε the emissivity, and d the diameter of the test specimen. The rate function of $\phi(T, \eta)$ is given by an Arrhenius form (Equation (2)) [18,19].

$$\phi(T, \eta) = \frac{\partial \eta}{\partial t} = K_0(1 - \eta) \exp\left(-\frac{E_a}{RT}\right) \qquad (2)$$

where η is the fraction of the reactant transformed into the product, K_0 the Arrhenius factor, and E_a the activation energy. The numerical scheme as well as the initial and boundary conditions were previously described [19]. The effective thermal conductivity of Equation (1) is essentially affected by the porosity of the test specimen. To account for the significant variation of k_{eff} with sample density [19,23], the ratio of k_{eff}/k_{bulk} between 0.002 and 0.05 was considered in the numerical model of this study.

3. Results and Discussion

3.1. Experimental Measurement and Analysis

A typical sequence of combustion images recorded from a powder compact of Ta:Si = 5:3 is illustrated in Figure 1, showing that a steady and nearly parallel combustion wave traveling in a self-sustaining fashion. Based upon the recorded films, the deduced flame-front propagation velocity (V_f) varying from 6.4 to 22.9 mm/s was reported in Figure 2 as a function of sample density and preheating temperature. As the sample density increased from 35% to 45% TMD, an appreciable increase in the flame-front velocity was observed. This was primarily caused by the enhancement of intimate contact between the reactant particles, which improved the layer-by-layer heat transfer and accelerated the combustion wave. In addition, owing to the increase of the reaction temperature with initial sample temperature, Figure 2 shows that the propagation rate of the combustion wave increases considerably with sample preheating temperature. It is useful to note that an increase in T_p from 25 to 300 °C almost doubles the flame-front velocity.

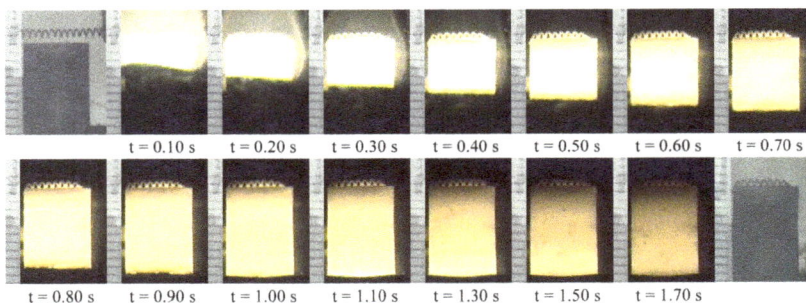

| t = 0.10 s | t = 0.20 s | t = 0.30 s | t = 0.40 s | t = 0.50 s | t = 0.60 s | t = 0.70 s |

| t = 0.80 s | t = 0.90 s | t = 1.00 s | t = 1.10 s | t = 1.30 s | t = 1.50 s | t = 1.70 s |

Figure 1. Solid state combustion propagating in a self-sustaining manner along a powder compact of Ta:Si = 5:3 with d_s = 35% TMD and T_p = 100 °C.

The dependence of the combustion front temperature (T_c) on sample density and preheating temperature is presented in Figure 3. According to the typical temperature profiles plotted in the insert of Figure 3, the abrupt rise in the temperature represents rapid arrival of the reaction front and the peak value signifies the combustion front temperature. Data points in Figure 3 signify the combustion front temperatures. In the insert, two temperature profiles measured from 45% TMD samples with T_p = 100 and 300 °C are presented. It is evident that the increase of preheating temperature increases the reaction temperature. As shown in Figure 3, for the samples without prior heating (T_p = 25 °C), the increase of sample density from 35% to 45% TMD significantly increased the peak combustion temperature from 1240 to 1530 °C. This could be attributed to a lesser amount of heat dissipated within the voids of the porous compact. As a result, the energy flux sustaining the combustion wave propagation was augmented. The combustion front temperature was also increased by preheating the sample. It was found that a more substantial increase of the combustion front temperature by preheating the sample was observed for the sample with a lower compaction density. Likewise, the influence of sample density was more pronounced for the sample with a lower preheating temperature. The highest combustion temperature reaching up to 1645 °C was measured from the sample of 45% TMD at T_p = 300 °C.

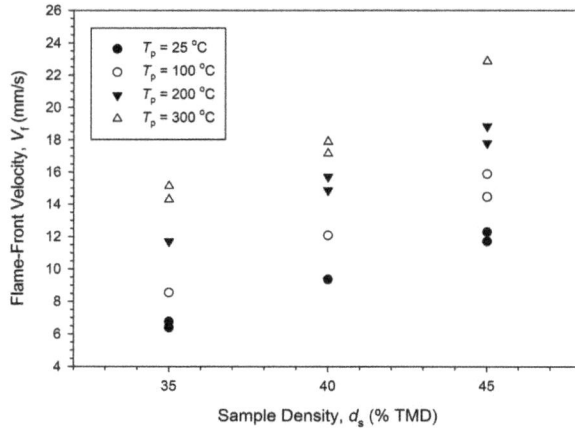

Figure 2. Effects of sample density and preheating temperature on flame-front velocity of elemental powder compacts of Ta:Si = 5:3.

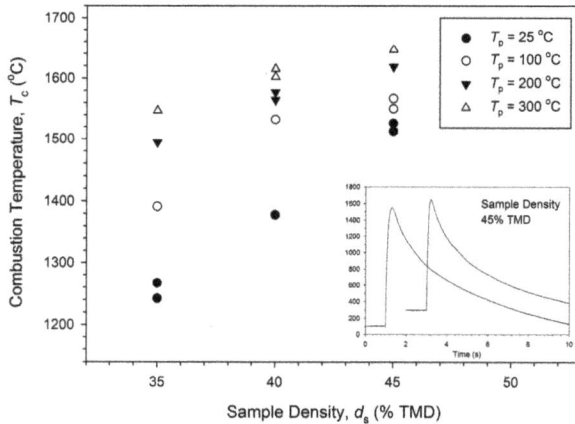

Figure 3. Effects of sample density and preheating temperature on combustion temperature of elemental powder compacts of Ta:Si = 5:3.

In order to determine the apparent activation energy (E_a) of the self-sustaining combustion reaction associated with formation of Ta_5Si_3, the following equation that relates the combustion wave velocity with the combustion temperature and the thermochemical parameters is considered [10,11].

$$\left(\frac{V_f}{T_c}\right)^2 = f(n)K\left(\frac{R}{E_a}\right)\exp\left(-\frac{E_a}{RT_c}\right) \qquad (3)$$

where f(n) is a function of the kinetic order of the reaction, R the universal gas constant, and K a constant which includes the heat capacity of the product, the thermal conductivity, and the heat of the reaction. Based upon Equation (3), a plot correlating $\ln(V_f/T_c)^2$ with $1/T_c$ is constructed in Figure 4. According to the slope of a best-fitted straight line for the data, $E_a \approx 108$ kJ/mol was obtained and employed in the reaction rate function of Equation (2) for the numerical simulation.

Figure 5 presents the XRD patterns of two synthesized products from samples of different densities and preheating temperatures. It is evident that Ta_5Si_3 was formed as the dominant silicide and a minor phase Ta_2Si was identified. A comparison between Figure 5a,b indicates that the content of Ta_2Si was reduced by increasing the sample compaction density and initial temperature. This implies that higher reaction temperature and better contact between the reactant powders could enhance the phase conversion.

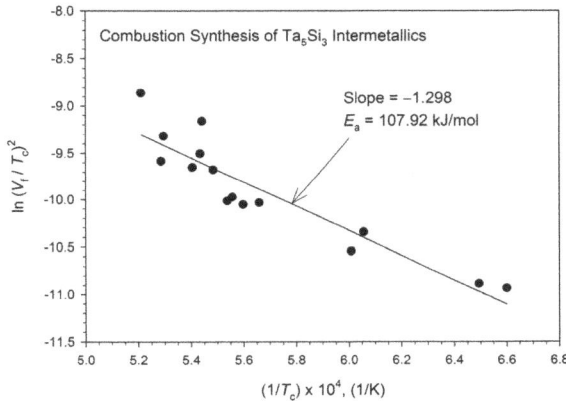

Figure 4. Activation energy (E_a) of combustion synthesis of Ta_5Si_3 determined from a relationship between combustion front velocity (V_f) and temperature (T_c).

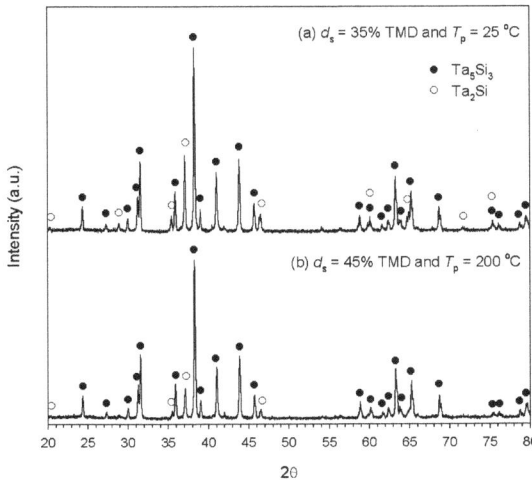

Figure 5. X-ray diffraction (XRD) patterns of SHS-derived products from Ta:Si = 5:3 powder compacts with (**a**) d_s = 35% TMD and T_p = 25 °C and (**b**) d_s = 45% TMD and T_p = 200 °C.

3.2. Numerical Simulation and Validation

The Arrhenius factor (K_0) of the rate function is essentially dissimilar for different reaction systems. According to the previous study [19], the magnitude of K_0 varies between 1.5×10^8 and 2.0×10^9 s^{-1} for combustion synthesis of intermetallic aluminides, CoAl and NiAl. It should be noted that very high combustion rates were detected in the Co–Al and Ni–Al combustion systems. Specifically, for the powder compacts with no prior heating, the flame-front velocity in the range of 11–37 mm/s was measured for the Co–Al system and 22–102 mm/s for the Ni–Al system [19]. In a series of numerical simulation performed by this study, the calculated results with $0.005 \leq k_{\text{eff}}/k_{\text{bulk}} \leq 0.02$ and and $T_p = 25\ °C$ showed that the adoption of $K_0 = 3.0 \times 10^9$ s^{-1} in the model yields $V_f = 18$–32 mm/s, $K_0 = 3.0 \times 10^8$ s^{-1} yields $V_f = 12$–20 mm/s, $K_0 = 3.0 \times 10^7$ s^{-1} yields $V_f = 8$–12 mm/s, and $K_0 = 3.0 \times 10^6$ s^{-1} yields $V_f = 4$–8 mm/s. The above results advised that the magnitude of K_0 on the order of 10^7 is the best fit for the Ta–Si combustion system. After a further refinement coupled with appropriate values of $k_{\text{eff}}/k_{\text{bulk}}$, it was found that $K_0 = 2.5 \times 10^7$ s^{-1} is the optimum to achieve the most agreeable results between measured and calculated combustion wave velocities for the 5Ta + 3Si combustion system.

Figure 6a,b plots the calculated instantaneous temperature profiles during the SHS process for the 5Ta + 3Si powder compacts with different effective thermal conductivities and preheating temperatures. The test specimen with a ratio of $k_{\text{eff}}/k_{\text{bulk}} = 1.0$ signifies a fully-dense sample, while that with $0 < k_{\text{eff}}/k_{\text{bulk}} < 1$ represents the powder compact with a certain degree of porosity. As depicted in Figure 6a, for the case of $k_{\text{eff}}/k_{\text{bulk}} = 0.005$ and $T_p = 25\ °C$, the combustion wave at $t = 1.8$ s is about to reach the bottom of the sample and the deduced flame-front velocity was 6.59 mm/s. Figure 6a also reveals the profiles featuring a comparable peak temperature of around 1260 °C, which stands for the combustion front temperature. For the powder compact with $k_{\text{eff}}/k_{\text{bulk}} = 0.02$ and $T_p = 200\ °C$, as shown in Figure 6b, the combustion front temperature rises to 1535 °C and the time required to complete the SHS process is noticeably shortened by a faster combustion front with $V_f = 20.22$ mm/s. The increase of k_{eff} is to numerically simulate the powder compact with a higher compaction density, which corresponds to the possession of a better degree of contact between the reactant powders. As a result, the rate of heat transfer from the combustion front to the unburned region is enhanced and the heat lost to the voids of the porous sample is reduced. This explains a faster combustion wave and a higher reaction front temperature for a denser powder compact. Preheating the sample also contributes to a higher reaction temperature, which favors the propagation of the combustion front. The calculated combustion velocities and temperatures of Figure 6 are within a reasonable range consistent with the experimental data, which confirms the reaction kinetics and exothermicity of the numerical model.

Figure 7 presents a comparison between measured and calculated flame-front velocities for combustion synthesis of Ta_5Si_3. The symbols plotted in Figure 7 are experimental data obtained from the 5Ta + 3Si samples and the continuous solid lines represent the computational results under $K_0 = 2.5 \times 10^7$ s^{-1} and $0.005 \leq k_{\text{eff}}/k_{\text{bulk}} \leq 0.016$. It is evident that for different preheating temperatures, the calculated flame speeds agree satisfactorily with experimental data of the samples with densities of 35%–45% TMD. The determination of the ratio of $k_{\text{eff}}/k_{\text{bulk}}$ not only has to yield the calculated combustion velocities in good agreement with experimental data, but should also be consistent with that reported by the previous study [19]. As mentioned above, k_{eff} of the Ni–Al powder compacts with relative densities of 50%–65% TMD falls within the extent of $k_{\text{eff}}/k_{\text{bulk}} = 0.016$–0.052. Therefore, the magnitudes of $k_{\text{eff}}/k_{\text{bulk}} = 0.005$–0.016 and $K_0 = 2.5 \times 10^7$ s^{-1} were justified for combustion synthesis of Ta_5Si_3 from elemental powder compacts of 35%–45% TMD.

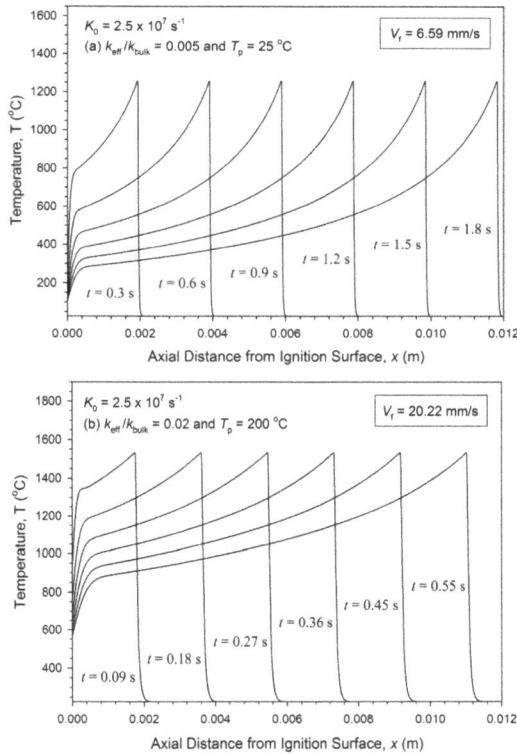

Figure 6. Calculated time histories of temperature profiles of Ta–Si powder compacts with (**a**) k_{eff}/k_{bulk} = 0.005 and T_p = 25 °C and (**b**) k_{eff}/k_{bulk} = 0.02 and T_p = 200 °C.

4. Conclusions

Formation of Ta_5Si_3 by combustion synthesis in the SHS mode was experimentally and numerically investigated by this study. Sample compacts with d_s = 35%, 40% and 45% TMD were prepared from an elemental powder mixture of Ta:Si = 5:3. Experiments were conducted with powder compacts at different preheating temperatures of T_p = 25–300 °C. It was found that both the measured combustion wave velocity ranging from 6.4 to 22.9 mm/s and combustion temperatures from 1240 to 1645 °C increased with increasing sample density and preheating temperature. The apparent activation energy, $E_a \approx 108$ kJ/mol, of the synthesis reaction was deduced from the dependence of combustion wave velocity on reaction temperature. Formation of Ta_5Si_3 as the dominant silicide phase from solid-state combustion was achieved. A minor phase Ta_2Si was reduced by increasing the sample density and combustion temperature.

In the numerical study, the Arrhenius factor of the rate function, $K_0 = 2.5 \times 10^7$ s^{-1}, was obtained to optimally simulate the combustion rate of the 5Ta + 3Si sample to form Ta_5Si_3. The effect of sample density on combustion wave kinetics was correlated with the variation of k_{eff} of the powder compact. Numerical simulations showed that the calculated combustion velocities and their dependent trend based upon $0.005 \leq k_{eff}/k_{bulk} \leq 0.016$ properly matched the experimental data of the samples with 35%–45% TMD. Moreover, the combustion exothermicity of the numerical model was well justified by the measured combustion temperatures.

9

Metals **2015**, *5*, 1580–1590

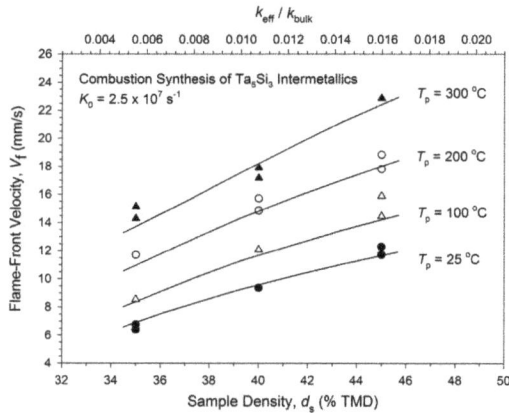

Figure 7. A comparison between measured and calculated flame-front velocities as functions of sample density and $k_{\text{eff}}/k_{\text{bulk}}$ for combustion synthesis of Ta_5Si_3.

Acknowledgments: This research was sponsored by the Ministry of Science and Technology, ROC under the grant of MOST 104-2221-E-035-057. The authors are grateful for the Precision Instrument Support Center of Feng Chia University in providing materials analytical facilities.

Author Contributions: C.-L. Yeh supervised the work, wrote the paper, and analyzed the experimental and numerical data; C.-C. Chou conducted the experiments and performed numerical simulation; and P.-W. Hwang contributed the numerical modeling.

Conflicts of Interest: The authors declare no conflict of interest.

References

1. Meschter, P.J.; Schwartz, D.S. Silicide-matrix materials for high temperature applications. *JOM* **1989**, *41*, 52–55. [CrossRef]
2. Petrovic, J.J.; Vasudevan, A.K. Key developments in high temperature structural silicides. *Mater. Sci. Eng. A* **1999**, *261*, 1–5. [CrossRef]
3. Tillard, M. The mixed intermetallic silicide $Nb_{5-x}Ta_xSi_3$ ($0 \leq x \leq 5$). Crystal and electronic structure. *J. Alloy. Compd.* **2014**, *584*, 385–392. [CrossRef]
4. Schlesinger, M.E. Thermodynamics of solid transition-metal silicides. *Chem. Rev.* **1990**, *90*, 607–628. [CrossRef]
5. Stoloff, N.S. An overview of powder processing of silicides and their composites. *Mater. Sci. Eng. A* **1999**, *261*, 169–180. [CrossRef]
6. Liu, G.; Li, J.; Chen, K. Combustion synthesis of refractory and hard materials: A review. *Int. J. Refract. Met. Hard Mater.* **2013**, *39*, 90–102. [CrossRef]
7. Gromov, A.A.; Chukhlomina, L.N. *Nitride Ceramics: Combustion Synthesis, Properties, and Applications*; Wiley-VCH: Weinheim, Germany, 2015.
8. Vicario, I.; Poulon-Quintin, A.; Lagos, M.A.; Silvain, J.F. Effect of material and process atmosphere in the preparation of Al–Ti–B grain refiner by SHS. *Metals* **2015**, *5*, 1387–1396. [CrossRef]
9. Yeh, C.L.; Chen, W.H. Combustion synthesis of $MoSi_2$ and $MoSi_2$–Mo_5Si_3 composites. *J. Alloy. Compd.* **2007**, *438*, 165–170. [CrossRef]
10. Yeh, C.L.; Chen, W.H.; Hsu, C.C. Formation of titanium silicides Ti_5Si_3 and $TiSi_2$ by self-propagating combustion synthesis. *J. Alloy. Compd.* **2007**, *432*, 90–95. [CrossRef]
11. Bertolino, N.; Anselmi-Tamburini, U.; Maglia, F.; Spinolo, G.; Munir, Z.A. Combustion synthesis of Zr–Si intermetallic compounds. *J. Alloy. Compd.* **1999**, *288*, 238–248. [CrossRef]
12. Yeh, C.L.; Chen, W.H. Preparation of Nb_5Si_3 intermetallic and Nb_5Si_3/Nb composite by self-propagating high-temperature synthesis. *J. Alloy. Compd.* **2005**, *402*, 118–123. [CrossRef]

13. Maglia, F.; Anselmi-Tamburini, U.; Milanese, C.; Bertolino, N.; Munir, Z.A. Field activated combustion synthesis of the silicides of vanadium. *J. Alloy. Compd.* **2001**, *319*, 108–118. [CrossRef]

14. Maglia, F.; Anselmi-Tamburini, U.; Bertolino, N.; Milanese, C.; Munir, Z.A. Field-activated combustion synthesis of Ta–Si intermetallic compounds. *J. Mater. Res.* **2001**, *16*, 534–544. [CrossRef]

15. Gras, C.; Gaffet, E.; Bernard, F. Combustion wave structure during the $MoSi_2$ synthesis by mechanically-activated self-propagating high-temperature synthesis (MASHS): *In situ* time-resolved investigations. *Intermetallics* **2006**, *14*, 521–529. [CrossRef]

16. Maglia, F.; Milanese, C.; Anselmi-Tamburini, U.; Doppiu, S.; Cocco, G.; Munir, Z.A. Combustion synthesis of mechanically activated powders in the Ta–Si system. *J. Alloy. Compd.* **2004**, *385*, 269–275. [CrossRef]

17. Li, H.P. The numerical simulation of the heterogeneous composition effect on the combustion synthesis of TiB_2 compound. *Acta Mater.* **2003**, *51*, 3213–3224. [CrossRef]

18. Li, H.P. An investigation of the ignition manner effects on combustion synthesis. *Mater. Chem. Phys.* **2003**, *80*, 758–767. [CrossRef]

19. Yeh, C.L.; Hwang, P.W.; Chen, W.K.; Li, J.Y. Modeling evaluation of Arrhenius factor and thermal conductivity for combustion synthesis of transition metal aluminides. *Intermetallics* **2013**, *39*, 20–24. [CrossRef]

20. Gennari, S.; Maglia, F.; Anselmi-Tamburini, U.; Spinolo, G. SHS (Self-sustained high-temperature synthesis) of intermetallic compounds: Effect of process parameters by computer simulation. *Intermetallics* **2003**, *11*, 1355–1359. [CrossRef]

21. Gennari, S.; Anselmi-Tamburini, U.; Maglia, F.; Spinolo, G.; Munir, Z.A. A new approach of the modeling of SHS reactions: Combustion synthesis of transition metal aluminides. *Acta Mater.* **2006**, *54*, 2343–2351. [CrossRef]

22. Yeh, C.L.; Lin, J.Z. Combustion synthesis of Cr–Al and Cr–Si intermetallics with Al_2O_3 additions from Cr_2O_3–Al and Cr_2O_3–Al–Si reaction systems. *Intermetallics* **2013**, *33*, 126–133. [CrossRef]

23. Miura, S.; Terada, Y.; Suzuki, T.; Liu, C.T.; Mishima, Y. Thermal conductivity of Ni–Al powder compacts for reaction synthesis. *Intermetallics* **2000**, *8*, 151–155. [CrossRef]

metals

MDPI

Article

Effect of Ceramic Content on the Compression Properties of TiB$_2$-Ti$_2$AlC/TiAl Composites

Shili Shu [1,2], Cunzhu Tong [2], Feng Qiu [1,3,*] and Qichuan Jiang [1,*]

[1] Key Laboratory of Automobile Materials (Ministry of Education), Department of Materials Science and Engineering, Jilin University, Changchun 130025, China; shushili@ciomp.ac.cn

[2] State Key Laboratory of Luminescence and Applications, Changchun Institute of Optics, Fine Mechanics and Physics, Chinese Academy of Sciences, Changchun 130012, China; tongcunzhu@ciomp.ac.cn

[3] Department of Mechanical Engineering, Oakland University, Rochester, MI 48309, USA

* Authors to whom correspondence should be addressed; qiufeng@jlu.edu.cn (F.Q.); jqc@jlu.edu.cn (Q.J.); Tel./Fax: +86-431-8509-5592 (F.Q.); +86-431-8509-4699 (Q.J.).

Academic Editor: Ana Sofia Ramos

Received: 29 September 2015; Accepted: 6 November 2015; Published: 25 November 2015

Abstract: *In situ* synthesized TiB$_2$-reinforced TiAl composites usually possess high strength. However, it is very expensive to use B powder to synthesize TiB$_2$ particles. Moreover, the strength enhancement of TiB$_2$/TiAl composite is generally at the cost of plasticity. In this study, *in situ* dual reinforcement TiB$_2$-Ti$_2$AlC/TiAl composites were fabricated by using B$_4$C powder as the B and C source, which greatly reduces the potential production cost. The 6 vol. % TiB$_2$-Ti$_2$AlC/TiAl composite fabricated by using the Ti-Al-B$_4$C system shows greatly improved compressive properties, *i.e.*, 316 MPa and 234 MPa higher than those of TiAl alloy and with no sacrifice in plasticity.

Keywords: TiAl; intermetallics; composite; compression properties

1. Introduction

Composite technology, *i.e.*, introducing stiff and hard particle reinforcements, is an effective approach to improve the strength of TiAl alloy [1–5]. TiB$_2$ and Ti$_2$AlC are the two most frequently-used reinforcing particles in TiAl matrix composites [4–6]. It has been reported that TiB$_2$ particles can effectively enhance the strength of TiAl alloy [7–10]. However, the strength enhancement of the TiAl matrix composite caused by the addition of TiB$_2$ particles is usually at the cost of plasticity [7–10]. In contrast to TiB$_2$, Ti$_2$AlC ceramic combines unusual properties of both metals and ceramics, that is it possesses both plasticity and strength [11–13]. Although the enhancement effect of Ti$_2$AlC is relatively weaker than that of TiB$_2$, Ti$_2$AlC particles could improve the strength of the TiAl alloy and, at same time, with no great plasticity damage [14]. Therefore, introducing TiB$_2$ and Ti$_2$AlC particles simultaneously into TiAl matrix is expected to fabricate the *in situ* dual reinforcement TiB$_2$-Ti$_2$AlC/TiAl composite with further improved strength and also excellent plasticity.

The most direct idea to fabricate *in situ* dual reinforcement TiB$_2$-Ti$_2$AlC/TiAl composite is using the Ti-Al-B-C system. However, B powder is very expensive, which limits its potential application in industrial practice. Thus, it is necessary to find another material to use as the B source instead of B powder. As is known, the cost of B$_4$C powder is at least 20-times less expensive than that of B powder. Meanwhile, B$_4$C powder can offer C to synthesize Ti$_2$AlC particles. Thus, it is a more cost-effective way to fabricate the *in situ* dual reinforcement TiB$_2$-Ti$_2$AlC/TiAl composite by using B$_4$C as the B and C source.

Recently, several methods have been applied to fabricate TiAl matrix composites. Van Meter *et al.* [7] fabricated 40 vol. % and 50 vol. % TiB$_2$/TiAl composites by the method of powder metallurgy (P/M). The compression strength of the composites was reported in the range of 2484–2866 MPa, and fracture strain was in the range of 0.4%–1.7%. Bohn *et al.* [9] fabricated Ti_5Si_3/TiAl composites by the method of hot isostatic pressing (HIP). Compression strength reached up to 2680 MPa for the sample with a mean grain size of 170 nm, while fracture strain was about 1.2%. Yang *et al.* [15] fabricated Ti_2AlC/TiAl composites by the method of spark plasma sintering (SPS). The compression strength of the composites reached 2058 MPa, and the fracture strain was about 0.16%. Compared with these above methods, the method of combustion synthesis and hot press consolidation represents an *in situ* processing technique for the preparation of composites, which takes advantage of the low energy requirement, cleaner particle-matrix interface, one-step forming process, density and high purity of the products. Consequently, the fabricated composites would possess better comprehensive properties.

In our previous work [16], we focused on the issues of the effect of B_4C size on the fabrication of *in situ* TiB_2-Ti_2AlC/TiAl composites. It was concluded that just when the size of B_4C particles is reduced to 3.5 µm, pure *in situ* TiB_2-Ti_2AlC/TiAl composites could be successfully fabricated. However, the effect of the content of synthesized ceramic particles on the compression properties of TiB_2-Ti_2AlC/TiAl composites was not discussed. Thus, in this work, under the basis of the mentioned research work, the B_4C particles with a size of 3.5 µm were directly used to fabricate TiB_2-Ti_2AlC/TiAl composites with different contents of the synthesized ceramic particles. The effect of the content of synthesized ceramic particles on the compression properties and work-hardening capacity of TiB_2-Ti_2AlC/TiAl composite was investigated. Moreover, the reinforcing effect of the TiB_2-Ti_2AlC particulates synthesized from Ti-Al-B_4C and Ti-Al-B-C systems was compared.

2. Experimental Section

Starting materials were made from commercial powders of Ti (99.5% purity, ~25 µm), Al (99% purity, ~74 µm), B_4C (99.5% purity, ~3.5 µm), B (98% purity, ~3 µm) and carbon black (99.9% purity). The powders of Ti, Al and B_4C (Ti-Al-B_4C system) corresponding to nominal 2, 4, 6 and 8 vol. % TiB_2-Ti_2AlC/TiAl composites were mixed sufficiently by ball milling for 8 h at a low speed (~35 rpm) in a conventional planetary ball-miller. Both the pot and balls were made of stainless steel, and the mass ratio of ball to powders was 20:1–25:1. Then, the mixed powders were cold pressed into cylindrical compacts using a stainless steel die. In addition, the powders of Ti, Al, B and carbon black (Ti-Al-B-C system) corresponding to nominal 6 vol. % TiB_2-Ti_2AlC/TiAl composite were also mixed and pressed into cylindrical compact, which was used to compare to the composites fabricated by using the Ti-Al-B_4C system. Unless specified otherwise, the composites were fabricated from the Ti-Al-B_4C system. The powder compact of 28 mm in diameter and approximately 36 mm in height was contained in a graphite mold, which was put into a self-made vacuum thermal explosion furnace. The heating rate of the furnace was about 30 K/min, and the temperature in the vicinity of the center of the compact was measured by Ni-Cr/Ni-Si thermocouples. When the temperature measured by the thermocouples suddenly rose rapidly, indicating that the sample should be ignited, the sample was quickly pressed just when it was still hot and soft. Pressure (~50 MPa) was maintained for 10 s, and then, the product was cooled down to ambient temperature at a cooling rate of ~10 K/min.

The phase constituents of composites were examined by X-ray diffraction (XRD, Rigaku D/Max 2500PC, Rigaku Corporation, Tokyo, Japan) with Cu Kα radiation. Microstructures were studied using scanning electron microscopy (SEM, Evo18, Carl Zeiss, Oberkochen, Germany) equipped with an energy-dispersive spectrometer (EDS, Oxford Instruments, Oxford, UK). The density of each sample was measured three times by Archimedes' water-immersion method, and average values are listed in Table 1. The cylindrical specimens with a diameter of 3 mm and a height of 6 mm were used for compression tests, and the loading surface was polished parallel to the other surface. Uniaxial compression tests were carried out under a servo-hydraulic materials' testing system (MTS, MTS 810, MTS Systems Corporation, Minneapolis, MN, USA) with a strain rate of $1 \times 10^{-4} \cdot s^{-1}$.

Metals **2015**, *5*, 2200–2209

3. Results and Discussion

3.1. Phase Constituents and Microstructures

In order to identify the synthesized ceramic particles, the phase constituents of the composites were examined by XRD. Figure 1 shows the X-ray diffraction results of the composites fabricated from the Ti-Al-B$_4$C system. Actually, due to the limitation of the detection capacity of XRD, no ceramic phases, but the γ-TiAl and α_2-Ti$_3$Al phases, were detected in the samples with 2, 4 and 6 vol. % nominal ceramic contents. In the sample with a high content of ceramic particles (the nominal ceramic content is 8 vol. %), the TiB$_2$ and Ti$_2$AlC phases were detected besides the γ-TiAl and α_2-Ti$_3$Al phases. The combustion synthesis reaction in the Ti-Al-B$_4$C system is complete, and no trace of residual B$_4$C particles was found, which indicated that TiB$_2$ and Ti$_2$AlC particles could be synthesized simultaneously in the Ti-Al-B$_4$C system and that *in situ* dual reinforcement TiB$_2$-Ti$_2$AlC/TiAl composites could be successfully fabricated by using B$_4$C as the B and C source.

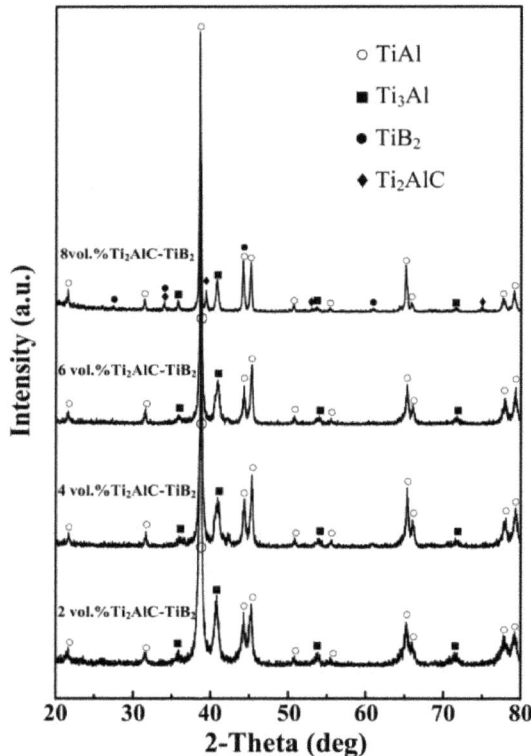

Figure 1. XRD patterns of *in situ* dual reinforcement TiB$_2$-Ti$_2$AlC/TiAl composites fabricated by using the Ti-Al-B$_4$C system.

Figure 2a–d shows the microstructures of the TiB$_2$-Ti$_2$AlC/TiAl composites fabricated from the Ti-Al-B$_4$C system. It can be seen from Figure 2a–c that when the contents of synthesized TiB$_2$ and Ti$_2$AlC particles increase from 2–6 vol. %, the ceramic particles distribute more homogeneously in the TiAl matrix. The increasing content of synthesized TiB$_2$ and Ti$_2$AlC particles results in more heat released by the reaction and, thus, a higher temperature. The increased temperature facilitates faster and more complete diffusion of the reactants during combustion synthesis, leading to a more uniform distribution of the ceramic particles in the TiAl matrix [17]. However, the ceramic particles begin to sinter together, as shown in Figure 2d, when too many ceramic particles formed in the composites (8 vol. % TiB_2–Ti_2AlC). In order to examine the distribution of TiB_2 and Ti_2AlC particles, EDS was used to distinguish the two ceramic particles from each other (as shown in Figure 2c). It can be seen that in 6 vol. % TiB_2-Ti_2AlC/TiAl composite, Ti_2AlC particles are rod-like in shape with sizes of 1–6 μm in length, while TiB_2 particles are in a near spherical shape with a size of less than 1 μm in diameter.

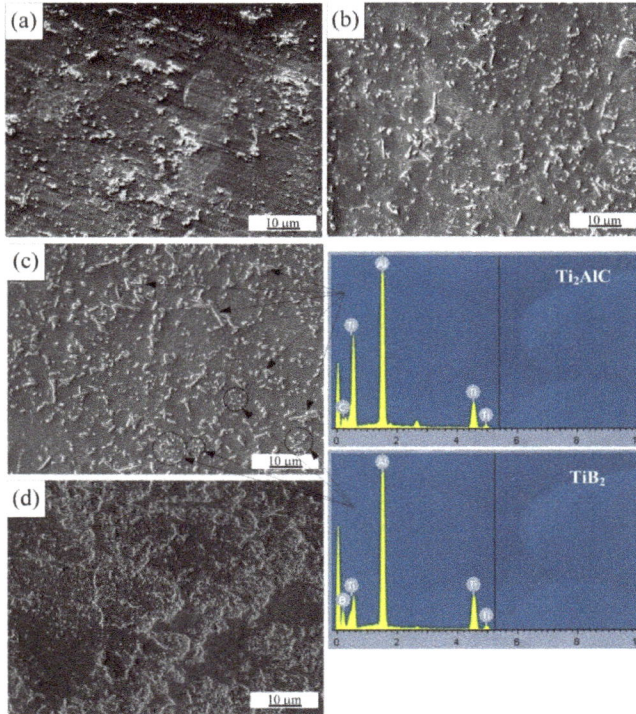

Figure 2. Microstructures of (**a**) 2 vol. % TiB_2-Ti_2AlC/TiAl; (**b**) 4 vol. % TiB_2-Ti_2AlC/TiAl; (**c**) 6 vol. % TiB_2-Ti_2AlC/TiAl and (**d**) 8 vol. % TiB_2-Ti_2AlC/TiAl composites fabricated by using the Ti-Al-B_4C system.

3.2. Compression Properties and Work-Hardening Capacity

Figure 3 shows true compression stress-strain curves of the TiAl alloy and the TiB_2-Ti_2AlC/TiAl composites fabricated from the Ti-Al-B_4C system. The compression properties are summarized in Table 1. The yielding strength (σ_y) and ultimate compression strength (σ_{ucs}) of TiB_2-Ti_2AlC/TiAl composites increase with increasing content of synthesized ceramic particles. The fracture strain (ε_f) of the composites does not change significantly when the content of synthesized TiB_2-Ti_2AlC particles increases from 2–6 vol. %. However, when the content of synthesized TiB_2-Ti_2AlC particles increases to 8 vol. %, ε_f decreases significantly, due to too many ceramic particles being synthesized and segregated together in the composite. This also can be confirmed by the fracture surface of 6 vol. % TiB_2-Ti_2AlC/TiAl and 8 vol. % TiB_2-Ti_2AlC/TiAl composites shown in Figure 4a,b. It can be compared to the fracture surface that the 6 vol. % TiB_2-Ti_2AlC/TiAl composite exhibits with a tear ridge, while the fracture surface of the 8 vol. % TiB_2-Ti_2AlC/TiAl composite is relatively flat. In 8 vol. % the TiB_2-Ti_2AlC/TiAl composite, the bad interface combination between aggregated particles and

TiAl matrix would play the role of crack initiator during deformation and facilitate the fracture of the composite.

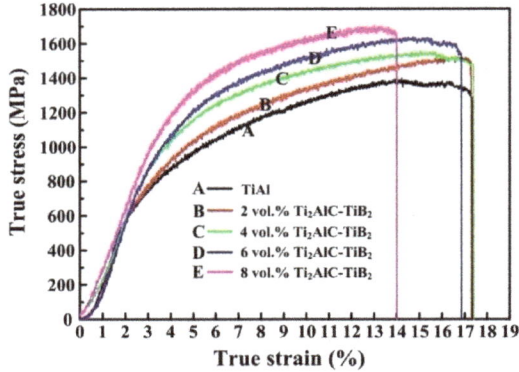

Figure 3. True compression stress-strain curves of TiAl alloy and the *in situ* dual reinforcement TiB_2-Ti_2AlC/TiAl composites fabricated by using the Ti-Al-B_4C system.

Table 1. Compression properties and work-hardening capacity (H_c) of TiAl alloy and the *in situ* dual reinforcement TiB_2-Ti_2AlC/TiAl composites fabricated by using the Ti-Al-B_4C system.

Ceramic Content	Measureddensity (g/cm^3)	σ_y (MPa)	σ_{ucs} (MPa)	ε_f (%)	H_c
TiAl	3.72 ± 0.03	465 ± 41	1415 ± 20	17.3 ± 0.0	2.07 ± 0.31
2 vol. %	3.74 ± 0.02	625 ± 24	1487 ± 32	16.8 ± 0.3	1.39 ± 0.15
4 vol. %	3.80 ± 0.02	711 ± 25	1562 ± 10	17.7 ± 0.3	1.20 ± 0.07
6 vol. %	3.82 ± 0.02	781 ± 25	1649 ± 12	16.6 ± 0.2	1.11 ± 0.05
8 vol. %	3.88 ± 0.05	865 ± 29	1695 ± 5	14.9 ± 0.9	0.96 ± 0.06

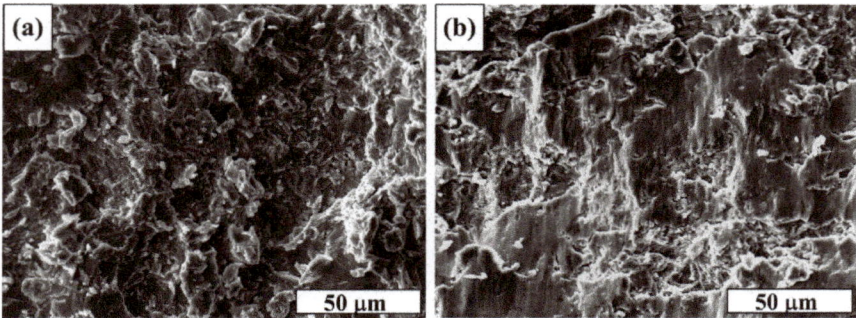

Figure 4. SEM images of the compression fractured surfaces of (**a**) 6 vol. % and (**b**) 8 vol. % TiB_2-Ti_2AlC/TiAl composites fabricated by using the Ti-Al-B_4C system.

The above results indicate that lower content (2–6 vol. %) TiB_2-Ti_2AlC particles could effectively improve the strength of TiAl alloy without sacrificing plasticity. The 6 vol. % TiB_2-Ti_2AlC/TiAl composite possesses the best compression properties. The average σ_y and σ_{ucs} of the 6 vol. % TiB_2-Ti_2AlC/TiAl composite fabricated by using the Ti-Al-B_4C system are 781 MPa and 1649 MPa, respectively, which are 316 MPa and 234 MPa higher than those of TiAl alloy. The enhancement of strength is mainly due to the reinforcing effect of stiff TiB_2 and Ti_2AlC particles. The uniform distribution of *in situ* TiB_2 and Ti_2AlC particles would be the reason for the maintenance of the high plasticity. Moreover, as discussed in our previous study [14], the metallic property of Ti_2AlC particles and the coherent interface between Ti_2AlC and TiAl would also contribute to the high plasticity.

It can be seen from the true stress-strain curves of the composites fabricated by using the Ti-Al-B_4C system shown in Figure 3 that these curves all show a clear work hardening. The work-hardening capacity of composites is calculated according to the formulary (H_c) ($H_c = (\sigma_{ucs} - \sigma_y)/\sigma_y$) [18]. The results are listed in Table 1. The H_c of TiB_2-Ti_2AlC/TiAl composite decreases with the increase in the content of synthesized ceramic particles. The onset of plastic deformation with an obvious strain hardening in composites represents a deformation mechanism of dislocation activity. The strain hardening of composites after yielding is mainly due to dislocation multiplication, accumulation and interaction [18–20]. As mentioned above, the σ_y of composites increases with increasing content of synthesized ceramic particles, which means that dislocation-nucleation threshold stress increases with the increase in ceramic content, that is the activation of dislocations becomes more difficult. Consequently, the dislocation interactions during plastic deformation would become weak with the increase in ceramic content, leading to a decrease in H_c. The stress-strain curves also exhibit a low modulus in the elastic region, which is similar to other researchers' work [21,22]. It is speculated that porosity might be playing a part in this. The porosities in the composites evaluated by the method of image analysis are approximately 1.9%, 1.5%, 1.8% and 1.4%, respectively. In addition, the 6 vol. % TiB_2-Ti_2AlC/TiAl composite was also fabricated from the Ti-Al-B-C system, which was used to compare to the composites fabricated from the Ti-Al-B_4C system. Figure 5 shows the true compression stress-strain curves of the *in situ* dual reinforcement 6 vol. % TiB_2-Ti_2AlC/TiAl composites fabricated by these two systems. The compression properties and their work-hardening capacity are summarized in Table 2. The results indicate that the compression properties and work-hardening capacity of the TiB_2-Ti_2AlC/TiAl composites fabricated from these two systems are similar. Thus, from the economic point of view, the TiB_2-Ti_2AlC/TiAl composite fabricated by using the Ti-Al-B_4C system could be widely used in practical production.

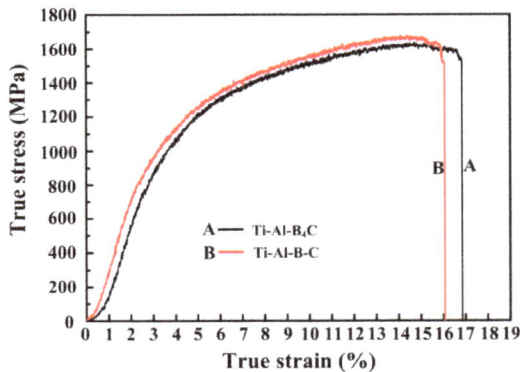

Figure 5. True compression stress-strain curves of the *in situ* dual reinforcement 6 vol. % TiB_2-Ti_2AlC/TiAl composites fabricated by using different systems.

Table 2. Compression properties and work-hardening capacity (H_c) of the *in situ* dual reinforcement TiB_2-Ti_2AlC/TiAl composites fabricated by using different systems.

System	Ceramic Content	σ_y (MPa)	σ_{ucs} (MPa)	ε_f (%)	H_c
Ti-Al-B_4C	6 vol. %	781 ± 25	1649 ± 12	16.6 ± 0.2	1.11 ± 0.05
Ti-Al-B-C	6 vol. %	771 ± 23	1677 ± 1	15.8 ± 0.3	1.18 ± 0.06

4. Conclusions

The content of TiB_2-Ti_2AlC particles significantly influences the compression properties and work-hardening capacity of TiAl matrix composites. With the increase in the content of synthesized TiB_2-Ti_2AlC particles, the σ_y and σ_{ucs} of TiB_2-Ti_2AlC/TiAl composites increase, while H_c decreases. When the content of synthesized ceramic particles is lower (2 vol. %–6 vol. %), ε_f does not change significantly. However, when the content of ceramic particles comes to 8 vol. %, ε_f decreases significantly. The synthesized TiB_2 and Ti_2AlC particles could effectively improve the σ_y and σ_{ucs} of TiAl alloy due to the reinforcing effect of stiff TiB_2 and Ti_2AlC particles. The uniform distribution of *in situ* TiB_2 and Ti_2AlC particles and the special characteristics of Ti_2AlC would be the reasons for the maintenance of the high plasticity. The compression properties of the TiB_2-Ti_2AlC/TiAl composites fabricated by using the Ti-Al-B_4C system are similar to those of the TiB_2-Ti_2AlC/TiAl composites fabricated by using the Ti-Al-B-C system. From the economic point of view, it is better to fabricate *in situ* dual reinforcement TiB_2-Ti_2AlC/TiAl composites by using the Ti-Al-B_4C system. The σ_y and σ_{ucs} of the 6 vol. % TiB_2-Ti_2AlC/TiAl composite fabricated by using the Ti-Al-B_4C system are 316 MPa and 234 MPa higher than those of the TiAl alloy, without sacrificing plasticity.

Acknowledgments: This work is supported by the National Natural Science Foundation of China (NNSFC, No. 51171071), the National Basic Research Program of China (973 Program, No. 2012CB619600), the NNSFC (No. 51501176), the Jilin Province Science and Technology Development Plan (No. 20140520127JH), the International Science Technology Cooperation Program of China (2013DFR00730), the Research Fund for the Doctoral Program of Higher Education of China (No. 20130061110037), the State Scholarship Fund of China Scholarship Council (201506175140), as well as by The Project 985-High Performance Materials of Jilin University.

Author Contributions: These authors accomplished this work together.

Conflicts of Interest: The authors declare no conflict of interest.

References

1. Kumar, K.S.; Bao, G. Intermetallic-matrix composites: An overview. *Compos. Sci. Technol.* **1994**, *52*, 127–150. [CrossRef]
2. Xiang, L.Y.; Wang, F.; Zhu, J.F.; Wang, X.F. Mechanical properties and microstructure of Al_2O_3/TiAl *in situ* composites doped with Cr_2O_3. *Mater. Sci. Eng. A* **2011**, *528*, 3337–3341. [CrossRef]
3. Rao, K.P.; Zhou, J.B. Characterization and mechanical properties of *in situ* synthesized Ti_5Si_3/TiAl composites. *Mater. Sci. Eng. A* **2003**, *356*, 208–218. [CrossRef]
4. Yang, C.H.; Wang, F.; Ai, T.T.; Zhu, J.F. Microstructure and mechanical properties of *in situ* TiAl/Ti_2AlC composites prepared by reactive hot pressing. *Ceram. Int.* **2014**, *40*, 8165–8171. [CrossRef]
5. Ai, T.T.; Wang, F.; Feng, X.M.; Ruan, M.M. Microstructural and mechanical properties of dual Ti_3AlC_2-Ti_2AlC reinforced TiAl composites fabricated by reaction hot pressing. *Ceram. Int.* **2014**, *40*, 9947–9953. [CrossRef]
6. Li, B.; Lavernia, E.J. Spray forming and co-injection of particulate reinforced TiAl/TiB_2 composites. *Acta Mater.* **1997**, *45*, 5015–5030. [CrossRef]
7. Van Meter, M.L.; Kampe, S.L.; Christodoulou, L. Mechanical properties of near-γ titanium aluminides reinforced with high volume percentages of TiB_2. *Scr. Mater.* **1996**, *34*, 1251–1256. [CrossRef]
8. Hirose, A.; Hasegawa, M.; Kobayashi, K.F. Microstructures and mechanical properties of TiB_2 particle reinforced TiAl composites by plasma arc melting process. *Mater. Sci. Eng. A* **1997**, *239–240*, 46–54.
9. Bohn, R.; Klassen, T.; Bormann, R. Room temperature mechanical behavior of silicon-doped TiAl alloys with grain sizes in the nano-and submicron-range. *Acta Mater.* **2001**, *49*, 299–311. [CrossRef]
10. Kim, S.H.; Chung, H.H.; Pyo, S.G.; Hwang, S.J.; Kim, N.J. Effect of B on the Microstructure and Mechanical Properties of Mechanically Milled TiAl Alloys. *Metall. Mater. Trans. A* **1998**, *29*, 2273–2283. [CrossRef]
11. Lin, Z.J.; Zhuo, M.J.; Zhou, Y.C.; Li, M.S.; Wang, Y.J. Microstructural characterization of layered ternary Ti_2AlC. *Acta Mater.* **2006**, *54*, 1009–1015. [CrossRef]
12. Chen, Y.L.; Yan, M.; Sun, Y.M.; Mei, B.C.; Zhu, J.Q. The phase transformation and microstructure of TiAl/Ti_2Al composites caused by hot pressing. *Ceram. Int.* **2009**, *35*, 1807–1812. [CrossRef]
13. Kulkarni, S.R.; Datye, A.V.; Wu, K.-H. Synthesis of Ti_2AlC by spark plasma sintering of TiAl-carbon nanotube powder mixture. *J. Alloys Compd.* **2010**, *490*, 155–159. [CrossRef]

14. Shu, S.L.; Qiu, F.; Lü, S.J.; Jin, S.B.; Jiang, Q.C. Phase transitions and compression properties of Ti$_2$AlC/TiAl composites fabricated by combustion synthesis reaction. *Mater. Sci. Eng. A* **2012**, *539*, 344–348. [CrossRef]
15. Yang, F.; Kong, F.T.; Chen, Y.Y.; Xiao, S.L. Effect of spark plasma sintering temperature on the microstructure and mechanical properties of a Ti$_2$AlC/TiAl composite. *J. Alloys Compd.* **2010**, *496*, 462–466. [CrossRef]
16. Shu, S.L.; Qiu, F.; Lin, Y.; Wang, Y.; Wang, J.; Jiang, Q. Effect of B$_4$C size on the fabrication and compression properties of *in situ* TiB$_2$-Ti$_2$AlC/TiAl composites. *J. Alloys Compd.* **2013**, *551*, 88–91. [CrossRef]
17. Alman, D.E. Reactive sintering of TiAl-Ti$_5$Si$_3$ *in situ* composites. *Intermetallics* **2005**, *13*, 572–579. [CrossRef]
18. Afrin, N.; Chen, D.L.; Cao, X.; Jahazi, M. Strain hardening behavior of a friction stir welded magnesium alloy. *Scr. Mater.* **2007**, *57*, 1004–1007. [CrossRef]
19. Sun, B.B.; Sui, M.L.; Wang, Y.M.; He, G.; Eckert, J.; Ma, E. Ultrafine composite microstructure in a bulk Ti alloy for high strength, strain hardening and tensile ductility. *Acta Mater.* **2006**, *54*, 1349–1357. [CrossRef]
20. Zhao, Y.H.; Bingert, J.F.; Liao, X.Z.; Cui, B.Z.; Han, K.; Sergueeva, A.V.; Mukherjee, A.K.; Valiev, R.Z.; Langdon, T.G.; Zhu, Y.T. Simultaneously increasing the ductility and strength of nano structured alloys. *Adv. Mater.* **2006**, *18*, 2949–2953. [CrossRef]
21. Calderon, H.A.; Garibay-Febles, V.; Umemoto, M.; Yamaguchi, M. Mechanical properties of nanocrystalline Ti-Al-X alloys. *Mater. Sci. Eng. A* **2002**, *329–331*, 196–205. [CrossRef]
22. Rao, K.P.; Vyas, A. Comparison of titanium silicide and carbide reinforced *in situ* synthesized TiAl composites and their mechanical properties. *Intermetallics* **2011**, *19*, 1236–1242. [CrossRef]

metals

MDPI

Article

Modeling of TiAl Alloy Grating by Investment Casting

Yi Jia [1,2], Shulong Xiao [1,2], Jing Tian [2], Lijuan Xu [2] and Yuyong Chen [1,2,3,*]

[1] National Key Laboratory of Science and Technology on Precision Heat Processing of Metals, Harbin Institute of Technology, Harbin 150001, China; samuel_jia@hotmail.com (Y.J.); xiaoshulong@hit.edu.cn (S.X.)

[2] School of Materials Science and Engineering, Harbin Institute of Technology, Harbin 150001, China; tianjing@hit.edu.cn (J.T.); xljuan@hit.edu.cn (L.X.)

[3] State Key Laboratory of Advanced Welding and Joining, Harbin Institute of Technology, Harbin 150001, China

* Author to whom correspondence should be addressed; yychen@hit.edu.cn; Tel./Fax: +86-451-8641-8802.

Academic Editor: Ana Sofia Ramos

Received: 19 October 2015; Accepted: 4 December 2015; Published: 9 December 2015

Abstract: The investment casting of TiAl alloys has become the most promising cost-effective technique for manufacturing TiAl components. This study aimed to investigate a series of problems associated with the investment casting of TiAl alloys. The mold filling and solidification of this casting model were numerically simulated using ProCAST. Shrinkage porosity was quantitatively predicted by a built-in feeding criterion. The results obtained from the numerical simulations were compared with experiments, which were carried out on Vacuum Skull Furnace using an investment block mold. The investment casting of TiAl grating was conducted for verifying the correctness and feasibility of the proposed method. The tensile test results indicated that, at room temperature, the tensile strength and elongation were approximately 675 MPa and 1.7%, respectively. The microstructure and mechanical property of the investment cast TiAl alloy were discussed.

Keywords: numerical simulation; TiAl alloys; investment casting; shrinkage porosity

1. Introduction

Energy as well as environmental issues have become the main obstacles for sustaining social and economic development. The substitution of lightweight materials for heavy ones is effective for solving this problem. Aviation and aerospace materials are developed with a predominant focus on the development of lightweight, high-strength materials. TiAl alloys exhibit excellent mechanical, oxidation, and corrosion resistance properties at elevated temperatures (greater than 600 °C), making them a possible replacement for traditional Ni-based superalloy components in the aircraft and automobile industry for increasing the thrust-to-weight ratio and efficiency while decreasing exhaust and noise pollution [1–3]. Because of the chemical heterogeneity and physical properties of TiAl alloys, numerous efforts have been focused on the introduction of titanium aluminide into the market, albeit with limited success. A limitation for the "mass market" manufacture of TiAl-based components is that TiAl exhibits very high chemical reactivity, high melting temperature, low ductility, and poor workability. In contrast, casting exhibits a significant advantage for complex-shaped components such as turbine blades, turbocharger rotators and exhaust valves. Because of these issues, investment casting, which can directly produce near-net-shaped components with a good surface finish and low production cost, is a subject of growing interest [4].

Different casting processes are employed for casting TiAl alloys, such as induction-skull melting (ISM), vacuum-arc remelting (VAR), counter-gravity low pressure atmosphere melting (CLIM), plasma-arc melting (PAM), *etc.* Although most of these processes produce high quality castings, the cost of the products is rather high.

Investment casting exhibits tremendous advantages for the production of quality cast components as it has the key benefits of accuracy, versatility and integrity. As a result, investment casting is one of the most economical methods for producing a wide range of metal castings.

Several efforts have been focused on the thermodynamic stability and mechanism of interaction between refractory materials such as CaO [5,6], Al_2O_3 [7,8], ZrO_2 [9,10] and Y_2O_3 [11,12] in contact with molten TiAl alloys. Such information is of great interest not only for the purpose of solidification studies but also for the induction melting of TiAl alloys and their investment casting in ceramic molds as well.

The casting process based on experience has the characteristics of high cost and long cycle. In addition, the pouring process is invisible. However, by numerical simulation, cost savings as well as a reduction in production cycles can be achieved. Signification promise and potential have been demonstrated by numerical simulations. This study aims to fabricate a grating with Ti–47Al–2.5V–1Cr (at. %) by investment casting and discusses the relevant microstructure and mechanical properties.

2. Experimental Section

The casting of the TiAl alloy grating was a disk with a diameter of 580 mm, which had a 180 mm hole in the center, and a thickness of 10 mm. Initially, we made a small test disk with a diameter of 400 mm, which had a 120 mm hole in the center, and a thickness of 10 mm. In the following text, we refer to full-size casting as the ultimate goal, and test casting as the minor one. All samples for characterization were cut from the test casting. The full-size casting only employed X-ray non-destructive inspection for the porosity in the casting.

2.1. Numerical Simulation

Experiment processes of TiAl casting were simulated by a finite element method (FEM) software ProCAST package (ESI Group, Paris, France). Tables 1 and 2 summarize the thermo-physical material properties of the casting and mold, respectively, which were implemented in the preprocessing procedure; the properties of TiAl were given by Sung [13,14], and the properties of ZrO_2 were given by ProCAST. The environment temperature in Table 2 was the temperature in the vacuum chamber during melting because it is too difficult to calculate the heat effect of the melting system to the environment. So, we ignored that, and determined it to be room temperature. The filling and solidification behavior was simulated by the calculation procedure. The filling behavior, temperature field and solidification parameters, with respect to the formation of shrinkage porosity, were analyzed during the post-processing procedure. The initial processing parameters used in the simulation were a pouring temperature of 1700 °C, and a filling time of 3 s.

Table 1. Thermo-physical properties of TiAl alloy varied with temperature.

Temperature/(°C)	Density/(kg·m^{-3})	Specific Heat/ (J·kg^{-1}·K^{-1})	Thermal Conductivity/ (W·m^{-1}·K^{-1})	Thermal Diffusion/ (m^2·s^{-1})
25	3857	598	13.2	-
200	-	630	16.7	-
400	-	667	20.2	-
600	-	703	23.1	-
700	-	799	25.4	8.4
800	-	769.3	26.8	8.8
900	-	900.1	28.	8.1
1000	-	975.9	27.9	7.7
1100	-	995.6	26.5	7.2
1200	-	1140.4	27.3	6.7
1600	3788	786	31	-
1800	3612	794	37	-

Table 2. Thermo-physical properties of TiAl alloy and ZrO_2 mold.

Thermophysical Properties	TiAl Alloy	Mold (ZrO_2)
Density/(kg·m^{-3})	3788	3970
Specific heat/(J·kg^{-1}·K^{-1})	598	777
Thermal conductivity/(W·m^{-1}·K^{-1})	13.2	39
Liquidus temperature/(°C)	1554	2323
Solidus temperature/(°C)	1478	-
Latent heat/(J·kg^{-1}·K^{-1})	435	-
Environment temperature/(°C)	25	25

2.2. Production of Castings

The alloy used in this study had a nominal composition of Ti–47Al–2.5V–1Cr (at. %, TiAl alloy hereafter). The castings were produced by VAM-150, ZXVAC (Shenyang, China). The charge material was pieces of TiAl alloy ingots produced by vacuum-arc remelting.

In this study, the conventional "lost wax" procedure was employed for fabricating ceramic shell molds. The replicated wax crowns were assembled on the numerically optimized runner and gating system.

The detailed manufacturing processes were described as follows. First, the wax patterns were dipped into the slurry, which was mixed with zirconia sol and ZrO_2 powders (diameter < 50 μm), stuccoed by zirconia sand and dried. After the primary coating, wax patterns were coated with a back-up slurry, which comprised alumina and silica sol. Back-up coating process was repeated several times for enhancing the strength of the ceramic molds. Finally, a sealing coat of the back-up slurry was applied. At the end of these processes, the total thickness of the ceramic molds was 7–10 mm. The de-waxing process of the ceramic molds was carried out at about 0.8 MPa and 150 °C in a steam autoclave. After de-waxing, the ceramic molds were sintered at 950 °C for 2 h.

Molds were kept in a refractory-filled can, the vacuum chamber was evacuated and backfilled with argon at a pressure of 8×10^4 Pa and the charge melted. Prior to the pouring process, the ceramic molds were first preheated, and then molten TiAl was poured into the preheated ceramic molds. After removing the ceramic molds, TiAl castings were obtained. There are some negative factors in TiAl casting, such as the static pressure head is low because of low density (3.8 g/cm^{-3}), narrow solidification interval (<80 °C), and bad fluidity. Hence, there is a tendency to employ a pouring temperature as high as possible for improving the quality of the TiAl casting. The machine can melt TiAl at the highest temperature 1700 °C; hence, it is chosen as the pouring temperature.

2.3. Characterization of Microstructure and Mechanical Property

Samples used for characterization were cut from the test casting. All microstructures had been ground and polished electrolytically (Perchloric 6% + Butoxyethanol 34% + Methanol 60%, −30 °C, 35 V, 9 mA). Metallographic specimens were etched by Kroll's reagent (a mixture of 10 mL HF, 5 mL HNO_3 and 85 mL H_2O) for observation under an optical microscope. Optical microscopy was performed with an Olympus microscope (Tokyo, Japan). The mechanical property of the castings was characterized by tensile deformation experiments at a constant strain rate (1×10^{-4} s^{-1}) performed at room temperature. A field emission gun scanning electron microscope (SEM, FEI, QUANTA 200F, Portland, OR, USA) was employed for characterizing the morphologies of the fracture surface.

3. Results and Discussion

3.1. Numerical Simulation for Test Casting

The three-dimensional (3D) model of test casting based on the first design is shown in Figure 1. The mold-filling process and solidification were calculated. All the casting gratings exhibited porosities under different initial conditions, as shown in Figure 2. The results indicated that casting defects are spread throughout the entire cast gratings, and the porosity is considerable. Casting defects are affected by the mold-filling process and the solidification of the cast grating. For a mold temperature of 25 °C and gravity casting, the worst results are obtained from the simulation (Figure 2a), and a large amount of shrinkage pores are spread throughout the entire cast grating. With the increase in the mold temperature to 800 °C, the situation improves slightly (Figure 2b). Moreover, when a centrifugal casting is employed, the situation significantly improves at rotation speeds of 200 rpm and 400 rpm (Figure 2c–f). However, with the increase in the rotation speed to 600 rpm, the amount of shrinkage pores increases (Figure 2g,h).

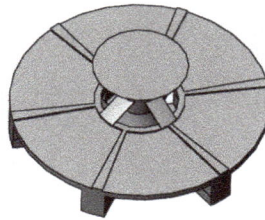

Figure 1. Three-dimensional drawing of runner system for test casting.

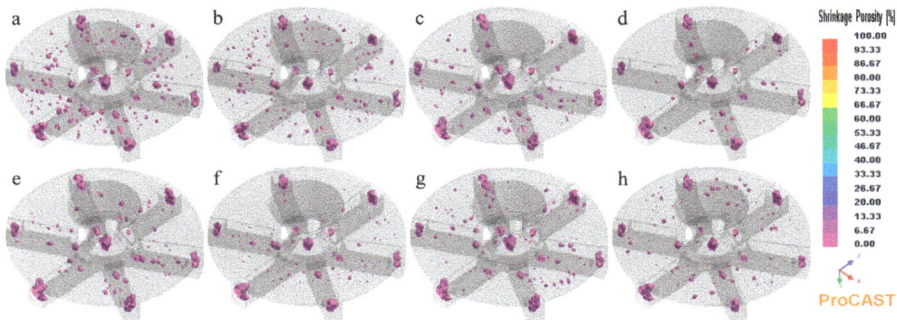

Figure 2. The predicted shrinkage porosity of test castings: (**a**) mold temperature of 25 °C and gravity casting (short for 25 °C, 0 rpm); (**b**) 800 °C, 0 rpm; (**c**) 25 °C, 200 rpm; (**d**) 800 °C, 200 rpm; (**e**) 25 °C, 400 rpm; (**f**) 800 °C, 400 rpm; (**g**) 25 °C, 600 rpm; (**h**) 800 °C, 600 rpm.

Because of the contact with the mold, the front-flowing molten metal loses a large amount of heat, which causes a sudden decrease in temperature, and the molten metal may no longer flow to the cold end of the mold. The possibility of shrinkage porosity slightly decreased with a high mold temperature. For a given pouring temperature and mold temperature, the molten TiAl alloy fills the mold at a rapid rate, it gets a long time to feed, and decreases the possible formation of cast defects. In the case when the pouring rate is limited by the casting equipment, the action of the centrifugal force results in the spreading of the molten metal. With this runner design, the liquid flow would break under excessive centrifugal force; hence, a high rotation speed causes deterioration instead (Figure 2g,h). When the temperature of the frontier molten metal rapidly decreases, a large number of dendrites would mix in with the molten alloy, which would increase the viscosity of the molten alloy and flow resistance, slowing down of the flow rate of the molten alloy. While centrifugal force stirs the molten alloy, dendrites mixed with the latter molten alloy with higher temperature for remelting. By comparing the two cases, a mold temperature of 25 °C and gravity casting and a mold temperature of 800 °C, 400 rpm, when the mold is completely filled, the solid fractions obtained are 11.3% and 0.2%, respectively.

For shrinkage porosity, Figure 2d showed best situation, Figure 2f was the second and Figure 2e was the third. However, Figure 3d,e exhibited voids, and Figure 3f was free of voids. In ProCAST, the voids predicted represent not only air bubbles but also oxide layers trapped in areas where fluid flow is restricted. Hot Isostatic Pressing (HIP), which is a necessary step in the foundry industry, can eliminate voids but does not remove micro-porosity, which means that voids are more serious than porosity. Hence, a mold temperature of 800 °C and a rotation speed of 400 rpm are probably good choices for the test casting, which exhibited porosity slightly more than that observed at a mold temperature of 800 °C and a rotation speed of 200 rpm, albeit free of voids.

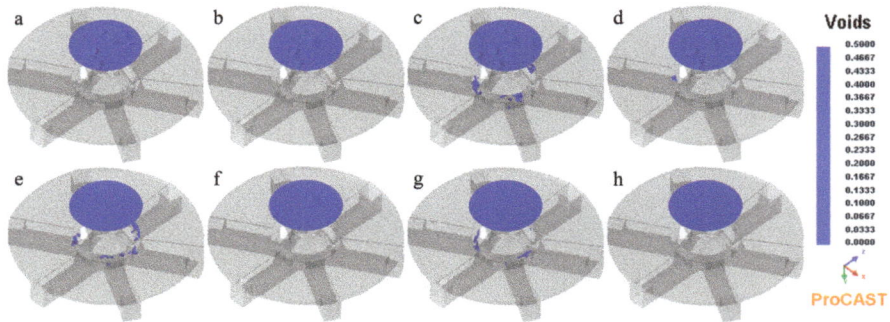

Figure 3. Predicted voids of test castings, (**a–h**), the same as the Figure 2.

3.2. The Quality of Test Casting

Figure 4a shows TiAl alloy test castings, which show visible pores on the surface. Figure 4b shows the location of the samples for metallographic observation and tensile test. As shown in Figures 5 and 6c,d, the orientation of the metallographic section is tangential, and Figure 6a shows the contact surface between the disk and runner.

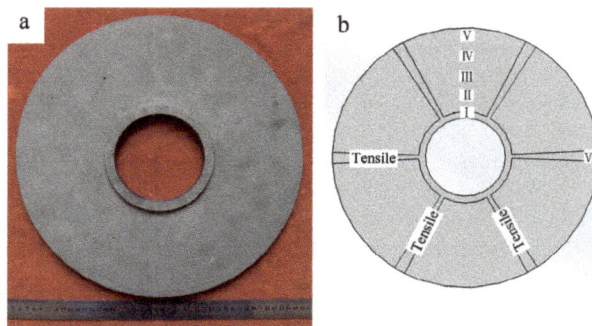

Figure 4. Test casting (**a,b**) showed the specimen locations: I, II, III, IV, and V, for Figure 5a–e, respectively; VI for Figure 6a; Tensile for the tensile test.

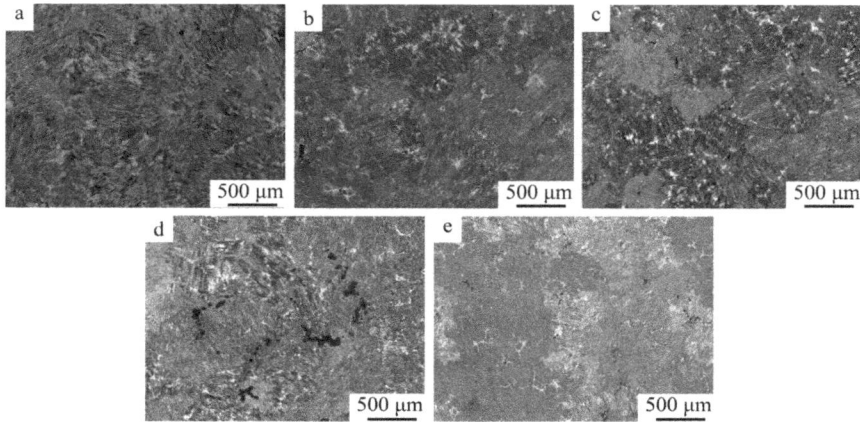

Figure 5. Optical microstructure of test casting (**a**–**e**) were from the center hole to the outer edge, and the interval between the two samples measured 20 mm.

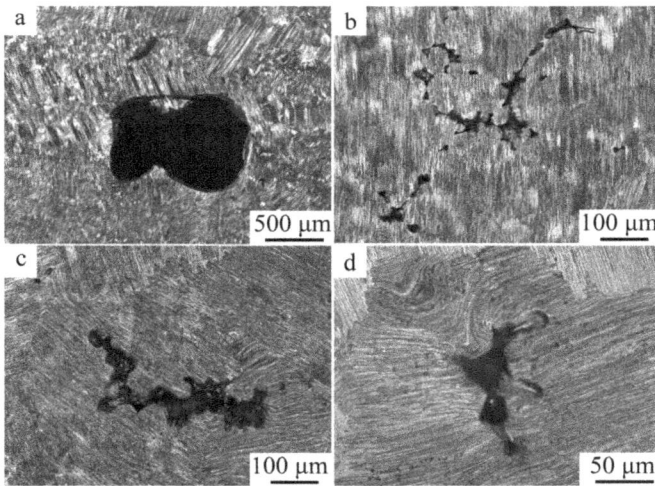

Figure 6. Micro-defects of test casting, (**a**) pore and (**b**–**d**) shrinkage.

Figure 5 shows a typical fully lamellar microstructure of as-cast TiAl, which contains a small amount of finely segregated γ-grains which are mainly located at the lamellar grain boundaries or in the interdendritic region as a result of the occurrence of double peritectic reactions during solidification. The microstructure does not significantly differ from the center hole to the outer edge; however, the casting defects are not the same everywhere. Figure 5a,b,e show micropores, and Figure 5c,d show extensive micro-porosity. Compared with the simulation data as shown in Figure 2d, the predicted shrinkage porosity of casting is in agreement with the experimental data, which show that the shrinkage porosity of casting enriches in the center of the casting and is absent near the center hole and outer edge. Figure 6a shows that a big hole appeared in runner, which is proved by Figure 7b. Figure 6a,b show that there is no effect of shrinkage on the fully lamellar microstructure; however, Figure 6c,d, show bending lamellar microstructure near the shrinkage. Figure 7a shows the slice view of the disk, and Figure 7b shows the slice view of the rib. As shown in Figure 7a, marginal shrinkage porosity is observed, and the amount of shrinkage porosity is significantly less than that in Figure 7b. The predicted shrinkage porosity by ProCAST is the probability, and Figure 5 meets the left half of Figure 7a—Figure 5a,b,e is almost porosity free, Figure 5c shows a few pores, and Figure 5d shows enrichment of shrinkage porosity.

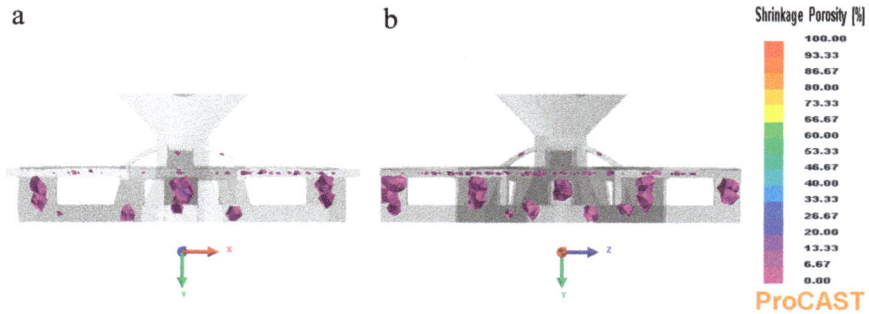

Figure 7. Slice view at a mold temperature of 800 °C and rotation speed of 400 rpm, (**a**) disk and (**b**) rib.

3.3. Mechanical Property for Test Casting

Figure 8a shows the mechanical property of the investment cast TiAl alloy at room temperature. The tensile strength and elongation at room temperature are about 675 MPa and 1.7%, respectively. The results obtained from this study are in good agreement with the data reported previously [4,15]. The mechanical property of a casting not only depends on its microstructure but also on the alloy composition, casting conditions and heat treatment. Figure 8b shows the fracture surface of an investment cast TiAl specimen at room temperature that presents an irregular and tortuous surface. It can be concluded that the main failure modes are inter-granular fracture in equiaxed γ-grains and trans-lamellar cracking in lamellar grains. The minor failure modes are trans-granular cracking in equiaxed γ-grains and local ductile failure in lamellar grains. It is suggested that the retained as-cast lamellar structure has a detrimental effect on tensile property, especially elongation. This is mainly related to the strong anisotropic flow stress behavior of the $\alpha2/\gamma$ lamellar [2].

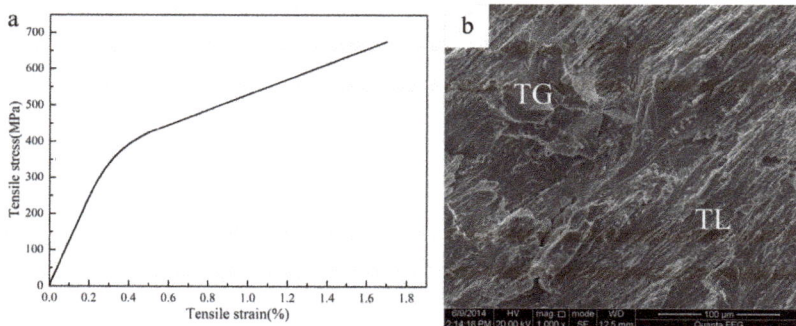

Figure 8. Tensile test stress-strain curve obtained at room temperature (**a**) and fracture surface (**b**) of as-cast TiAl specimen, transgranular (TG) and translamellar (TL).

3.4. Numerical Simulation for Full-Size Casting

From simulation results and the microstructure of the test casting, the thickness of the disk is too small to be compared with its diameter. Hence, the mold is hardly full with molten alloy in a gravity cast. However, when a centrifugal force is employed, the molten alloy can be easily torn in the flat disk. In this case, the advantage of centrifugal force is weakened. Hence, the main ideas of improvement with respect to the runner system are to provide a sufficient molten alloy to fill the mold and keep the fluid stable. Therefore, it is decided to moderately increase the gate size, as shown in Figure 9.

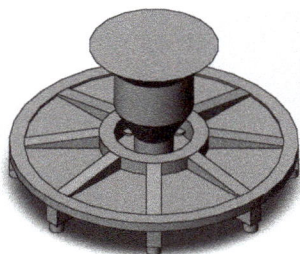

Figure 9. Three-dimensional drawing of runner system for full-size casting.

The simulation results obtained for the test casting show that mold temperature is not a factor that decides the quality of casting below 800 °C. On the other hand, high mold temperature causes a heavy interfacial reaction [13], hence the preheat temperature of the mold for the full-size casting decreases.

Figure 10 shows the shrinkage porosity in the full-size castings: casting defects in the optimized design significantly decrease more when compared to those in the test casting. Because of the enlarged filling gate, the centrifugal effects exhibit dramatic improvement of the mold filling (Figure 10b–e).

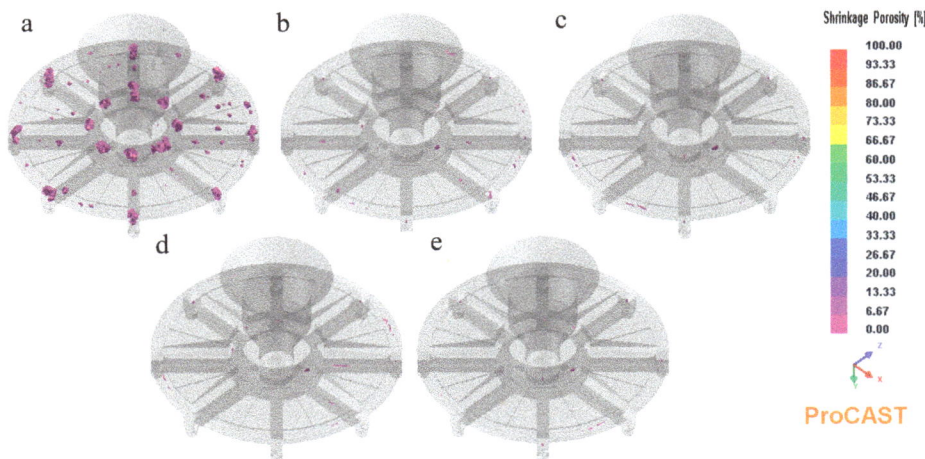

Figure 10. The predicted shrinkage porosity of full-size castings, (**a**) mold temperature of 600 °C and gravity casting (short for 600 °C, 0 rpm); (**b**) 200 °C, 200 rpm; (**c**) 400 °C, 200 rpm; (**d**) 600 °C, 200 rpm; (**e**) 600 °C, 400 rpm.

Shrinkage porosity is the result of failure of feeding to operate effectively. In this case, centrifugal force can effectively decrease shrinkage porosity; however, it would produce voids, which are caused by gas trapped in the liquid alloy. Figure 11 shows the voids of the full-size castings. The 10-mm-thick disk is the required component; hence, the voids in the rib and platform are acceptable. The simulation data show that a mold temperature of 600 °C and rotation speed of 200 rpm are the best choice for the full-size casting. Although it is only slightly better than the other three plans (Figure 11b–e), considering that there are some harmful effects not included in the simulation such as sand hoppers, this marginal advantage is still worth consideration.

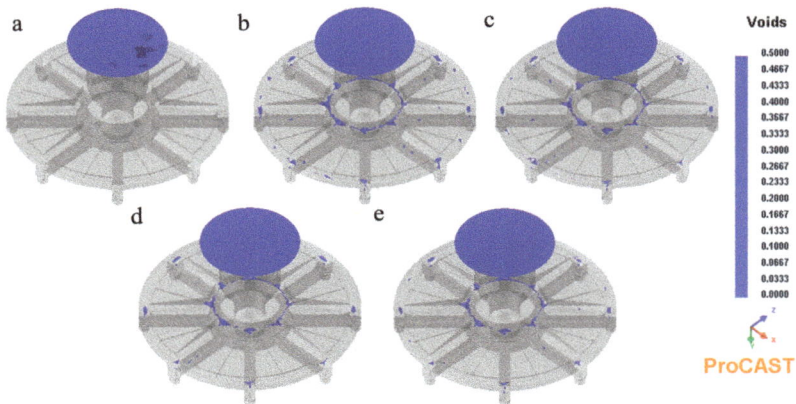

Figure 11. Predicted voids of full-size castings, (**a–e**) the same as the Figure 10.

As shown in Figure 12a, few pores are observed in the rib, and the surface of the disk is defect-free. Radiography revealed major internal defects present (Figure 12b). As shown in Figure 12c, some micro-porosity is also observed in the disk. The pores exhibit a round, smooth edge; hence, they are blowholes. At a mold temperature of 600 °C and rotation speed of 200 rpm, marginal shrinkage porosity and voids are observed in the simulation (Figures 10d and 11d). We speculate that their presence is attributed to the fact that the permeability of the mold is not sufficient. There is an ideal condition for ProCAST, but not for the experiment. There is also an indication of shrinkage in the junction between the rib and the outer ring of the disk (Figure 12d), which is a hot spot because storage is present under it.

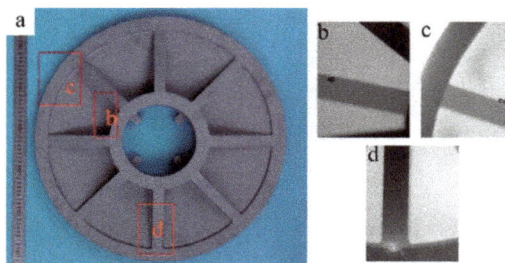

Figure 12. X-ray nondestructive inspection results of full-size casting, (**a**) the grating casting and (**b–d**) correspond to b, c and d areas on (**a**), respectively.

4. Conclusions

The mold-filling and solidification process of TiAl grating by centrifugal investment casting were simulated. The following principal conclusions were drawn from this study:

(1)   Gratings with diameters of 400 and 580 mm were successfully produced.

(2)   The casting parameters for the test casting were a pouring temperature of 1700 °C, a mold preheated temperature of 800 °C, and a rotation speed of 400 rpm.

(3)   The optimal casting parameters for full-size casting were a pouring temperature of 1700 °C, a mold preheated temperature of 600 °C, and a rotation speed of 200 rpm.

(4)   The specimens showed a typical fully lamellar microstructure, which exhibits finely segregated γ-grains. TiAl as-cast specimens exhibited a moderate mechanical property. At room temperature, the tensile strength and elongation were about 675 MPa and 1.7%, respectively.

Acknowledgments: This research was financially supported by the National Natural Science Foundation of China (No. 51001040) and the National Natural Science Foundation of China (No. 51371064).

Author Contributions: Yi Jia and Yuyong Chen conceived and designed the experiments; Yi Jia and Shulong Xiao performed the experiments; Yi Jia, Lijuan Xu and Jing Tian analyzed the data; Yi Jia wrote the paper.

Conflicts of Interest: The authors declare no conflict of interest.

References

1.   Kim, Y.W. Gamma-titanium aluminides: Their status and future. *J. Miner.* **1995**, *47*, 39–41. [CrossRef]

2.   Yamaguchi, M.; Inui, H.; Ito, K. High-temperature structural intermetallics. *Acta Mater.* **2000**, *48*, 307–322. [CrossRef]

3.   Varin, R.A.; Gao, Q. The effect of chromium on the microstructure and micromechanical properties of TiAl-base alloys. *Mater. Manuf. Process.* **1996**, *11*, 381–410. [CrossRef]

4.   Kuang, J.P.; Harding, R.A.; Campbell, J. Microstructures and properties of investment castings of γ-titanium aluminide. *Mater. Sci. Eng. A* **2002**, *329*, 31–37. [CrossRef]

5.   Gomes, F.; Barbosa, J.; Ribeiro, C.S. Induction melting of γ-TiAl in CaO crucibles. *Intermetallics* **2008**, *16*, 1292–1297. [CrossRef]

6.   Tsukihashi, F.; Tawara, E.; Hatta, T. Thermodynamics of calcium and oxygen in molten titanium and titanium-aluminum alloy. *Metall. Mater. Trans. B* **1996**, *27*, 967–972. [CrossRef]

7.   Barbosa, J.; Ribeiro, C.S.; Monteiro, A.C. Influence of superheating on casting of γ-TiAl. *Intermetallics* **2007**, *15*, 945–955. [CrossRef]

8.   Kuang, J.P.; Harding, R.A.; Campbell, J. Investigation into refractories as crucible and mould materials for melting and casting γ-TiAl alloys. *Mater. Sci. Technol.* **2000**, *16*, 1007–1016. [CrossRef]

9.   Jia, Q.; Cui, Y.Y.; Yang, R. Intensified interfacial reactions between γ-titanium aluminide and CaO stabilised ZrO$_2$. *Int. J. Cast Met. Res.* **2004**, *17*, 23–27. [CrossRef]

10.   Nowak, R.; Lanata, T.; Sobczak, N.; Ricci, E.; Giuranno, D.; Novakovic, R.; Holland-Moritz, D.; Egry, I. Surface tension of γ-TiAl-based alloys. *J. Mater. Sci.* **2010**, *45*, 1993–2001. [CrossRef]

11.   Cui, R.J.; Gao, M.; Zhang, H.; Gong, S.K. Interactions between TiAl alloys and yttria refractory material in casting process. *J. Mater. Process. Technol.* **2010**, *210*, 1190–1196.

12.   Teodoro, O.; Barbosa, J.; Naia, M.D.; Moutinho, A.M.C. Effect of low level contamination on TiAl alloys studied by SIMS. *Appl. Surf. Sci.* **2004**, *231*, 854–858. [CrossRef]

13.   Sung, S.Y.; Kim, Y.J. Modeling of titanium aluminides turbo-charger casting. *Intermetallics* **2007**, *15*, 468–474. [CrossRef]

14.   Fu, P.X.; Kang, X.H.; Ma, Y.C.; Liu, K.; Li, D.Z.; Li, Y.Y. Centrifugal casting of TiAl exhaust valves. *Intermetallics* **2008**, *16*, 130–138. [CrossRef]

15.   Yang, R.; Cui, Y.Y.; Dong, L.M.; Jia, Q. Alloy development and shell mould casting of γ-TiAl. *J. Mater. Process. Technol.* **2003**, *135*, 179–188. [CrossRef]

Article

Effect of Laser Surface Treatment on the Corrosion Behavior of FeCrAl-Coated TZM Alloy

Jeong-Min Kim [1,*], Tae-Hyung Ha [1], Joon-Sik Park [1] and Hyun-Gil Kim [2]

[1] Department of Advanced Materials Engineering, Hanbat National University, 125 Dongseo-daero, Yuseong-gu, Daejeon 34158, Republic of Korea; htman15@naver.com (T.-H.H.); jsphb@hanbat.ac.kr (J.-S.P.)
[2] Light Water Reactor Fuel Technology Division, Korea Atomic Energy Research Institute, 989-111 Daedeok-daero, Yuseong-gu, Daejeon 34057, Republic of Korea; hgkim@kaeri.re.kr
* Correspondence: jmk7475@hanbat.ac.kr; Tel.: +82-42-821-1235; Fax: +82-42-821-1592

Academic Editor: Ana Sofia Ramos
Received: 6 December 2015; Accepted: 27 January 2016; Published: 29 January 2016

Abstract: The current study involves the coating of Titanium-Zirconium-Molybdenum (TZM) alloy with FeCrAl through plasma thermal spraying which proved effective in improving the oxidation resistance of the substrate. A post-laser surface melting treatment further enhanced the surface protection of the TZM alloy. Oxidation tests conducted at 1100 °C in air indicated that some Mo oxides were formed at the surface but a relatively small amount of weight reduction was observed for FeCrAl-coated TZM alloys up to 60 min of treatment. The post-laser surface treatment following the plasma thermal spray process apparently delayed the severe oxidation process and surface spalling of the alloy. It was suggested that the slow reduction in weight in the post-laser-treated specimen was related to fewer defects in the coating layer. It was also found that a surface reaction layer formed through the diffusion of Fe into the Mo alloy substrate at high temperature. The layer mainly consisted of Fe-saturated Mo and FeMo intermetallic compounds. In order to observe the corrosion behavior of the laser-treated alloy in 3.5% NaCl solution, electrochemical characteristics were also investigated. A proposed equivalent circuit model for the specimen indicated localized corrosion of coated alloy with some permeable defects in the coating layer.

Keywords: Mo alloy; FeCrAl; plasma thermal spray; laser treatment

1. Introduction

Due to excellent high temperature properties, Mo alloys are used in a large number of elevated temperature applications in the aerospace and nuclear industries [1]. However, a primary drawback of these alloys is susceptibility to high temperature oxidation because volatile MoO_3 is formed in the oxidizing environment at a high temperature such as 873 °C [2]. Therefore, some protective coatings are often necessary to protect the surface of Mo alloys from oxidation [3–5]. To this end, FeCrAl alloy can be applied as a coating layer to the surface of Mo alloys since it can withstand a temperature as high as 1400 °C [6,7]. The thermal spraying method has been sucessfully employed to improve the oxidation resistance of substrate alloys, and the process has several advantages, such as low cost and simple operation [4,5]. However, thermal spray techniques are often implicated in introducing defects in the coating, such as pores and weak interfacial bondings [7].

Post-treatments such as laser surface melting following the thermal spray can be useful to enhace the interface stability and resultant oxidation resistance. It has been reported that relatively high porosity, as well as poor adhesion in various coating/substrate systems, could be modified by such post-treatments. The laser surface melting process could significantly enhance the adhesion of the metallic coating layer to a substrate and reduce microdefects in the coating [8–13].

In the current study, we primarily focused on increasing the stability of the protective FeCrAl coating layer by at least a few hours at high temperatures above 1000 °C, which would be highly significant from an industrial standpoint. Even if the treatment time at high temperatures is not long, some interdiffusion occurs between the coating and the substrate. Such a diffusion process may cause the formation of MoFe intermetallic compounds, thereby resulting in a change in the microstructure of the FeCrAl/Mo system at the high temperature. This issue has also been addressed in the current work. Furthermore, the general corrosion behavior of FeCrAl-coated Mo alloy after the laser treatment has been also investigated.

2. Experimental Section

In the current research TZM Mo alloy was used as the substrate. FeCrAl alloy with a composition of Fe-20 wt. % Cr-5 wt. % Al was deposited on the surface of sand-blasted and ultrasonic-cleaned TZM substrate via air plasma thermal spray process. A mixture of Ar and H_2 was used as the plasma gas and argon was used as the powder carrier gas. The plasma thermal spraying process parameters employed in this study are as follows: 400 A, Ar gas pressure of 100 MPa, H_2 gas pressure of 6 MPa, and spray distance of 100 mm. Some of the as-sprayed FeCrAl coatings were surface-treated by using a continuous wave (CW) diode laser with a maximum power of 300 W (PF-1500F model; HBL Co., Daejeon, Korea) and a powder supplier (Pwp14Y04K model; Yesystem Co., Daejeon, Korea). The laser process parameters, such as laser power, scanning speed, and powder injection, were set up based on a previous work [14]. The optimized applied power for the current research was 180 W, the scanning speed was 8 mm/s, and powder injection rate was 6.5 g/min. To prevent specimens from oxidation during the process, argon was continuously supplied into the melting zone.

Oxidation behaviors of FeCrAl-coated specimens were investigated in air at 1100 °C. The specimens were 10 × 10 × 7 mm in size and were isothermally heated in an electric resistance furnace (PCAM Korea Co., Daejeon, Korea). The oxidation resistance was evaluated by measuring the average weight changes with respect to the holding time for the oxidized specimens. In order to observe the microstructural changes of the laser-treated coating layer during the high temperature oxidation, the coating specimen was isothermally heated at 1100 °C for 5 h under argon atmosphere. Microstructural analyses were performed using a scanning electron microscope (SEM, JEOL, JSM-5610, Tokyo, Japan), equipped with energy dispersive X-ray spectrometer (EDS, JEOL, Tokyo, Japan), and X-ray diffractometer (XRD, Rigaku, Smartlab, Tokyo, Japan). Electrochemical measurements were performed using a potentiostat and an electro impedance spectroscopy (EIS) analysis software (ZIVE SP1, WONATECH, Seoul, Korea). Saturated calomel electrode was used as the reference electrode and platinum plate was used as the counter-electrode. NaCl solution (3.5 wt. %) was used as the electrolyte and the potential scan rate for polarization tests was 1 mV/s.

3. Results and Discussion

3.1. Oxidation Resistance of FeCrAl-Coated TZM Alloy at 1100 °C

Figure 1 shows cross-sectional SEM images of plasma thermal sprayed FeCrAl coatings on TZM alloys before and after the laser surface treatment. Comparatively sound coating layers could be observed in both the as-sprayed and the post-laser treated specimens. A somewhat nonuniform thickness of the FeCrAl coating layer was found after the laser surface melting; however, the interface appeared to be a little more compact in that case. The darker area, which was aluminum-rich based on the SEM-EDS analyses, seen in the FeCrAl coating layers with and without the post-treatment, implied that some segregation occurred during the coating processes. Although the area looks a little darker in the as-sprayed specimen, the darker area was found also to be aluminum-rich.

The oxidation resistance of the FeCrAl-coated specimens at 1100 °C was measured as the percent weight loss with respect to the reference, as compared in Figure 2. It appears that the coating layers on the surface of the Mo alloys were consistently detached from the substrates, especially after a long exposure time. Initially, the amount of weight reduction for both the specimens remained small up to 60 min, after which significant weight loss was observed. This indicated that the FeCrAl coatings, regardless of preparation route, clearly protected the substrate because significant oxidation and spalling of the substrate TZM alloys without any coating are known to occur almost instantly at high temperatures [15]. In our samples, the oxidation behavior appeared to be dependent on whether the post-laser treatment was conducted or not for long exposure times. In comparison with the plasma thermal spray, the post-laser surface treatment following the spray showed a slower weight reduction rate. When the FeCrAl-coated Mo alloys are exposed to a high temperature, for instance 1100 °C, the initial oxidation will take place mainly through the defect areas, such as pores and cracks in the coating layers. Therefore, it can be concluded that the whole oxidation process was relatively slower in the laser-treated specimen with reduced porosity in the FeCrAl coating. Another reason can be related to the bonding characteristics of the interface. The interface between the as-sprayed coating and the substrate was mechanically bonded, while the interfacial bonding has been reported to become stronger metallurgically after the laser treatment [8]. Quite a few research results showed that laser treatments reduced the porosity and strengthened the coating adhesion in many coating/substrate systems [8–13].

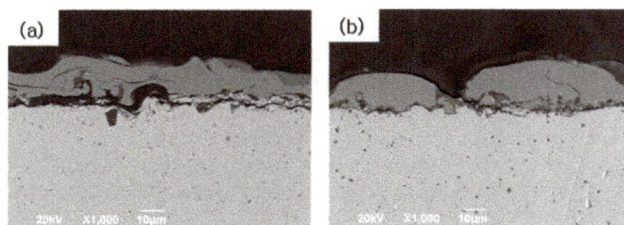

Figure 1. Cross-sectional microstructures of FeCrAl plasma thermal spray-coated TZM alloy: (**a**) as-sprayed; (**b**) after the laser surface melting process.

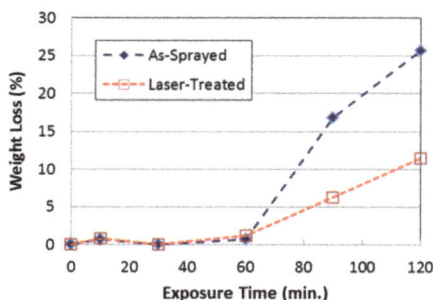

Figure 2. Oxidation behaviors of FeCrAl plasma thermal spray-coated TZM alloys before and after the laser surface treatment.

Figure 3 shows the typical surface appearance of oxidized specimens prepared with and without the post-laser surface treatment. Severely damaged surfaces with a lot of MoO_3 phases were observed in the as-sprayed samples, while a relatively smooth surface, except for some small portions of damaged areas, was present in the post-laser-treated specimen. These observations implied that the oxidation of Mo substrate proceeded preferentially through the defect areas of the coating layer. If Mo oxides are formed on the surface of a substrate, the oxidized layer will expand and volatilize. Once some cracks are formed at the coating/substrate interface, the oxidation process becomes more facilitated.

The XRD analysis results (Figure 4) showed various oxides, such as MoO_3, Fe_2O_3, and Cr_2O_3. Cr and Al oxides are generally known to protect metallic materials effectively from oxidation [16]. Al oxides were not clearly found, possibly because of the relatively small amount of Al additions. It is noteworthy that MoO_3 phases were also formed on the surface of the laser-treated specimen even though its surface morphology appeared to be quite different from that of the as-sprayed ones. Since fewer defect areas were observed in the coating layer on the surface of the laser-treated alloy, it could be harder for oxygen to diffuse through the coating layer. However, the laser-treated coating layers were also found to contain some defects.

Figure 3. Surface morphologies of FeCrAl plasma thermal spray-coated TZM alloys: (**a,b**) as-sprayed; (**c,d**) after the laser treatment.

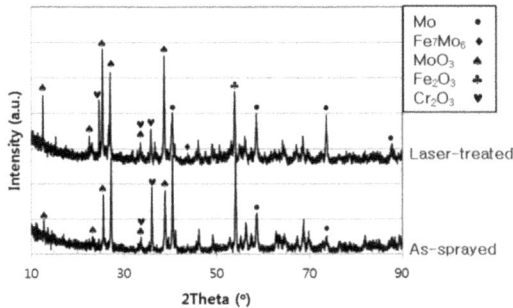

Figure 4. XRD analysis results of surfaces of oxidation test specimens.

3.2. Microstructural Variations of FeCrAl-Coated TZM Alloy at 1100 °C

The microstructure of the coating layer is expected to change when it is exposed to high temperatures which in turn should affect the oxidation resistance of FeCrAl-coated alloy. As previously mentioned, the initial oxidation process depended mainly on the defect areas of the coating layers. However, after a certain period of time, the Mo substrate would become exposed to air because of not only the severe spalling of surface coating layers that occurs but also the interdiffusion that takes place between the Mo alloy substrate and the FeCrAl coating layer. Then Mo oxides are formed in large quantities, resulting in severely damaged surfaces and spalling. In order to understand the microstructural changes of the FeCrAl coating layers during high temperature oxidation, the laser-treated specimen was isothermally heated at the same temperature (1100 °C) under Ar atmosphere.

As shown in Figure 5, some interfacial reactions occurred in the isothermally heated specimen. Diffusion of iron into the Mo matrix apparently occurred during the isothermal heating at 1100 °C, and SEM-EDS analyses indicated that the reaction layer consisted of mainly two different areas: low Fe and high Fe regions. The dominant low Fe area is represented as the Mo phase containing about 10% Fe. The maximum solubility of Fe in Mo is about 31 at. % at 1611 °C, but it is sharply reduced with a decrease in temperature and the solubility at 1100 °C is about 10%. Some MoFe intermetallic compounds might have formed from the supersaturated Mo matrix either during the heating or the cooling. It has been reported that μ-phase and σ-phase can form between the Fe coating layer and the Mo substrate [17]. Since the isothermal heating was carried out at 1100 °C, the high Fe area was postulated to be Fe_7Mo_6 μ-phase. According to the SEM-EDS analysis results it is suggested that Fe diffused into the Mo substrate nonuniformly, mainly through the less dense areas. When Fe exceeded its solubility limit in the Mo matrix at a certain temperature, the Fe_7Mo_6 intermetallic compound was finally formed from the supersaturated matrix. Figure 6 also indicates that the MoFe phases were formed among the low Fe-containing Mo phases. Cr was uniformly distributed in the reacted surface region, which was attributed to the fact that Mo and Cr show miscibility in both solid and liquid states.

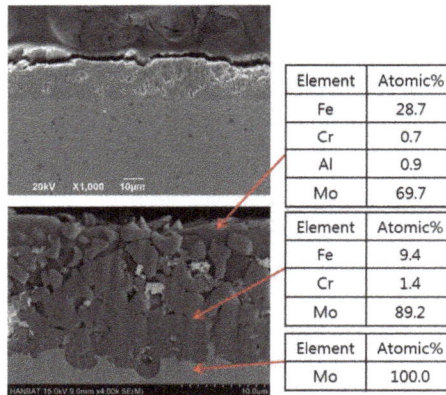

Element	Atomic%
Fe	28.7
Cr	0.7
Al	0.9
Mo	69.7
Element	Atomic%
Fe	9.4
Cr	1.4
Mo	89.2
Element	Atomic%
Mo	100.0

Figure 5. Cross-sectional microstructure of the post-laser-treated TZM alloy that was isothermally heated at 1100 °C.

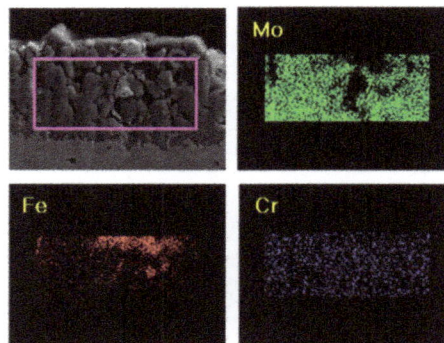

Figure 6. SEM micrograph and EDS mapping of the post-laser-treated TZM alloy that was isothermally heated at 1100 °C.

3.3. Corrosion Behavior of FeCrAl-Coated TZM Alloys in 3.5% NaCl

Although the oxidation behavior of FeCrAl-coated Mo alloys at high temperature is focused on in the present research, the corrosion resistance of the alloys in 3.5% NaCl solution is also industrially important. As shown in Figure 7, the $E_{corr.}$ for the post-laser-treated specimen was a little more noble than that for the as-sprayed specimen. The corrosion current density for both specimens was found to be similar, even though the accuracy may not be high due to a lack of linearity in the plots. Because of the lower corrosion potential and the similar corrosion current density of the as-sprayed specimen obtained from the curves, a little lower corrosion resistance can be considered for the as-sprayed than the post-laser-treated specimen.

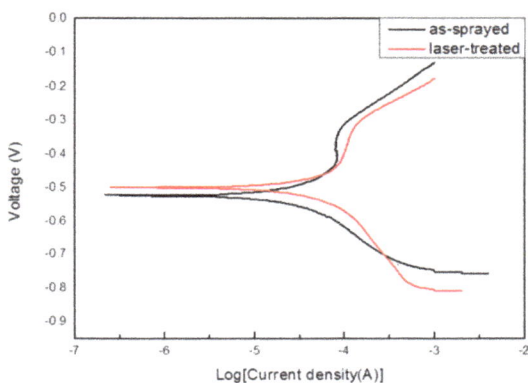

Figure 7. Potentiodynamic polarization scans for FeCrAl plasma spray-coated Mo alloys in 3.5% NaCl solution with and without the post-laser treatment.

These polarization curves were obtained after seven days' immersion in 3.5% NaCl solution and some variations in the potential were observed. The variations in potential were reported to be due to irregular repassivation and depassivation processes [18]. Since $E_{corr.}$ for the substrate TZM alloy was measured at about -443 mV, the substrate was apparently more noble than the FeCrAl coating layers, regardless of whether the post-laser treatment was perfomed or not. This means that the substrate itself will be protected even if it is exposed to the solution through defect areas in the coating layers.

Figure 8 indicates a proposed equivalent circuit model for the laser-treated specimen immersed in 3.5% NaCl solution for seven days. An equivalent model typically represents localized corrosion of the coated alloy containing some permeable defects in the coating layer [19–21]. Rs corresponds to the solution resistance of the electrolyte between the reference and the working electrodes. $Q1$ is the coating capacitance and $R1$ is the pore electrical resistance to the ionic current through the pores. $Q2$ and $R2$ represent the polarization resistance and capacitance of the Mo alloy, respectively. A constant phase element (CPE) is often known to be a practical way to describe the aforementioned capacitance [19]. The values of the parameters for the equivalent circuit in Figure 8 are listed in Table 1.

Table 1. Electrochemical parameters obtained from EIS spectra of the post-laser-treated TZM alloy immersed in 3.5% NaCl solution for seven days.

Parameters	$R_s (\Omega \cdot cm^2)$	CPE1 $(\mu F \cdot cm^{-2})$	$R1 (\Omega \cdot cm^2)$	CPE2 $(\mu F \cdot cm^{-2})$	$R2 (\Omega \cdot cm^2)$
Values	2.56	0.038	6.86	2.52	6.38×10^9

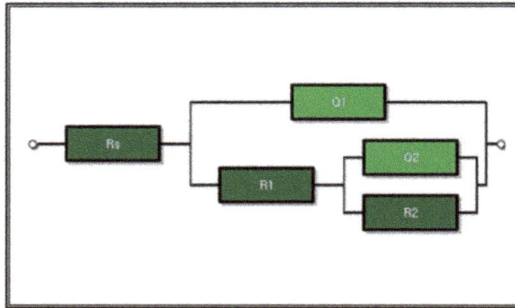

Figure 8. Equivalent circuit model representing the corrosion behavior of the post-laser-treated TZM alloy immersed in 3.5% NaCl solution for seven days.

4. Conclusions

In order to enhance the oxidation resistance of TZM alloy at high temperatures above 1000 °C, FeCrAl coating was deposited on TZM alloy with and without the post-laser surface melting process, and the influence of the laser treatment on the corrosion behavior was investigated. The oxidation tests conducted at 1100 °C in air indicated that oxidation and spalling took place even for the FeCrAl-coated specimens, regardless of whether the post-laser treatment was executed or not, at high temperature. However, the laser process significantly reduced the overall rate of weight loss due to spalling of the coating layer from the substrate.

Meanwhile, diffusion of coating elements occurred from the laser-treated coating layer into the Mo substrate at high temperature, resulting in a reaction layer near the surface. The surface reaction layer mainly consisted of the Fe-saturated Mo phase and the Fe_7Mo_6 intermetallic compound. Electrochemical corrosion potentials and current densities for both the as-sprayed and the laser-treated specimens in 3.5% NaCl solution were observed to be similar. Finally, an equivalent model for the laser-treated specimen implied that some permeable defects existed in the coating layer.

Acknowledgments: Acknowledgments: This work was supported by the National Research Foundation of Korea (NRF) grant funded by Korea government (MSIP) (NRF-2012M2A8A5025822).

Author Contributions: Author Contributions: J.-M.K. designed the research and wrote the manuscript with help from the other authors; T.-H.H. performed the experiments; J.-M.K., J.-S.P. and H.-G.K. analyzed the data.

Conflicts of Interest: The authors declare no conflict of interest.

References

1. Alur, A.P.; Chollacoop, N.; Kumar, K.S. High-temperature compression behavior of Mo-Si-B alloys. *Acta Mater.* **2004**, *52*, 5571–5587. [CrossRef]
2. Smolik, G.R.; Petti, D.A.; Schuetz, S.T. Oxidation and volatilization of TZM alloy in air. *J. Nucl. Mater.* **2000**, *283*, 1458–1462. [CrossRef]
3. Park, J.S.; Kim, J.M.; Cho, S.H.; Son, Y.I.; Kim, D. Oxidation of MoSi$_2$-coated and uncoated TZM (Mo-0.5Ti-0.1Zr-0.02C) alloys under high temperature plasma flame. *Mater. Trans.* **2013**, *54*, 1517–1523. [CrossRef]
4. Chakraborty, S.P. Studies on the development of TZM alloy by aluminothermic coreduction process and formation of protective coating over the alloy by plasma spray technique. *Int. J. Refract. Met. Hard Mater.* **2011**, *29*, 623–630. [CrossRef]
5. Wang, Y.; Wang, D.; Yan, J.; Sun, A. Preparation and characterization of molybdenum disilicide coating on molybdenum substrate by air plasma spraying. *Appl. Surf. Sci.* **2013**, *284*, 881–888. [CrossRef]

6. Naumenko, D.; Le-Coze, J.; Wessel, E.; Fischer, W.; Quadakkers, W.J. Effect of trace amounts of carbon and nitrogen on the high temperature oxidation resistance of high purity FeCrAl alloys. *Mater. Trans.* **2002**, *43*, 168–172. [CrossRef]

7. Hao, S.; Zhao, L.; He, D. Surface microstructure and high temperature corrosion resistance of arc-sprayed FeCrAl coating irradiated by high current pulsed electron beam. *Nucl. Instrum. Methods B Beam Interact. Mater. Atoms* **2013**, *312*, 97–103. [CrossRef]

8. Li, C.; Wang, Y.; Wang, S.; Guo, L. Laser surface remelting of plasma-sprayed nanostructured Al_2O_3-13 wt. % TiO_2 coatings on magnesium alloy. *J. Alloys Compd.* **2010**, *503*, 127–132. [CrossRef]

9. Qian, M.; Li, D.; Liu, S.B.; Gong, S.L. Corrosion performance of laser-remelted Al-Si coating on magnesium alloy AZ91D. *Corros. Sci.* **2010**, *52*, 3554–3560. [CrossRef]

10. Sova, A.; Grigoriev, S.; Okunkova, A.; Smurov, I. Cold spray deposition of 316L stainless steel coatings on aluminium surface with following laser post-treatment. *Surf. Coat. Technol.* **2013**, *235*, 283–289. [CrossRef]

11. Marrocco, T.; Hussain, T.; McCartney, D.G.; Shipway, P.H. Corrosion performance of laser posttreated cold sprayed titanium coatings. *J. Therm. Spray Technol.* **2011**, *20*, 909–917. [CrossRef]

12. Ciubotariu, C.R.; Frunzaverde, D.; Marginean, G.; Serban, V.A.; Birdeanu, A.V. Optimization of the Laser Remelting Process for HVOF-Sprayed Stellite 6 Wear Resistant Coatings. *Optics Laser Technol.* **2016**, *77*, 98–103. [CrossRef]

13. Gao, Y.; Xiong, J.; Gong, D.; Li, J.; Ding, M. Improvement of Solar Absorbing Property of Ni-Mo Based Thermal Spray Coatings by Laser Surface Treatment. *Vacuum* **2015**, *121*, 64–69. [CrossRef]

14. Kim, H.G.; Kim, I.H.; Jung, Y.I.; Park, D.J.; Park, J.Y.; Koo, Y.H. High-Temperature Oxidation Behavior of Cr-Coated Zirconium. In Proceedings of the LWR Fuel Performance Meeting, Charlotte, NC, USA, 15–19 September 2013; p. 840.

15. Yang, F.; Wang, K.S.; Hu, P.; He, H.C.; Kang, X.Q.; Wang, H.; Liu, R.Z.; Volinsky, A.A. La doping effect on TZM alloy oxidation behavior. *J. Alloys Compd.* **2014**, *593*, 196–201. [CrossRef]

16. Leyens, C. Oxidation and Protection of Titanium Alloys and Titanium Aluminides. In *Titanium and Titanium Alloys*, 1st ed.; Leyens, C., Peters, M., Eds.; Wiley-VCH: Weinheim, Germany, 2003; pp. 187–230.

17. Rajkumar, V.B.; Kumar, K.C.H. Thermodynamic modeling of the Fe-Mo system coupled with experiments and *ab initio* calculations. *J. Alloys Compd.* **2014**, *611*, 303–312. [CrossRef]

18. Verdian, M.M.; Raeissi, K.; Slehi, M. Corrosion performance of HVOF and APS thermally sprayed NiTi intermetallic coatings in 3.5% NaCl solution. *Corros. Sci.* **2010**, *52*, 1052–1059. [CrossRef]

19. Liu, C.; Bi, Q.; Leyland, A.; Matthews, A. An electrochemical impedance spectroscopy study of the corrosion behavior of PVD coated steels in 0.5 N NaCl aqueous solution: Part I. Establishment of equivalent circuits for EIS data modelling. *Corros. Sci.* **2003**, *45*, 1243–1256. [CrossRef]

20. Kim, W.J.; Ahn, S.H.; Kim, H.G.; Kim, J.G.; Ozdemir, I.; Tsunekawa, Y. Corrosion performance of plasma-sprayed cast iron coatings on aluminum alloy for automotive components. *Surf. Coat. Technol.* **2005**, *200*, 1162–1167. [CrossRef]

21. Liu, C.; Bi, Q.; Matthews, A. EIS comparison on corrosion performance of PVD TiN and CrN coated mild steel in 0.5 N NaCl aqueous solution. *Corros. Sci.* **2001**, *43*, 1953–1961. [CrossRef]

metals

MDPI

Article

Simulation and Experimental Investigation for the Homogeneity of Ti$_{49.2}$Ni$_{50.8}$ Alloy Processed by Equal Channel Angular Pressing

Diantao Zhang [1,2], Mohamed Osman [1,2,*], Li Li [1,2], Yufeng Zheng [3] and Yunxiang Tong [1,2,*]

[1] Key Laboratory of Superlight Material and Surface Technology, Ministry of Education, Harbin Engineering University, Harbin 150001, China; zzzzzz9025@sina.com (D.Z.); lili_heu@hrbeu.edu.cn (L.L.)
[2] Center for Biomedical Materials and Engineering, College of Materials Science and Chemical Engineering, Harbin Engineering University, Harbin 150001, China
[3] Department of Materials Science and Engineering, College of Engineering, Peking University, Beijing 100871, China; yfzheng@pku.edu.cn
* Correspondence: drosman1975@gmail.com (M.O.); tongyx@hrbeu.edu.cn (Y.T.); Tel.: +86-451-8251-8173 (Y.T.); Fax: +86-451-8251-8644 (Y.T.)

Academic Editor: Ana Sofia Ramos
Received: 12 January 2016; Accepted: 15 February 2016; Published: 25 February 2016

Abstract: Ti$_{49.2}$Ni$_{50.8}$ shape memory alloy (SMA) was processed by equal channel angular pressing (ECAP) for eight passes at 450 °C. The deformation homogeneity was analyzed on various planes across the thickness by Deform-3D software. Strain standard deviation (SSD) was used to quantify deformation homogeneity. The simulation result shows that the strain homogeneity is optimized by the third pass. Deformation homogeneity of ECAP was analyzed experimentally using microhardness measurements. Experimental results show that the gradual evolution of hardness with increasing numbers of passes existed and the optimum homogeneity was achieved after three passes. This is in good agreement with simulation results.

Keywords: TiNi shape memory alloy; equal channel angular pressing; finite element method; strain homogeneity; microhardness

1. Introduction

TiNi-based shape memory alloys (SMAs) are considered one of the most promising materials for engineering and biomedical applications due to their unique shape memory effect and superelasticity [1,2]. In order to further improve the functional properties of TiNi SMAs, severe plastic deformation (SPD) methods have been employed to refine the microstructure of alloys [3]. The used techniques include high pressure torsion (HPT) [4] and equal channel angular pressing (ECAP) [5]. From an engineering application point of view, ECAP has the advantage of large sample size. Therefore, ECAP processing of TiNi SMAs receives more attention. In 2002, Pushin and his coworkers carried out the first ECAP of TiNi alloys and the grain size was reduced from 50–80 µm to 0.2–0.3 µm [5].

Until now, the principle of grain refinement resulting from ECAP has been well understood [6]. During conventional ECAP processing, an inhomogeneous microstructure may be achieved due to die geometry [7], friction [8] and strain hardenability of material [9]. For TiNi SMAs, martensitic transformation behavior and shape recovery properties are sensitive to microstructure. It has been reported that the transformation temperatures decrease with decreasing of grain size [10–12]. The dislocations introduced during cold working also might suppress martensitic transformation and improve the shape memory effect to some extent [13–15]. Therefore, deformation homogeneity is of critical importance for understanding processing-microstructure relationship and providing guideline on the optimization of shape recovery properties.

Finite element method (FEM) has been regarded as one of the important approaches to understand the deformation behavior and estimate the developed strain in the ECAP process [16–18]. However, to date, no report is available on the deformation homogeneity along different longitudinal planes of TiNi alloys. In the present work, FEM was used to analyze the deformation homogeneity by Deform-3D software. The simulation results were further experimentally validated by microhardness measurements.

2. Simulation Models

Deform-3D Version 6.1 (Scientific Forming Technologies Corporation, Columbus, OH, USA) was used to carry out the simulation of ECAP processing. In order to obtain the data of flow stress, the solution-treated $Ti_{49.2}Ni_{50.8}$ alloy samples was compressed using Gleeble 3500 machine (Dynamic Systems Inc, Poestenkill, NY, USA) at different strain rates. The compress tests were performed at 450 °C, at which most of the ECAP processing of intermetallic TiNi-based SMAs were carried out [4,19].

Figure 1 shows the FEM model for numerical simulation, including pressing ram, billet and ECAP die with $\varphi = 120°$ and $\psi = 60°$. The intersectional angle was selected because it gave the highest strain dispersal uniformity [17]. Route Bc was used since it was the optimum one for producing an ultrafine structure [6]. The billet used for analysis was cylindrical in shape with a diameter of 10 mm and a length of 60 mm. The simulation conditions, including geometry and process parameters, the FEM elements number, meshing method as well as the physical properties of $Ti_{49.2}Ni_{50.8}$ are summarized in Table 1. Figure 2 shows the plane sections taken through the processed ECAP billet in the simulation.

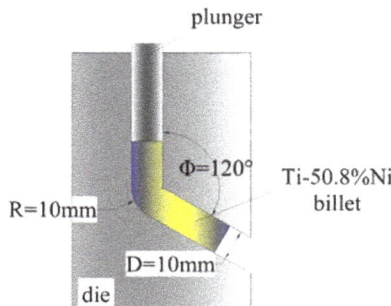

Figure 1. FEM model for ECAP process.

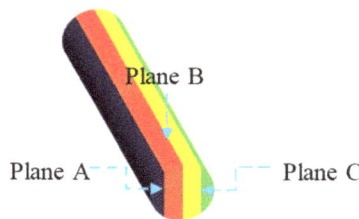

Figure 2. Plane sections taken through the processed ECAP billet in the simulation and experimental work.

Table 1. Simulation parameters and physical properties of $Ti_{49.2}Ni_{50.8}$ sample.

Parameter	Value
Billet length (mm)	60
Billet Diameter (mm)	10
Initial billet temperature (°C)	450
Initial tooling temperature (°C)	450
Temperature range for flow stress (°C)	450
Strain rate range for flow stress (s^{-1})	0.001–10
Punch speed (mm/s)	10
Friction coefficient between Die and billet	0.25
Friction coefficient between Die and ram	0.25
Total number of mesh elements	12,000
Minimum size of mesh element (mm)	0.7
Mesh density type	relative
Relative interference depth	0.7
Density ($g \cdot cm^{-3}$)	6.45

3. Homogeneity Calculation

The degree of strain distribution homogeneity was calculated from the simulation model by a mathematical coefficient called strain standard deviation (SSD) [20]:

$$SSD = \sqrt{\frac{\sum_{i=1}^{n} \left(\varepsilon_i - \varepsilon_{avg} \right)^2}{n}} \tag{1}$$

where ε_i is the plastic strain magnitude in point i and ε_{avg} is the average plastic strain from 300 points at each plane section in the billet. The smaller the SSD value is, the better homogeneity of strain distribution [17,21].

4. Experimental Details

Before processing, the samples were annealed at 850 °C for 1 h, followed by water quenching. The samples were processed at the same parameters as described in the simulation model. The rod was kept at 450 °C for 10 min in a furnace prior to each pass, transferred to the pre-heated ECAP die as quickly as possible and then pressed at a rate of 10 mm/s. A graphitic lubricant was used to reduce the friction effect between the die and sample. The ECAPed billet was cut to the same three planes as in the simulation model (Figure 2) using the low speed diamond saw cutting machine to avoid the possible change of microstructure.

Vickers microhardness test was conducted on the sectioned planes with an applied load of 200 g and a dwell time of 10 s. The values of Vickers hardness (HV) were recorded using a HX-1000TM digital hardness tester (Shanghai Zhaoyi Photonics Co., Ltd, Shanghai, China) equipped with a Vickers diamond indenter. In order to investigate HV hardness distributions, 120 reading at every section were measured, as shown in Figure 3. The three narrow bands represent the different zones in the same plane. The width of the indentation is about 50 μm. In order to ensure the accuracy of the microhardness results, the distance between each test points is at least 300 μm.

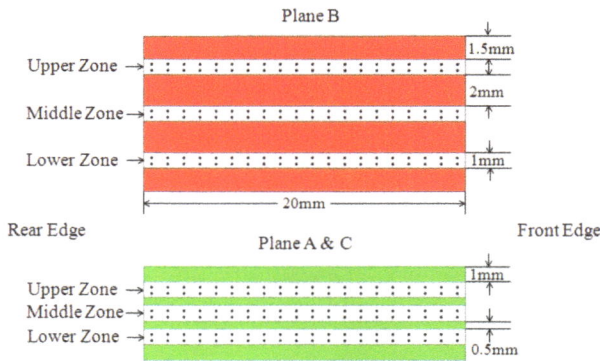

Figure 3. Schematic illustrations of the positions of hardness measurement on planes.

5. Results and Discussion

In order to investigate strain homogeneity at each plane for the processed billet, it is necessary to measure the average strain at every plane. As an example, Figure 4 shows the examined zones in each plane. Hundred points were taken to measure the strain variation at every zone in each plane which is sufficient to track strain variance. The length of examined zone is 20 mm along the deformed sample. The middle zone is the center of plane C. The upper zone and the lower zone are symmetric, and the distance is 0.6 mm from the edge of plane C.

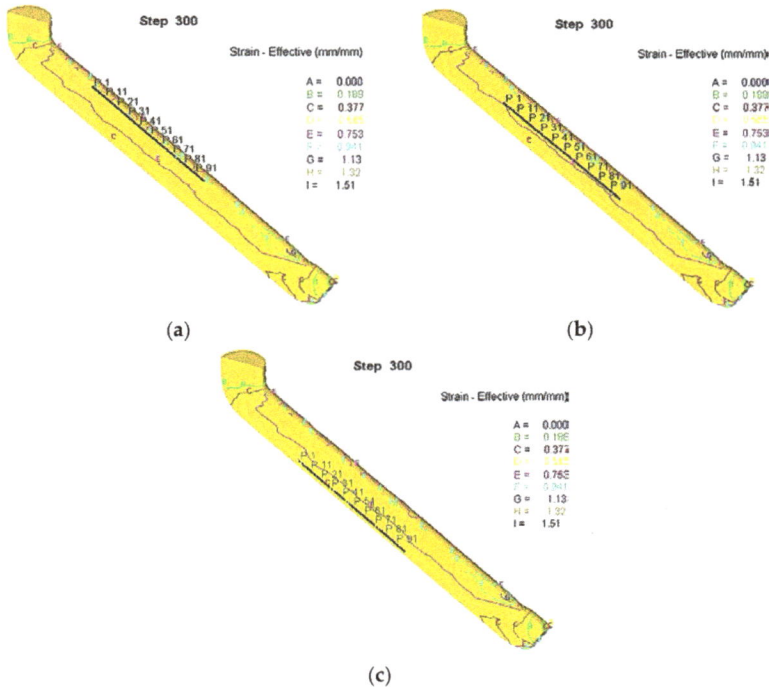

Figure 4. Variable point tracking of effective strain at different positions in plane C: (**a**) upper zone; (**b**) middle zone; and (**c**) lower zone.

The effective strain was calculated using Equation (2)

$$\varepsilon_{eq} = \left[\cfrac{2 \left[\varepsilon_x^2 + \varepsilon_y^2 + \varepsilon_z^2 + \cfrac{\gamma_{xy}^2 + \gamma_{yz}^2 + \gamma_{zx}^2}{2} \right]}{3} \right]^{1/2} \tag{2}$$

where ε_x, ε_y and ε_z are normal strain in x, y and z direction, respectively; and γ_{xy}, γ_{yz} and γ_{zx} are shearing strain for x-y planes, y-z planes and z-x planes, respectively. The obtained effective strain distribution across various planes at 1, 3, 4 and 8 passes are presented in Figure 5. Generally speaking, the average strain increases as the pass number increases. Every plane shows the strain heterogeneity for each pass, but strain variance between planes is reduced to a great extent at the third pass rather than other passes. Figure 6 shows the deformation homogeneity using SSD quantifier as functions of pass number and plane section. Usually, increasing pass number results in an increase of strain distribution uniformity [18]. Thus, it is expected that SSD values will decrease with increasing pass number. SSD measurement shows that homogeneity is optimized at the third pass rather than other passes, which is consistent with the reported results [21]. Next, the microhardness of different planes was measured to experimentally validate the simulation results.

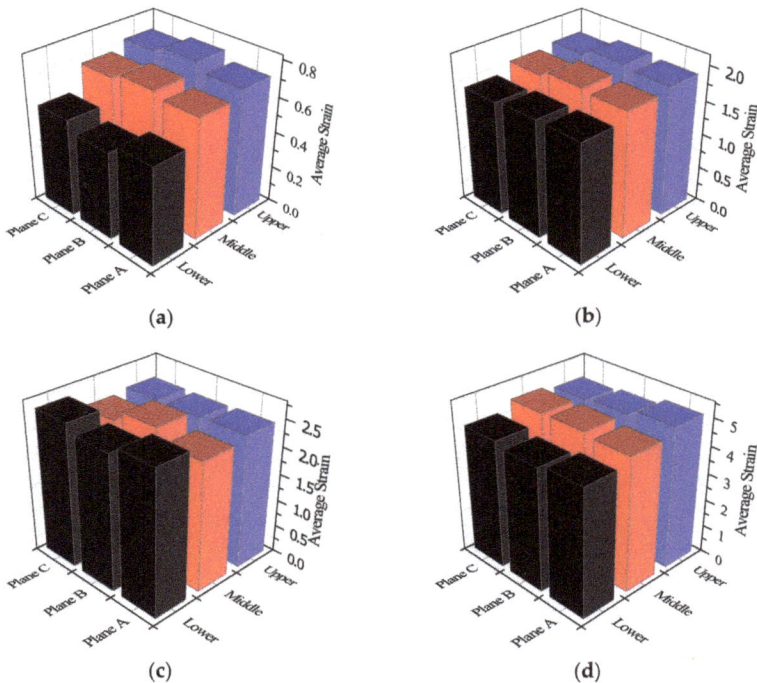

Figure 5. Average strain distribution in the different planes of Ti$_{49.2}$Ni$_{50.8}$ alloy processed by ECAP for different passes: (**a**) 1 pass; (**b**) 3 passes; (**c**) 4 passes; and (**d**) 8 passes.

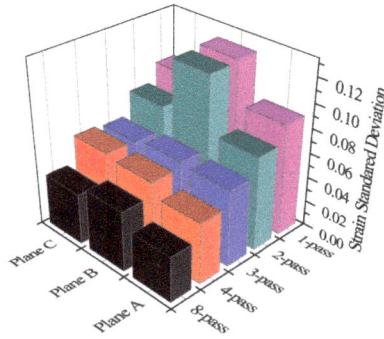

Figure 6. The effect of pass number on SSD for ECAPed Ti$_{49.2}$Ni$_{50.8}$ alloy.

The homogeneity was first analyzed in terms of mapping contours, which were constructed by the software of Tecplot. The microhardness measurements were plotted with the position as shown in Figure 3. The blank region in Figure 3 was filled by the software using interpolation method. Figure 7 shows the individual microhardness plotted against the position on various planes of the solution-treated sample. The individual measurements are plotted in the form of color-coded contour maps to provide a direct and visual representation of the data. The significance of the colors is shown by the color scale given above the drawings. It is seen that solution-treated sample shows a uniform distribution of microhardness irrespective of the position. The average value of microhardness was determined to be 265 HV. This means that the homogeneous microstructure presents in the solution-treated sample. Figure 8 shows the individual microhardness measurements for the ECAPed samples with different pass numbers. It should be noted that the values of HV are plotted within two ranges in incremental step of 10, the first range from 260 to 330 for the samples with one and three passes, and the second range from 300 to 360 for the samples with four and eight passes. These maps are constructed with the vertical axis in the Y direction where $Y = 0$ denotes the lower rear point for each plane along the X-axis and the horizontal axis in the X direction where $X = 0$ and 20 mm mean the rear and front positions of each plane, respectively.

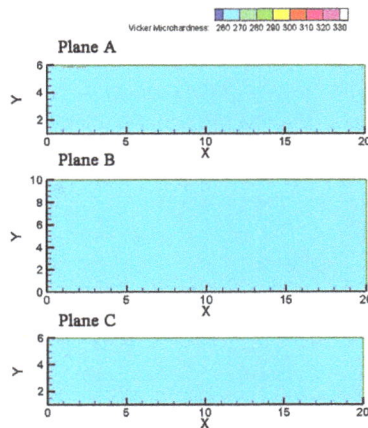

Figure 7. Color-coded contour maps showing Vickers microhardness distributions at solution treated sample before deformation along various planes.

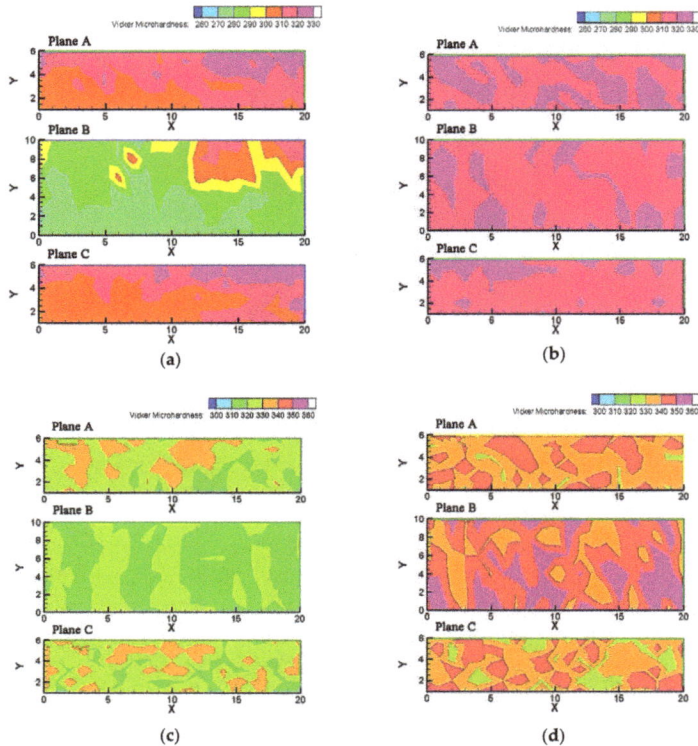

Figure 8. Color-coded contour maps showing Vickers microhardness distributions along various planes for the samples processed after different pass numbers: (**a**) 1 pass; (**b**) 3 passes; (**c**) 4 passes; and (**d**) 8 passes.

Figure 8a shows the contour maps of the sample processed for one pass. Two important features can be observed. First, the microhardness is significantly higher than that of the solution-treated sample. This increase occurs over the entire planes due to the strengthening effect resulting from grain refinement and high-density dislocations introduced during ECAP. Since the sample is axis-symmetric, microhardness distribution of plane A and plane C are nearly the same. Second, an obvious inhomogeneity is revealed. The distribution of microhardness is less homogeneous than that of the solution-treated sample. The microhardness in the vicinity of lower surface are obviously lower than upper surface at each plane. The average hardness of bottom surface is reduced by 15–25 as compared to the remainder region at each plane. This inhomogeneous region is confined to within a width of about 2 mm from the bottom surface. The occurrence of the lower hardness region in the bottom of the sample, which is the characteristics of the ECAP die with outer corner radius, is attributed to the faster flow of the outer part compared with the inner part in the main deformation zone [22–24].

Figure 8b shows the contour maps of microhardness along entire planes after three passes. It is seen that microhardness increases. Furthermore, the homogeneity is improved as compared to that of the sample processed for one pass. Figure 8c,d shows the contour maps of microhardness along various planes for the samples processed after four and eight passes, respectively. The microhardness increases with increasing pass number to 8 passes. This is a general tendency and consistent with the simulation results in which the deformation strain increases with increasing pass number. As compared to the results shown in Figure 8b, the differences between microhardness values at different zones and different planes increase, as evidenced by the different color regions. For example, for the results of the eight-passed sample shown in Figure 8d, there are three kinds of color regions at each plane, respectively, and four kinds of color regions in the sample. This indicates that the inhomogeneous distribution of microhardness becomes more obvious with increasing pass number. In order to provide a clear comparison, Table 2 shows the detailed microhardness values with standard deviation. It is seen that the three-passes sample is more homogeneous among various planes.

Table 2. Average microhardness values at different planes for the solution-treated and ECAP processed samples.

Process	Pass No.	Position at the Billet		
		Plane A	Plane B	Plane C
Solution treated	0	265 ± 2	265 ± 2	265 ± 2
ECAPed	1	313 ± 14	285 ± 20	310 ± 15.5
	3	319 ± 6	317 ± 7.5	318 ± 6.5
	4	326 ± 11	320 ± 7.5	327 ± 11.5
	8	338 ± 10	346 ± 9	337 ± 12

6. Conclusions

Higher accumulative strain is obtained by increasing ECAP pass number, which leads to more homogeneity, but heterogeneity still exists between planes. The simulation results show that the optimum homogeneity can be obtained after three passes at 450 °C, which is experimentally confirmed by the microhardness measurement. The results may provide the guideline on the optimization of microstructure and shape memory properties by ECAP processing.

Acknowledgments: This work was supported by the Fundamental Research Funds for the Central Universities and Harbin City Innovative Talents Research Special Program (2015RAXXJ033).

Author Contributions: All authors contributed equally to this work. Mohamed Osman and Yunxiang Tong conceived and designed the experiments; Diantao Zhang performed the experiments; Li Li analyzed the data; Yufeng Zheng contributed reagents/materials/analysis tools; and Diantao Zhang wrote the paper.

Conflicts of Interest: The authors declare no conflict of interest. The founding sponsors had no role in the design of the study; in the collection, analyses, or interpretation of data; in the writing of the manuscript, and in the decision to publish the results.

Abbreviations

The following abbreviations are used in this manuscript:

ECAP	Equal channel angular pressing
SMA	Shape memory alloy
SSD	Strain standard deviation
FEM	Finite element method
SPD	Severe plastic deformation
HPT	High pressure torsion

References

1. Otsuka, K.; Kakeshita, T. Science and technology of shape-memory alloys: New developments. *MRS Bull.* **2002**, *27*, 91–100. [CrossRef]
2. Otsuka, K.; Ren, X. Physical metallurgy of Ti-Ni-based shape memory alloys. *Prog. Mater. Sci.* **2005**, *50*, 511–678. [CrossRef]
3. Valiev, R.Z.; Mukherjee, A.K. Nanostructures and unique properties in intermetallics, subjected severe plastic deformation. *Scripta Mater.* **2001**, *44*, 1747–1750. [CrossRef]

4. Prokofiev, E.; Gunderov, D.V.; Lukyanov, A.; Pushin, V.; Valiev, R.Z. Mechanical behavior and stress-induced martensitic transformation in nanocrystalline $Ti_{49.4}Ni_{50.6}$ alloy. *Mater. Sci. Forum.* **2008**, *584*, 470–474. [CrossRef]

5. Valiev, R.Z.; Langdon, T.G. Principles of equal-channel angular pressing as a processing tool for grain refinement. *Prog. Mater. Sci.* **2006**, *51*, 881–981. [CrossRef]

6. Waitz, T.; Kazykhanov, V.; Karnthaler, H.P. Martensitic phase transformations in nanocrystalline NiTi studied by TEM. *Acta Mater.* **2004**, *52*, 137–147. [CrossRef]

7. Kočiško, R.; Kvačkaj, T.; Kováčová, A.; Zemko, M. The influence of ECAP geometry on the effective strain distribution. *Adv. Mater. Res.* **2015**, *1127*, 135–141. [CrossRef]

8. Quang, P.; Nghiep, D.M.; Kim, H.S. Simulation of the effective of friction on the deformation in equal channel angular pressing (ECAP). *Key Eng. Mater.* **2015**, *656–657*, 526–531. [CrossRef]

9. Medeiros, N.; Moreira, L.P. Upper-bound analysis of die corner gap formation for strain-hardening materials in ECAP process. *Comput. Mater. Sci.* **2014**, *91*, 350–358. [CrossRef]

10. Waitz, T.; Tsuchiya, K.; Antretter, T.; Fischer, F.D. Phase transformations of nanocrystalline martensitic materials. *MRS Bull.* **2009**, *34*, 814–821. [CrossRef]

11. Tong, Y.X.; Liu, Y.; Miao, J.M.; Zhao, L.C. Characterization of a nanocrystalline NiTiHf high temperature shape memory alloy thin film. *Scripta Mater.* **2005**, *52*, 983–987. [CrossRef]

12. Kuranova, N.N.; Makarov, V.V.; Pushin, V.G.; Uksusnikov, A.N. Thermo- and deformation induced martensitic transformations in binary TiNi-based alloys, subjected severe plastic deformation. *Mater. Sci. Forum.* **2013**, *738–739*, 530–534. [CrossRef]

13. Lin, H.C.; Wu, S.K.; Chou, T.S.; Kao, H.P. The Effects of cold rolling on the martensitic transformation of an equiatomic TiNi alloy. *Acta Metall. Mater.* **1991**, *39*, 2069–2080. [CrossRef]

14. Miller, D.A.; Lagoudas, D.C. Influence of cold work and heat treatment on the shape memory effect and plastic strain development of NiTi. *Mater. Sci. Eng. A* **2001**, *308*, 161–175. [CrossRef]

15. Zhao, L.C.; Zheng, Y.F.; Cai, W. Study of deformation micromechanism in cold-deformed TiNi based alloys. *Intermetallics* **2005**, *13*, 281–288. [CrossRef]

16. Lu, S.K.; Liu, H.Y.; Yu, L.; Jiang, Y.L.; Su, J.H. 3D FEM simulations for the homogeneity of plastic deformation in aluminum alloy HS6061-T6 during ECAP. *Procedia Eng.* **2011**, *12*, 35–40. [CrossRef]

17. Djavanroodi, F.; Omranpour, B.; Ebrahimi, M.; Sedighi, M. Designing of ECAP parameters based on strain distribution uniformity. *Prog. Nat. Sci.* **2012**, *22*, 452–460. [CrossRef]

18. Mahallawy, N.E.; Shehata, F.A.; Hameed, M.A.E.; Aal, M.I.A.E.; Kim, H.S. 3D FEM simulations for the homogeneity of plastic deformation in Al-Cu alloys during ECAP. *Mater. Sci. Eng. A* **2010**, *527*, 1404–1410. [CrossRef]

19. Tong, Y.X.; Guo, B.; Chen, F.; Tian, B.; Li, L.; Zheng, Y.F.; Prokofiev, E.A.; Gunderov, D.V.; Valiev, R.Z. Thermal cycling stability of ultrafine-grained TiNi shape memory alloys processed by equal channel angular pressing. *Scripta Mater.* **2012**, *67*, 1–4. [CrossRef]

20. Zaïri, F.; Aour, B.; Gloaguen, J.M.; Naït-Abdelaziz, M.; Lefebvre, J.M. Numerical modelling of elastic-viscoplastic equal channel angular extrusion process of a polymer. *Comput. Mater. Sci.* **2006**, *38*, 202–216. [CrossRef]

21. Zhang, X.; Hua, L.; Liu, Y. FE simulation and experimental investigation of ZK60 magnesium alloy with different radial diameters processed by equal channel angular pressing. *Mater. Sci. Eng. A* **2012**, *535*, 153–163. [CrossRef]

22. Kim, H.S. Finite element analysis of equal channel angular pressing using a round corner die. *Mater. Sci. Eng. A* **2001**, *315*, 122–128. [CrossRef]

23. Kim, H.S. On the effect of acute angles on deformation homogeneity in equal channel angular pressing. *Mater. Sci. Eng. A* **2006**, *430*, 346–349. [CrossRef]

24. Kim, H.S.; Seo, M.H.; Hong, S.I. On the die corner gap formation in equal channel angular pressing. *Mater. Sci. Eng. A* **2000**, *291*, 86–90. [CrossRef]

metals

MDPI

Article

Structural Origin of the Enhanced Glass-Forming Ability Induced by Microalloying Y in the ZrCuAl Alloy

Gu-Qing Guo, Shi-Yang Wu and Liang Yang *

College of Materials Science and Technology, Nanjing University of Aeronautics and Astronautics,
Nanjing 210016, China; guoguqing@nuaa.edu.cn (G.-Q.G.); shiyangwu0914@gmail.com (S.-Y.W.)
* Correspondence: yangliang@nuaa.edu.cn; Tel.: +86-25-5211-2903; Fax: +86-25-5211-2626

Academic Editor: Ana Sofia Ramos
Received: 16 January 2016; Accepted: 17 March 2016; Published: 23 March 2016

Abstract: In this work, the structural origin of the enhanced glass-forming ability induced by microalloying Y in a ZrCuAl multicomponent system is studied by performing synchrotron radiation experiments combined with simulations. It is revealed that the addition of Y leads to the optimization of local structures, including: (1) more Zr-centered and Y-centered icosahedral-like clusters occur in the microstructure; (2) the atomic packing efficiency inside clusters and the regularity of clusters are both enhanced. These structural optimizations help to stabilize the amorphous structure in the ZrCuAlY system, and lead to a high glass-forming ability (GFA). The present work provides an understanding of GFAs in multicomponent alloys and will shed light on the development of more metallic glasses with high GFAs.

Keywords: metallic glasses; microalloying; glass-forming ability; synchrotron radiation; microstructure; reverse Monte-Carlo simulation

1. Introduction

In principle, liquid cooled as rapidly as possible can transform into a solid having a glassy structure. For the metallic glass (MG) systems, the minimum cooling rate required for glass forming ranges from 10^{-1} K/s up to 10^6 K/s [1], which sets the good glass forming ability (GFA) systems apart from the marginal systems. Searching for optimized compositions with good GFA, particularly in multicomponent MGs, has been attracting a lot of effort over decades, but so far it is still largely based on a strategy of trial and error in a multidimensional composition space without the effective guidance of any general theory [2–9]. It has been realized that the GFA is extremely sensitive to composition [10,11]. Based on this phenomenon, a practical empirical rule called "microalloying" [12,13] has been widely used to facilitate the development of new MGs with improved GFA [14–17]. Nevertheless, the mechanism of the microalloying effect on GFA remains elusive.

Understanding the GFA of amorphous alloys from the structural perspective has been pursued for decades [18–20]. Although drawing the explicit overall structural picture of MGs has not been solved, local atomic and/or cluster structural models have been proposed by building and stacking clusters to fill space efficiently, including the hard-sphere random-packing model [21], the stereochemically designed model [22], the cluster packing model [23], and the quasi-equivalent clusters model [24]. These structural models can statistically describe the so-called short-to-medium range order in MGs with simple compositions very well. Experiments revealed that the microalloying-enhanced GFA of multicomponent alloys usually has a local maximum in a pinpoint composition [11]. This is associated with fine structural changes in such pinpoint composition [25] which however could not been well described by using the existing theoretical structure models. Therefore, further studies are required to address this issue.

In this work, a feasible scheme for addressing this issue is developed by performing a series of state-of-the-art synchrotron radiation-based experiments combined with simulations to investigate the microstructures of amorphous alloys. We choose the representative CuZrAlY quaternary alloy system as a research prototype. In this system, 5 at. % yttrium addition to CuZrAl mother alloy dramatically increases the critical casting size of the alloy from 3 to 10 mm [14], indicating the microalloying Y-induced increase of GFA.

2. Experimental Section

$Cu_{46}Zr_{47}Al_7$ ternary and corresponding $Cu_{46}Zr_{42}Al_7Y_5$ quaternary alloy ingots were prepared by arc melting mixtures of Cu (99.9 wt. %), Zr (99.9 wt. %), Al (99.9 wt. %), and Y (99.9 wt. %) in Ti-gettered high-purity argon atmosphere. Amorphous ribbons with a cross section of 0.04×2 mm^2 were produced from these ingots via single-roller melt spinning at a wheel surface velocity of 40 m/s in purified Ar atmosphere. Firstly, X-ray diffraction (Cu K$_\alpha$, radiation) was performed to confirm the amorphous state of the as-prepared samples. Subsequently, room temperature X-ray diffraction (XRD) measurement was performed using a high-energy synchrotron radiation monochromatic beam (about 100 keV) on beam line BW5 in Hasylab, Germany. Two-dimensional diffraction data was collected by a Mar345 image plate and then was integrated to Q-space by subtracting the background using the program Fit2D [26]. The output data was normalized by PDFgetX software to obtain structure factor $S(Q)$ according to the Faber-Ziman equation [27]. Furthermore, extended X-ray absorption fine structure (EXAFS) measurements for Zr, Cu, and Y K-edge were carried out using transmission mode at beam lines BL14W1 in the Shanghai Synchrotron Radiation Facility of China and U7C in the National Synchrotron Radiation Laboratory of China. Because of the experimentally-inaccessible energy value of the Al K-edge (1.560 keV), EXAFS did not allow the measurement of the local structure around the Al atoms. These EXAFS raw data were normalized via a standard data-reduced procedure, employing the Visual Processing in EXAFS Researches [28].

In order to obtain the atomic structural information as reliably as possible, both the normalized diffraction and EXAFS data were simulated simultaneously under the framework of reverse Monte-Carlo (RMC) [29]. Cubic boxes we used in the RMC simulation contained 40,000 atoms, matching $Cu_{46}Zr_{42}Al_7Y_5$ and $Cu_{46}Zr_{47}Al_7$ compositions. During RMC simulation, atoms move randomly within a determined time interval. The experimental data are compared to the simulation with an iterative calculation [30]:

$$\delta^2 = \frac{1}{\varepsilon^2}\sum_n (S_m(Q_n) - S_{exp}(Q_n))^2 + \frac{1}{\varepsilon_i^2}\sum_n (\chi_{m,El}(\kappa_n) - \chi_{exp,El}(\kappa_n))^2 \tag{1}$$

where δ^2 represents the deviation between the experimental and simulation data, ε parameters regulate the weight of the data set given in the fitting procedure, E_i denotes Ni, Nb, and Zr elements, and $S(Q)$ and $\chi(k)$ parameters are the XRD structural factor in Q space and the EXAFS signal, respectively. The subscripts "m" and "exp" represent the simulations and the experiments, respectively. The theoretical EXAFS signal, $\chi(k)$, of the ith element is calculated from the following equation:

$$\chi_{m,i}(\kappa) = \sum_j 4\pi c_j\rho \int r^2 \gamma_{ij}(r,\kappa)g_{ij}(r)dr \tag{2}$$

where c_j is the concentration of the jth element and γ_{ij} can be calculated by:

$$\gamma_{ij}(r,\kappa) = A_{ij}(\kappa,r)\sin(2\kappa r + \Phi_{ij}(\kappa r)) \tag{3}$$

where A_{ij} and Φ_{ij} denote the amplitude and the phase shift, respectively. They can be obtained by using the FEFF 8.1 code [30].

It should be noted that although during RMC simulation, the simulated experimental data do not include the Al K-edge EXAFS, we can still get reliable structural information, for the following reasons: (1) EXAFS is an element-specific method available for measuring the surroundings of each kind of atom. In other words, all the neighbor atoms around each atom can be distinguished when EXAFS data is fitted or simulated [31]. Since Zr, Cu, and Y EXAFS data can reflect all of their neighborhood information containing Zr, Cu, Y, and Al atoms, how the Al atoms distribute around the Zr (Cu or Y) centers can be determined accordingly; (2) four independent sets of experimental data (one set of XRD and three sets of EXAFS (Zr, Cu, and Y) were simultaneously simulated in this work. During the RMC simulation, all of these experimental data should fit their counterparts calculated theoretically from the same structural model. Such constraint can eliminate the computational randomness. Additionally, the RMC-simulated atomic structural models were further analyzed by the Voronoi tessellation method [32].

3. Results and Discussion

Compared with routine lab experiments (such as X-ray powder diffraction measurement), the synchrotron radiation-based XRD measurements can provide high-resolution data, which are more reliable for detecting the fine structures in materials, in particular in amorphous alloys. The original two-dimensional diffraction patterns of both $Cu_{46}Zr_{42}Al_7Y_5$ and $Cu_{46}Zr_{47}Al_7$ samples are plotted in Figure 1a,b. To obtain more accessible structural information, these two-dimensional patterns are transformed into one-dimensional curves. Figure 1c,d shows the corresponding one-dimensional structural factor. The amorphous nature of both samples can be confirmed because there are no circle lines or dots in the two-dimensional diffraction patterns and no sharp Bragg peaks behind the first strong peak in the structural factor curves, while these features are usually observed in the diffraction data of polycrystals or single crystals [33].

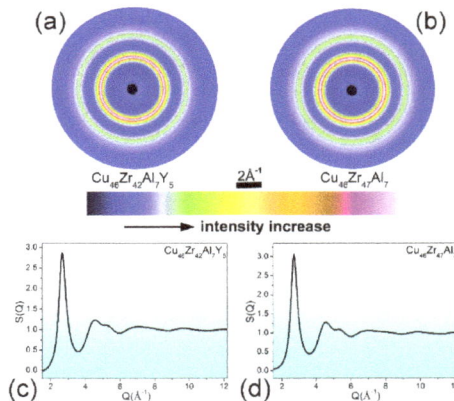

Figure 1. The two-dimensional X-ray diffraction (XRD) patterns in (**a**) $Cu_{46}Zr_{42}Al_7Y_5$ and (**b**) $Cu_{46}Zr_{47}Al_7$ amorphous samples; and the deduced data, including: the structural factors ($S(Q)$s) of (**c**) $Cu_{46}Zr_{42}Al_7Y_5$ and (**d**) $Cu_{46}Zr_{47}Al_7$. To highlight oscillations in low Q region of $S(Q)$, the Q region here was shortened to about 12 Å^{-1}.

Figure 2a–d shows the XRD and EXAFS experimental data as well as their corresponding RMC simulated curves. To ensure the proper interpretation of all the structural information during normalization, the EXAFS data for Zr, Cu, and Y K-edge were weighted by κ^3 values. This does not reduce the reliability of RMC simulation, because all the simulated Zr, Cu, and Y K-edge EXAFS spectra also are strictly weighted by κ^3 values, so that no systematic errors will be generated [28].

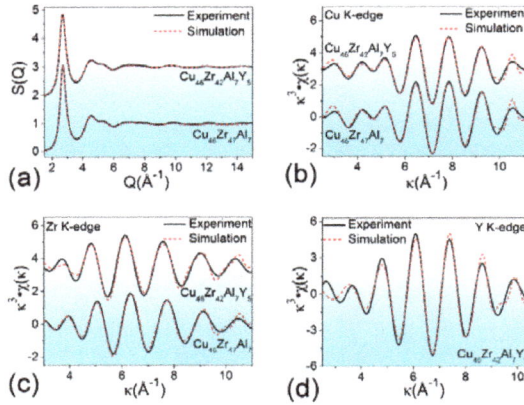

Figure 2. XRD and extended X-ray absorption fine structure (EXAFS) experimental data as well as their corresponding simulated curves, including (**a**) $S(Q)$; (**b**) Cu K-edge, (**c**) Zr K-edge, and (**d**) Y K-edge EXAFS data. Experimental and simulated data are plotted with the solid and the dashed lines, respectively. All the experimental and the simulated signals were weighted by κ^3. To highlight the $S(Q)$ difference between the experimental and the simulated data for both samples, the Q region here was shortened to about 15 Å^{-1}.

The good matching between all the experiment/simulation pairs confirms the success of the RMC simulations. Based on the simulated structural models, atomic-level structural information can be deduced. The coordination numbers (CNs) around Zr, Cu, Al, and Y center atoms, as well as all kinds of atomic-pair distances, are listed in Table 1.

Table 1. The first-shell atomic-pair information deduced from the reverse Monte-Carlo (RMC)-simulated structural models of $Cu_{46}Zr_{42}Al_7Y_5$ and $Cu_{46}Zr_{47}Al_7$ samples, including coordination numbers (CNs) and atomic-pair distances (R). Note that M denotes Cu, Zr, Al, or Y. The Goldschmidt radii of Zr, Cu, Y, and Al are 1.60 Å, 1.28 Å, 1.80 Å, and 1.43 Å, respectively.

Atomic Pairs	$Cu_{46}Zr_{47}Al_7$			$Cu_{46}Zr_{42}Al_7Y_5$		
	R(Å) $\pm$ 0.01	CNs	Uncertainty of CNs	R(Å) $\pm$ 0.01	CN	Uncertainty of CNs
Cu-Cu	2.59	5.0	$\pm$0.2	2.60	5.3	$\pm$0.3
Cu-Zr	2.88	5.4	$\pm$0.3	2.89	4.2	$\pm$0.3
Cu-Al	2.61	0.7	+0.3,−0.1	2.60	0.7	+0.2, −0.1
Cu-Y	-	-	-	2.98	0.9	-
Cu-M	-	11.1	-	-	11.1	-
Zr-Cu	2.88	5.4	$\pm$0.2	2.89	5.5	$\pm$0.2
Zr-Zr	3.20	5.9	$\pm$0.2	3.20	5.3	$\pm$0.3
Zr-Al	2.82	0.8	+0.2,−0.1	2.79	0.7	+0.3,−0.1
Zr-Y	-	-	-	3.54	0.5	+0.1
Zr-M	-	12.1	-	-	12.0	-
Al-Cu	2.61	4.9	$\pm$0.2	2.60	5.5	$\pm$0.3
Al-Zr	2.82	5.7	$\pm$0.3	2.79	4.9	$\pm$0.2
Al-Al	2.70	0.1	+0.1	2.70	0.1	+0.1
Al-Y	-	-	-	3.29	0.2	+0.2
Al-M	-	10.7	-	-	10.7	-
Y-Cu	-	-	-	2.98	7.7	$\pm$0.4
Y-Zr	-	-	-	3.54	4.5	$\pm$0.3
Y-Al	-	-	-	3.29	0.2	+0.2
Y-Y	-	-	-	3.76	0.3	+0.1
Y-M	-	-	-	-	12.7	-

For CN values, it is shown that there are obvious CN decreases in Cu-Zr and Al-Zr atomic pairs from $Zr_{47}Cu_{46}Al_7$ to $Zr_{42}Cu_{46}Y_5Al_7$ (5.4 to 4.2 and 5.7 to 4.9). In previous work, separation between Y and Zr was suggested due to the positive heat of mixing between and Y and Zr [34]. Therefore, Y and Zr atoms are prone to avoid each other, and Cu and Al atoms are expected to favor more Y atoms and less Zr atoms. In addition, it seems that Y addition does not induce a change of the total CNs of Zr, Cu, and Al centers. In both samples, the CN of Zr centers is about 12, while the CNs around Cu and Al centers are only 11 and 10.7, respectively. The CN around Al centers is a relatively small value. In previous work [35], it has been revealed that there is a bond shortening for Al-connected atomic pairs. This leads to fewer neighbor atoms around the Al centers. For atomic-pair distances, Zr-Zr, Zr-Cu, Cu-Cu, and some other atomic pairs have almost the same values in both samples. Nevertheless, it is found that the Zr-Y distance (3.54 Å) is obviously larger than the sum of Zr and Y Goldschmidt radii (1.60 + 1.80 = 3.40 Å), indicating the relatively weak interaction between the Zr-Y pair. From the atomic-level structural information mentioned above, we can conclude that Y doping does tune the local structures, but how Y doping affects the GFA needs to be further studied.

We can obtain cluster-level structural information via the Voronoi-tessellation approach. In other words, Voronoi clusters (VCs) can be extracted from the RMC simulated structural models, and indexed based on their geometrical features. The major VCs centered with Zr, Cu, Y, and Al atoms are plotted in Figure 3a–d. The ideal icosahedral cluster (<0,0,12,0>) and the icosahedral-like VCs (<0,2,8,2>, <0,3,6,3>, and <0,4,4,4>) whose CN are 12 have been proved to be the favorite structural units in the microstructures of MGs [36–38]. As shown in Figure 3, the major Cu-, Zr-, Al-, and Y-centered VCs (whose fractions are larger than 2% are selected) have broad CN distributions ranging from 9 to 13, 11 to 15, 8 to 12, and 10 to 14, respectively. Take Figure 3b for example, the popular Zr-centered VCs are 12-CN ones (<0,2,8,2> and <0,3,6,3>) and 13-CN ones (<0,1,10,2> and <0,3,6,4>), while none of them has a fraction higher than 16%. This indicates that various clusters co-existing to form the microstructure is the intrinsic structural feature in glassy alloys [39], and icosahedral clusters are the preferred but not the only clusters favored in the glassy structure, because only stacking icosahedral clusters can not fill space completely [39]. In addition, the Zr-centered icosahedral-like VCs in both samples have high weights, while fractions of the Cu-centered icosahedral-like counterparts are relatively small, let alone the Al-centered ones. This implies that the Zr-centered icosahedral-like VCs rather than Cu- or Al-centered counterparts contribute to the glass formation in $Cu_{46}Zr_{42}Al_7Y_5$ and $Cu_{46}Zr_{47}Al_7$ compositions. In addition, compared with $Cu_{46}Zr_{47}Al_7$, there are higher fractions of these icosahedral-like VCs (such as <0,2,8,2> and <0,3,6,3>) and lower weights of non-icosahedral VCs (such as <0,1,10,2> and <0,3,6,4>) in $Cu_{46}Zr_{42}Al_7Y_5$. Especially, the Zr-centered icosahedral clusters in $Cu_{46}Zr_{42}Al_7Y_5$ have a total fraction about 30%. This illustrates that when 5 at. % Y is added, Zr-centered icosahedral-like local structures became more popular in the microstructure. Furthermore, it is worth noting that the Y-centered icosahedral-like VCs possess relatively high proportions, which are even higher than the Zr-centered counterparts. Since Y atoms are the substitutes of Zr ones, it implies that the ZrCuAlY quaternary MG contains more icosahedral-like VCs than the corresponding ZrCuAl ternary sample, contributing to the higher GFA in the former.

Changing the configuration of one cluster while keeping its indexed character can apparently change the packing of atoms inside this cluster. In our previous work, it has been pointed out that the atomic packing efficiency inside clusters strongly relates to the GFA in binary alloys [25]. This is consistent with the widely-accepted dense packing principle [40,41]. The atomic packing efficiency can be calculated by

$$APE = V_a/V_u \tag{4}$$

where V_a and V_u denote the volume of atoms contained in clusters and the total volume of clusters themselves, respectively. V_u can be obtained by summing the volumes of all the tetrahedra in VCs because each VC is built by stacking tetrahedra with a shared vertex, located at the site of the VC's center atom. Because each atom embedded in the cluster is truncated as a cone ball, V_a can be calculated by summing the volumes of all the cone balls [25]. Because bond shortening around Al atoms in ZrCuAl MG was found in previous work [35], the atomic radius of the Al atom (1.26 Å) is estimated by

$$r_{Al} = \frac{CN_{Al\text{-}Cu} \times (d_{Cu\text{-}Al} - r_{Cu}) + CN_{Al\text{-}Zr} \times (d_{Zr\text{-}Al} - r_{Zr})}{CN_{Al\text{-}Cu} + CN_{Al\text{-}Zr}} \tag{5}$$

where r_{Al} is the Al atomic radius, CN_{Al-Cu} and CN_{Zr-Cu} stand for the numbers of Zr and Cu atoms around Al centers, respectively. In addition, d_{Cu-Al} and d_{Al-Zr} denote the distances of Al-Cu and Al-Zr pairs, respectively. For Zr, Cu, and Y atoms, because their neighbors are almost Cu and Zr atoms, and there are not strong bondings in Zr-Zr, Zr-Cu, Zr-Y, Cu-Y pairs, so that Zr, Cu, and Y atomic radii are the half of Zr-Zr, Cu-Cu, and Y-Y bond lengths, respectively (*i.e.*, 1.60 Å, 1.30 Å, and 1.78 Å). The average atomic packing efficiency values of both samples are plotted in Figure 4. It is obvious that the atomic packing efficiency of $Cu_{46}Zr_{42}Al_7Y_5$ has a higher value than that of $Cu_{46}Zr_{47}Al_7$. This indicates that Y addition not only changes the geometrical index of clusters, but also tunes the atomic packing inside these clusters, leading to a denser packing.

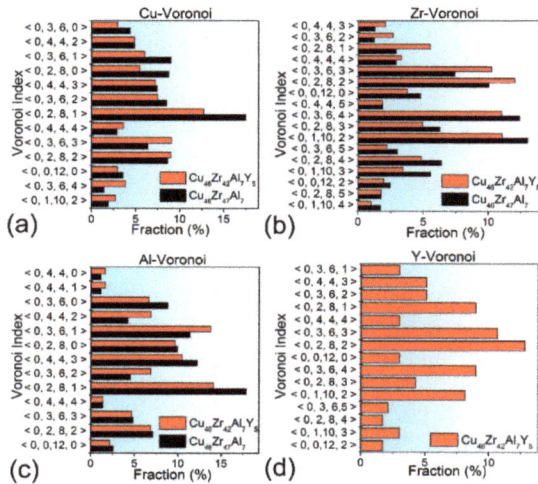

Figure 3. Distribution of the major Voronoi clusters (VCs), including: (**a**) Cu-centered VCs; (**b**) Zr-centered VCs; (**c**) Al-centered VCs; and (**d**) Y-centered VCs. Only those whose fractions are larger than 2% are selected. The CN value denotes the number of shell atoms of the corresponding VC; *i.e.*, the CN around the center atom.

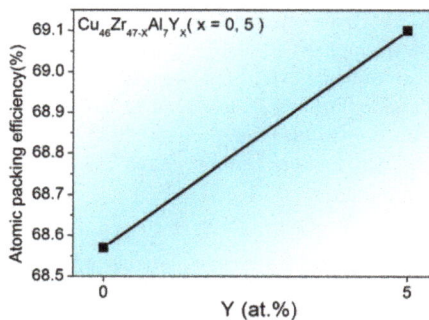

Figure 4. Atomic packing efficiency values of $Cu_{46}Zr_{42}Al_7Y_5$ and $Cu_{46}Zr_{47}Al_7$ samples.

Furthermore, the VC's shape strongly depends on the "stoichiometry" of its containing atoms, due to the atomic radial diversity among heterogeneous atoms [42]. As a result, there are different structural regularities in a number of VCs having the same index, let alone VCs with different indexes. For example, the regularity of a $Cu_6Zr_5Y_1Al_1$ <0,2,8,2> VC is obviously different from that of another $Cu_6Zr_5Y_1Al_1$ <0,2,8,2> counterpart, as shown in Figure 5. In previous work, it was suggested that the regularity of clusters may influence the glass formation in alloys [25,43,44]. Thus, we should know whether or not the regularity of VCs contributes to the GFA of $Cu_{46}Zr_{42}Al_7Y_5$ and $Cu_{46}Zr_{47}Al_7$.

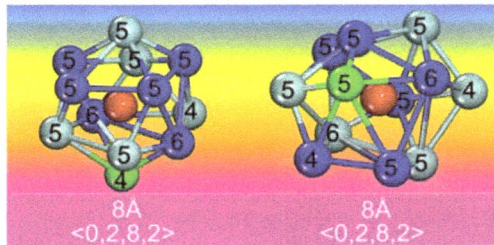

Figure 5. Configurations of two VCs extracted from the RMC-simulated model of $Cu_{46}Zr_{42}Al_7Y_5$, with the same <0,2,8,2> index. The number labeled on each shell atom stands for the number of its neighbor (connected) shell atoms, also indicates the *i*-fold symmetry. The blue, celadon, red, and green balls stand for Cu, Zr, Y, and Al atoms, respectively.

Here, a structural parameter (T) is adopted for indicating the regularity of clusters. T is the differences of all the edge lengths in a tetrahedron. Each VC can be formed by piling up a set of Delaunay tetrahedra which share a common vertex at the center atom of this VC [32], so that the regularity of their containing tetrahedra can be examined and extrapolated to any given VCs. T can be calculated by

$$T = \sum_{i \neq j} (e_i - e_j)^2 / 15 <e>^2 \tag{6}$$

where e_i denotes the length of the *i*th edge on the triangle face of a given tetrahedron, and $<e>$ is the corresponding mean value. For a regular tetrahedron where all edges are the same, the value of T should be zero. Thus, if a tetrahedron has a smaller T value, it has a higher regularity. The average T values in both samples are plotted in Figure 6. It is worth noting that $Cu_{46}Zr_{42}Al_7Y_5$ has a lower T value than $Cu_{46}Zr_{47}Al_7$. This indicates that when microalloying Y in the ZrCuAl alloy, the configurations of clusters are modified due to the denser atomic packing, exhibiting a higher degree of regularity.

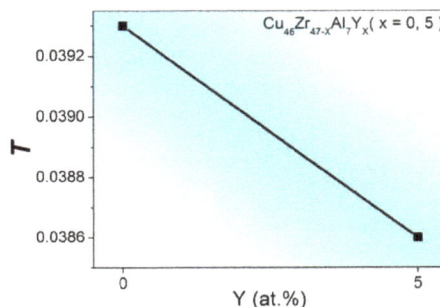

Figure 6. Variance of the edge lengths (T) of all the tetrahedra forming VCs in $Cu_{46}Zr_{42}Al_7Y_5$ and $Cu_{46}Zr_{47}Al_7$. This value determines the regularity of VCs.

In previous work, the GFA dependence on Y addition has been studied from thermodynamic or kinetic aspects [14,45]. In this work, we try to address this issue from the microstructure aspect. According to the numerous atomic- and cluster-level structural features and factors deduced above, we can analyze the reason why minor Y addition significantly affects the GFA in the ZrCuAl alloy system. It has been revealed in previous work [35] that in CuZrAl ternary composition, local structures centered with Zr and Cu solvents (major VCs) establish the structural basis, and Al solutes mainly play the role of connecting these major VCs to fill the space. In particular, the Zr atoms are apt to enter the center site of icosahedra, whereas the Cu-centered icosahedral-like VCs have a relatively low fraction.

When minor Y atoms (5 at. % in our case) are doped to replace Zr atoms, although the average CN around Zr atoms does not change, the fraction of icosahedral-like VCs with Zr centers increases. In addition, Y atoms are more likely to enter the center sites of icosahedral-like VCs. Therefore, more icosahedral-like local structures exist in the $Cu_{46}Zr_{42}Al_7Y_5$ quarternary MG. In previous work, it has been revealed that icosahedral local structures contribute to the formation of MGs [46,47], because stacking clusters with abundant five-fold point symmetrical features such as icosahedra can result in the exclusion of structural periodicity, which is required in crystals [39]. In this sense, increase of the icosahedral-like clusters leads to stabilization of the amorphous microstructure in the $Cu_{46}Zr_{42}Al_7Y_5$ alloy, greatly enhancing the GFA. Moreover, Y addition tunes the connections between heterogeneous atoms, leading to higher atomic packing inside clusters and a higher degree of regularity of clusters. Such optimization of local structures also contributes to the stabilization of the amorphous microstructure.

In one recent article studying the microalloying effect on the glass formation and the mechanical properties of MGs [48], the authors present a theory built around the experimental evidence that the microalloying elements organize a neighborhood around them that differs from both the crystalline and the glassy phases of the material in the absence of the additional elements. They also claim that a minute amount of foreign atoms (the so-called pinpoint effect [11]) influencing the GFA can be predicted by their theory. In our work studying the microalloying effect on the GFA in a ZrCuAl system from the microstructure aspect, the microstructural parameters proposed to contribute to glass formation (such as fraction of icosahedral clusters, the atomic packing efficiency, and the regularity of clusters) surely do not increase their values monotonously when adding more Y atoms in a ZrCuAl system. This also can explain why there is a minute amount of (Y) foreign atoms leading to enhancement of the GFA. This indicates that both the theoretical work [48] and our experimental work can shed light on the microalloying effect on the GFAs in multicomponent alloy systems.

4. Conclusions

The microstructures of $Cu_{46}Zr_{42}Al_7Y_5$ and $Cu_{46}Zr_{47}Al_7$ MGs are investigated by calculations based on the data obtained from synchrotron radiation-based XRD and EXAFS experiments. It is revealed that microalloying Y in the ZrCuAl alloy not only increases the fraction of Zr-centered icosahedral-like local structures, but also enhances the atomic packing efficiency and the regularity of clusters. This stabilizes the glassy-state structure, contributing significantly to the great enhancement of GFA. This study provides an in-depth understanding on how fine structures tune the glass formation in a mass of multicomponent MGs.

Acknowledgments: The authors would like to thank the HASYLAB in Germany, the Shanghai Synchrotron Radiation Facility in China, and the National Synchrotron Radiation Laboratory of China for the use of the advanced synchrotron radiation facilities. Financial supports from the National Natural Science Foundation of China (Grant No. U1332112 and 51471088), the Fundamental Research Funds for the Central Universities (Grant No. NE2015004), the Funding for Outstanding Doctoral Dissertation in NUAA (Grant No. BCXJ12-08), the Funding of Jiangsu Innovation Program for Graduate Education (Grant No. CXLX13-152), and the project funded by the Priority Academic Program Development (PAPD) of Jiangsu Higher Education Institutions are gratefully acknowledged.

Author Contributions: G.-Q.G. performed simulation work upon the experimental data. L.Y. performed analysis of this work and wrote this article. S.-Y.W contributed to the experimental research work.

Conflicts of Interest: The authors declare no conflict of interest.

References

1. Inoue, A. Stabilization of metallic supercooled liquid and bulk amorphous alloys. *Acta Mater.* **2000**, *48*, 279–306. [CrossRef]
2. Turnbull, D. Under what conditions can a glass be formed? *Contemp. Phys.* **1969**, *10*, 473–488. [CrossRef]
3. Greer, A.L. Confusion by Design. *Nature* **1993**, *366*, 303–304. [CrossRef]
4. Johnson, W.L. Bulk glass-forming metallic alloys: Science and technology. *MRS Bull.* **1999**, *24*, 42–56. [CrossRef]
5. Lu, Z.P.; Tan, H.; Li, Y.; Ng, S.C. The correlation between reduced glass transition temperature and glass forming ability of bulk metallic glasses. *Scr. Mater.* **2000**, *42*, 667–673. [CrossRef]
6. Lu, Z.P.; Liu, C.T. Glass formation criterion for various glass-forming systems. *Phys. Rev. Lett.* **2003**. [CrossRef] [PubMed]
7. Wang, W.H.; Dong, C.; Shek, C.H. Bulk metallic glasses. *Mater. Sci. Eng. R* **2004**, *44*, 45–89. [CrossRef]
8. Schroers, J.; Johnson, W.L. Ductile bulk metallic glass. *Phys. Rev. Lett.* **2004**. [CrossRef] [PubMed]
9. Ashby, M.F.; Greer, A.L. Metallic glasses as structural materials. *Scr. Mater.* **2006**, *54*, 321–326. [CrossRef]
10. Na, J.H.; Demetriou, M.D.; Floyd, M.; Hoff, A.; Garrett, G.R.; Johnson, W.L. Compositional landscape for glass formation in metal alloys. *Proc. Natl. Acad. Sci. USA* **2014**, *111*, 9031–9036. [CrossRef] [PubMed]
11. Ma, D.; Tan, H.; Wang, D.; Li, Y.; Ma, E. Strategy for pinpointing the best glass-forming alloys. *Appl. Phys. Lett.* **2005**. [CrossRef]
12. Lu, Z.P.; Liu, C.T. Role of minor alloying additions in formation of bulk metallic glasses: A review. *J. Mater. Sci.* **2004**, *39*, 3965–3974. [CrossRef]
13. Wang, W.H. Roles of minor additions in formation and properties of bulk metallic glasses. *Prog. Mater. Sci.* **2007**, *52*, 540–596. [CrossRef]
14. Xu, D.H.; Duan, G.; Johnson, W.L. Unusual glass-forming ability of bulk amorphous alloys based on ordinary metal copper. *Phys. Rev. Lett.* **2004**. [CrossRef] [PubMed]
15. Wang, D.; Tan, H.; Li, Y. Multiple maxima of GFA in three adjacent eutectics in Zr-Cu-Al alloy system-A metallographic way to pinpoint the best glass forming alloys. *Acta Mater.* **2005**, *53*, 2969–2979. [CrossRef]
16. Zhang, Q.S.; Zhang, W.; Inoue, A. Preparation of $Cu_{36}Zr_{48}Ag_8Al_8$ bulk metallic glass with a diameter of 25 mm by copper mold casting. *Mater. Trans.* **2007**, *48*, 629–631. [CrossRef]
17. Chen, L.Y.; Fu, Z.D.; Zhang, G.Q.; Hao, X.P.; Jiang, Q.K.; Wang, X.D.; Cao, Q.P.; Franz, H.; Liu, Y.G.; Xie, H.S.; *et al.* New class of plastic bulk metallic glass. *Phys. Rev. Lett.* **2008**. [CrossRef] [PubMed]
18. Egami, T.; Waseda, Y. Atomic size effect on the formability of metallic glasses. *J. Non-Cryst. Solids* **1984**, *64*, 113–134. [CrossRef]
19. Hirata, A.; Guan, P.F.; Fujita, T.; Hirotsu, Y.; Inoue, A.; Yavari, A.R.; Sakurai, T.; Chen, M.W. Direct observation of local atomic order in a metallic glass. *Nat. Mater.* **2011**, *10*, 28–33. [CrossRef] [PubMed]
20. Wu, Z.W.; Li, M.Z.; Wang, W.H.; Liu, K.X. Hidden topological order and its correlation with glass-forming ability in metallic glasses. *Nat. Commun.* **2015**. [CrossRef] [PubMed]
21. Bernal, J.D. A geometrical approach to the structure of liquids. *Nature* **1959**, *183*, 141–147.
22. Gaskell, P.H. A new structural model for transition metal-metalloid glasses. *Nature* **1978**, *276*, 484–485. [CrossRef]
23. Miracle, D.B. A structural model for metallic glasses. *Nat. Mater.* **2004**, *3*, 697–702. [CrossRef] [PubMed]
24. Sheng, H.W.; Luo, W.K.; Alamgir, F.M.; Bai, J.M.; Ma, E. Atomic packing and short-to-medium-range order in metallic glasses. *Nature* **2006**, *439*, 419–425. [CrossRef] [PubMed]
25. Yang, L.; Guo, G.Q.; Chen, L.Y.; Huang, C.L.; Ge, T.; Chen, D.; Liaw, P.K.; Saksl, K.; Ren, Y.; Zeng, Q.S.; *et al.* Atomic-Scale Mechanisms of the Glass-Forming Ability in Metallic Glasses. *Phys. Rev. Lett.* **2012**. [CrossRef] [PubMed]
26. Hammersley, A.P.; Svensson, S.O.; Hanfland, M.; Fitch, A.N.; Häusermann, D. Two-dimensional detector software: From real detector to idealised image or two-theta scan. *High Press. Res.* **1996**, *14*, 235–248. [CrossRef]
27. Qiu, X.; Thompson, J.W. PDFgetX2: A GUI-driven program to obtain the pair distribution function from X-ray powder diffraction data. *J. Appl. Crystallogr.* **2004**, *37*, 110–116. [CrossRef]

28. Klementev, K.V. Extraction of the fine structure from X-ray absorption spectra. *J. Phys. D* **2001**, *34*, 209–217. [CrossRef]

29. McGreevy, R.L. Reverse Monte Carlo modelling. *J. Phys. Condens. Matt.* **2001**, *13*, R877–R913. [CrossRef]

30. Saksl, K.; Jovari, P.; Franz, H.; Zeng, Q.S.; Liu, J.F.; Jiang, J.Z. Atomic structure of $Al_{89}La_6Ni_5$ metallic glass. *J. Phys. Condens. Matter.* **2006**, *18*, 7579–7592. [CrossRef] [PubMed]

31. Ravel, B.; Newville, M. Athena, Artemis, Hephaestus: Data analysis for X-ray absorption spectroscopy using IFEFFIT. *J. Synch. Rad.* **2005**, *12*, 537–541. [CrossRef] [PubMed]

32. Medvedev, N.N. The algorithm for three-dimensional Voronoi polyhedral. *J. Comput. Phys.* **1986**, *67*, 223–229. [CrossRef]

33. Zeng, Q.S.; Sheng, H.W.; Ding, Y.; Wang, L.; Yang, W.G.; Jiang, J.Z.; Mao, W.L.; Mao, H.K. Long-range topological order in metallic glass. *Science* **2011**, *332*, 1404–1406. [CrossRef] [PubMed]

34. Park, E.S.; Kim, D.H. Phase separation and enhancement of plasticity in Cu-Zr-Al-Y bulk metallic glasses. *Acta Mater.* **2006**, *54*, 2597–2604. [CrossRef]

35. Yang, L.; Guo, G.Q.; Chen, L.Y.; Wei, S.H.; Jiang, J.Z.; Wang, X.D. Atomic structure in Al-doped multicomponent bulk metallic glass. *Scr. Mater.* **2010**, *63*, 879–882. [CrossRef]

36. Wang, S.Y.; Kramer, M.J.; Xu, M.; Wu, S.; Hao, S.G.; Sordelet, D.J.; Ho, K.M.; Wang, C.Z. Experimental and *ab initio* molecular dynamics simulation studies of liquid $Al_{60}Cu_{40}$ alloy. *Phys. Rev. B* **2009**. [CrossRef]

37. Yang, L.; Guo, G.Q. Preferred clusters in metallic glasses. *Chin. Phys. B* **2010**. [CrossRef]

38. Fang, H.Z.; Hui, X.; Chen, G.L.; Liu, Z.K. Al-centered icosahedral ordering in $Cu_{46}Zr_{46}Al_8$ bulk metallic glass. *Appl. Phys. Lett.* **2009**. [CrossRef]

39. Guo, G.Q.; Wu, S.Y.; Luo, S.; Yang, L. Detecting Structural Features in Metallic Glass via Synchrotron Radiation Experiments Combined with Simulations. *Metals* **2015**, *5*, 2093–2108. [CrossRef]

40. Li, Y.; Guo, Q.; Kalb, J.A.; Thompson, C.V. Matching glass-forming ability with the density of the amorphous phase. *Science* **2008**, *322*, 1816–1819.

41. Greer, A.L.; Ma, E. Bulk metallic glasses: At the cutting edge of metals research. *Mater. Res. Bull.* **2007**, *32*, 611–615. [CrossRef]

42. Fujita, T.; Konno, K.; Zhang, W.; Kumar, V.; Matsuura, M.; Inoue, A.; Sakurai, T.; Chen, M.W. Atomic-scale heterogeneity of a multicomponent bulk metallic glass with excellent glass forming ability. *Phys. Rev. Lett.* **2009**. [CrossRef] [PubMed]

43. Xi, X.K.; Li, L.L.; Zhang, B.; Wang, W.H.; Wu, Y. Correlation of atomic cluster symmetry and glass-forming ability of metallic glass. *Phys. Rev. Lett.* **2007**. [CrossRef] [PubMed]

44. Xi, X.K.; Sandor, M.T.; Liu, Y.H.; Wang, W.H.; Wu, Y. Structural changes induced by microalloying in $Cu_{46}Zr_{47-x}Al_7Gd_x$ metallic glasses. *Scr. Mater.* **2009**, *61*, 967–969. [CrossRef]

45. Zhang, Y.; Chen, J.; Chen, G.L.; Liu, X.J. Glass formation mechanism of minor yttrium addition in CuZrAl alloys. *Appl. Phys. Lett.* **2006**. [CrossRef]

46. Saida, J.; Matsushita, M.; Inoue, A. Direct observation of icosahedral cluster in $Zr_{70}Pd_{30}$ binary glassy alloy. *Appl. Phys. Lett.* **2001**. [CrossRef]

47. Saksl, K.; Franz, H.; Jovari, P.; Klementiev, K.; Welter, E.; Ehnes, A.; Saida, J.; Inoue, A.; Jiang, J.Z. Evidence of icosahedral short-range order in $Zr_{70}Cu_{30}$ and $Zr_{70}Cu_{29}Pd_1$ metallic glasses. *Appl. Phys. Lett.* **2003**. [CrossRef]

48. Hentschel, H.G.E.; Moshe, M.; Procaccia, I.; Samwer, K.; Sharon, E. Microalloying and the mechanical properties of amorphous solids. 2015, arXiv:1510.03108.

Article

Intermetallic Reactions during the Solid-Liquid Interdiffusion Bonding of $Bi_2Te_{2.55}Se_{0.45}$ Thermoelectric Material with Cu Electrodes Using a Sn Interlayer

Chien-Hsun Chuang [1],*, Yan-Cheng Lin [2] and Che-Wei Lin [2]

[1]   Wire Technology Co., LTD, Taichung 432, Taiwan
[2]   Institute of Materials Science and Engineering, National Taiwan University, Taipei 106, Taiwan;
      d02527013@ntu.edu.tw (Y.-C.L.); r98527029@ntu.edu.tw (C.-W.L.)
*   Correspondence: josh604@hotmail.com; Tel./Fax: +886-2-2369-6171

Academic Editor: Ana Sofia Ramos
Received: 26 February 2016; Accepted: 19 April 2016; Published: 22 April 2016

Abstract: The intermetallic compounds formed during the diffusion soldering of a $Bi_2Te_{2.55}Se_{0.45}$ thermoelectric material with a Cu electrode are investigated. For this bonding process, $Bi_2Te_{2.55}Se_{0.45}$ was pre-coated with a 1 μm Sn thin film on the thermoelectric element and pre-heated at 250 °C for 3 min before being electroplated with a Ni barrier layer and a Ag reaction layer. The pre-treated thermoelectric element was bonded with a Ag-coated Cu electrode using a 4 μm Sn interlayer at temperatures between 250 and 325 °C. The results indicated that a multi-layer of $Bi–Te–Se/Sn–Te–Se–Bi/Ni_3Sn_4$ phases formed at the $Bi_2Te_{2.55}Se_{0.45}/Ni$ interface, ensuring sound cohesion between the $Bi_2Te_{2.55}Se_{0.45}$ thermoelectric material and Ni barrier. The molten Sn interlayer reacted rapidly with both Ag reaction layers to form an Ag_3Sn intermetallic layer until it was completely exhausted and the Ag/Sn/Ag sandwich transformed into a $Ag/Ag_3Sn/Ag$ joint. Satisfactory shear strengths ranging from 19.3 and 21.8 MPa were achieved in $Bi_2Te_{2.55}Se_{0.45}/Cu$ joints bonded at 250 to 300 °C for 5 to 30 min, dropping to values of about 11 MPa for 60 min, bonding at 275 and 300 °C. In addition, poor strengths of about 7 MPa resulted from bonding at a higher temperature of 325 °C for 5 to 60 min.

Keywords: $Bi_2Te_{2.55}Se_{0.45}$ thermoelectric material; diffusion soldering; intermetallic compounds; bonding strength

1. Introduction

$Bi_2Te_{2.55}Se_{0.45}$ intermetallic compound has been widely used as an *N*-Type thermoelectric (TE) material. For the manufacturing of thermoelectric modules, the TE elements are traditionally soldered with metallic electrodes [1]. A typical example was demonstrated by Chien *et al.* for the soldering of Bi_2Te_3/Cu couples using a $Sn_3Ag_{0.5}Cu$ alloy [2]. Although satisfactory joints in TE modules can be attained by conventional soldering, they cannot endure temperatures higher than the melting point of the solder alloy. To solve this problem, an additional water heat exchanger is usually required for cooling at the hot end of soldered TE modules industrially applied as power generators or waste heat recyclers. Ritzer *et al.* further reported that excessive molten solder can wick up the sides of TE pellets and cause electrical shorts between TE couples [3]. Another method for the bonding of TE elements with metallic electrodes is brazing, which uses a filler metal with a melting point higher than 400 °C. The manufactured thermoelectric modules can be operated at temperatures higher than that of soldered joints [4]. However, cracking at the brazed interfaces can occur due to the high thermal stress induced by the solidified filler metal. In addition, the liquid filler metal can strongly diffuse into the thermoelectric element, leading to the degradation of its TE efficiency. Neither the wicking of molten metal nor the risk of short circuits during the soldering process can be prevented in the brazing of TE modules.

An alternate diffusion soldering technique (also called solid-liquid interdiffusion bonding) uses a thin film solder (LT) inserted between the high-melting metallic work pieces or the metallization on certain substrates (HT1 and HT2) that are to be bonded. The LT interlayer, which is molten at low temperatures and acts as a transient liquid phase material, reacts rapidly with the HT1 and HT2 metals to form intermetallic phases. After a short period of solid-liquid interfacial reaction during the diffusion soldering process, the thin film solder (LT) is exhausted and has completely transformed into intermetallic compounds. The melting point of the newly formed intermetallics is much higher than that of the original LT interlayer, so the resulting joints can withstand considerably high temperatures during the operation of the manufactured thermoelectric modules. In fact, diffusion soldering has been applied in the past few decades to the manufacturing of microwave packages, high power devices, thick-film resistors, GaAs/Si wafer packages, and even gold jewelry, as reported by Jacobson and Humpston [5]. Such a novel bonding technique has also been employed by Chuang *et al.* to join Si chips with ceramic substrates [6,7]. In addition, certain advanced applications, such as those for micro electro mechanical systems (MEMS) packaging [8], semiconductor packaging [9], hybrid joining [10], and hermetic package sealing, have also been reported [11]. The mechanism of the intermetallic reactions was intensively studied by Bader, Gust, and Hieber [12].

Recently, diffusion soldering has been used to bond thermoelectric elements with metallic electrodes. In the studies of Yang *et al.*, $Bi_{0.5}Sb_{1.5}Te_3$ and GeTe thermoelectric materials were bonded with Cu electrodes using the diffusion soldering process at temperatures ranging from 250 to 325 °C with an additional thin-film Sn interlayer [13,14]. In both cases, satisfactory joints with sufficient bonding strengths were obtained. A sound bonding effect was also reported by Chuang *et al.* for the diffusion soldering of (Pb,Sn)Te thermoelectric elements with Cu electrodes using a Sn interlayer [15]. They further lowered the bonding temperatures for manufacturing, such as the (Pb,Sn)Te thermoelectric module, to a range of 170 °C to 250 °C by changing the Sn thin film interlayer to In [16]. The use of this bonding technique for manufacturing a $Bi_2Te_{2.55}Se_{0.45}$ module with a Cu electrode using a Sn interlayer will be further verified in this work. This study focuses on the intermetallic reactions occurring at the various interfaces of the multilayers in the $Bi_{0.5}Sb_{1.5}Te_3$/Cu joint, diffusion-soldered under various joining conditions and the resultant bonding strengths.

2. Experimental Section

$Bi_2Te_{2.55}Se_{0.45}$ thermoelectric material with an average composition (at. %) of Bi:Te:Se = 40:49:11 was vacuum-melted at 750 °C and then zone-refined with a speed of 1 mm/min. For the preparation of diffusion-soldered $Bi_2Te_{2.55}Se_{0.45}$ specimens with Cu electrodes, the $Bi_2Te_{2.55}Se_{0.45}$ ingot was cut into TE elements with a size of $3 \times 3 \times 3$ mm and ground with 4000 Grit SiC paper. The bonding surfaces of these $Bi_2Te_{2.55}Se_{0.45}$ TE specimens were pre-coated with a 1 μm Sn thin film and pre-heated at 250 °C for 3 min, after which they were electroplated with a ~4 μm Ni diffusion barrier layer and a 10 μm Ag reaction layer. The Cu electrodes were also electroplated with a ~4 μm Ag layer and a 4 μm Sn interlayer. The pre-treated $Bi_2Te_{2.55}Se_{0.45}$ thermoelectric element and Cu electrode were assembled in a vacuum furnace of 5.3×10^{-4} Pa and subsequently heated for the diffusion soldering process under an external pressure of 3 MPa, as shown in Figure 1. The $Bi_2Te_{2.55}Se_{0.45}$ /Cu assemblies were bonded at temperatures ranging from 250 to 325 °C for various times between 5 and 60 min. The diffusion-soldered $Bi_2Te_{2.55}Se_{0.45}$/Cu joints were cross-sectioned, ground with 4000 Grit SiC paper, and polished with 1 and 0.3 μm Al_2O_3 powders. The morphologies and compositions of the intermetallic compounds that formed at the interfaces were analyzed via scanning electron microscopy (SEM) (JEOL JSM-6510, JEOL Ltd., Tokyo, Japan) and energy dispersive X-ray spectroscopy (EDX) (Oxford Instruments, Abingdon, UK). In addition, the growth of interfacial intermetallics was directly estimated from the images in SEM. The average value of a minimum of 30 measurements for each diffusion-soldered specimen was determined to signify the intermetallic thickness (X).

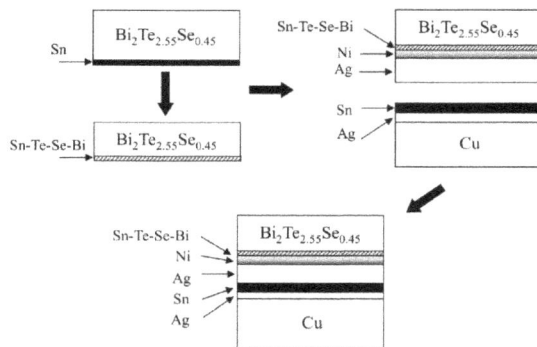

Figure 1. The schematic presentation of the diffusion soldering process for the bonding of $Bi_2Te_{2.55}Se_{0.45}$ thermoelectric materials with Cu electrodes using Sn thin film as a transient liquid phase interlayer.

The bonding strengths of various $Bi_2Te_{2.55}Se_{0.45}$/Cu joints were shear-tested with a DAGE 4000 Bond Tester (Nordson DAGE, Aylesbury, UK) at a speed of 0.3 mm/s. The fractured surfaces of the shear-tested specimens were observed via SEM. For the measurements of intermetallic thicknesses and shear strengths, at least 3 specimens were employed for each bonding condition. The standard deviations (σ) for all measurements were calculated from the average values (μ) of measured data (x_i) and plotted as the error bars of the experimental quantitative results:

$$\sigma = \sqrt{\frac{1}{N}\sum_{i=1}^{N}(x_i - \mu)^2} \tag{1}$$

where N is the number of measurements.

3. Results and Discussion

Figure 2 shows that, after pre-coating the Sn layer (Figure 2a) and directly heating it at 250 °C for 3 min, the $Bi_2Te_{2.55}Se_{0.45}$ thermoelectric material reacted with the thin-film Sn layer to form a mixed phase (shown in Figure 2b) consisting of many Bi-rich particles with a composition (at. %) of Bi:Te:Se = 55:40:5 embedded in the Sn-rich matrix of Sn:Te:Se = 52:44:4.

The compositions of the Bi-rich particles and Sn-rich matrix corresponded to the BiTe and SnTe intermetallic phases. It is obvious that the Te and Se elements in the $Bi_2Te_{2.55}Se_{0.45}$ thermoelectric material reacted with the pre-coated Sn film to form a continuous Sn–Te–Se layer. The result is consistent with the report of Chiu *et al.* that a SnTe intermetallic phase with a B1 crystal structure mainly formed in a Bi–Sn–Te ternary system [17]. In this study, certain Te lattices in the SnTe phase were replaced by Se atoms. The Bi atoms in the $Bi_2Te_{2.55}Se_{0.45}$ thermoelectric material remained after the reaction, appearing as Bi-rich islands in Figure 2b. For comparison, the $Bi_2Te_{2.55}Se_{0.45}$ thermoelectric material was also coated with a 1 μm Sn thin film and a ~4 μm Ni barrier layer and then heated at 250 °C for 3 min. Figure 2c shows that a thick Ni_3Sn_4 intermetallic compound appeared between the Ni and Sn–Te–Se layers in this case. It is obvious that the formation of Ni_3Sn_4 intermetallics is preferential to that of Sn–Te–Se phase during the interfacial reaction between Ni and $Bi_2Te_{2.55}Se_{0.45}$ thermoelectric materials.

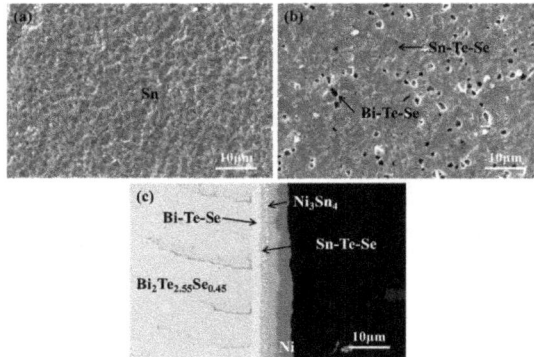

Figure 2. Surface of $Bi_2Te_{2.55}Se_{0.45}$ thermoelectric material after (**a**) electroplating with 1 μm Sn film, (**b**) subsequent heating at 250 °C for 3 min, and (**c**) pre-coated with Sn/Ni and heated 250 °C for 3 min.

To prevent the over-consumption of the Ni barrier layer, a pre-treatment process for $Bi_2Te_{2.55}Se_{0.45}$ thermoelectric material before diffusion soldering was selected. A 1 μm Sn thin film was electroplated on the material, and the composite was directly heated at 250 °C for 3 min. The pre-treated $Bi_2Te_{2.55}Se_{0.45}$ thermoelectric material was then coated with an additional Ni barrier layer and Ag reaction layer with thicknesses of about 4 μm and 10 μm, respectively. The composite was subsequently bonded with a Ag-coated Cu electrode using a Sn interlayer in a vacuum furnace of 5.3×10^{-4} Pa at temperatures ranging from 250 °C to 325 °C for various times between 5 and 60 min under an external pressure of 3 MPa. Micrographs of the interfacial reactions during such a diffusion soldering process at 250 °C for various times are shown in Figure 3. It can be observed that a Ni_3Sn_4 intermetallic layer appeared before the Ni barrier layer. In addition, a very thin layer of Bi–Te–Se phase (white in color) with a composition (at. %) of Bi:Te:Se = 52:36:12 formed on the surface of the $Bi_2Te_{2.55}Se_{0.45}$ thermoelectric material. Between the Ni_3Sn_4 intermetallic layer and the Bi–Te–Se phase, a thick Sn-rich layer (gray in color) embedded with Bi-rich islands (white in color) can be observed. The compositions (at. %) were Sn:Te:Se:Bi = 43:36:10:11 and Bi:Te:Se:Sn = 54:38:5:3, respectively. Increasing the bonding times, the Ni barrier layer reacted with the Sn-rich phase to form a Ni_3Sn_4 intermetallic layer, improving the cohesion between Ni barrier and $Bi_2Te_{2.55}Se_{0.45}$ thermoelectric material. Underneath the Ni barrier layers, the pre-coated Ag reaction layers remained after the diffusion soldering process, while the Sn thin film interlayer was exhausted and reacted with part of the Ag layer to transform into a Ag_3Sn intermetallic layer. Since the Ag_3Sn intermetallic phase has a melting point of 480 °C, the diffusion-soldered $Bi_2Te_{2.55}Se_{0.45}$ TE modules in this study can be applied at temperatures much higher than that for conventional soldered modules. Figure 3 also reveals that a Cu_3Sn intermetallic phase formed at the interface between the Ag_3Sn intermetallic layer and the Cu electrode.

It is worth mentioning that sound interfaces were obtained in the $Bi_2Te_{2.55}Se_{0.45}$/Cu joints bonded in this study, contrary to the report of Bader *et al.*, which voids frequently occurred at the interface of diffusion-soldered Ni/Sn/Ni and Cu/Sn/Cu specimens [12]. This difference may be explained by the external pressure of 3 MPa employed in this work, which is higher than that used in the work of Bader *et al.* In addition, it is postulated that fewer voids occur inherently during the intermetallic reaction at the Ag/Sn/Ag interface than during reactions at Ni/Sn/Ni and Cu/Sn/Cu interfaces. The result provides a beneficial effect of eliminating voids at the diffusion-soldered Ni/Sn/Ni or Cu/Sn/Cu interfaces through the insertion of an Ag_3Sn intermetallic layer between the Ni-Sn and Cu-Sn intermetallic compounds [18].

From Figure 3, it is obvious that the Sn–Te–Se–Bi, Ni_3Sn_4, and Ag_3Sn intermetallic compounds in the $Bi_2Te_{2.55}Se_{0.45}$/Cu joints, bonded with diffusion soldering at 250 °C, grew with bonding times of 5 min to 30 min. Increasing the bonding temperature to 300 °C further increased the thickness of these intermetallic compounds, as shown in Figure 4. Similar growth of the intermetallics can be observed in Figure 5, which shows the results of the diffusion soldering of $Bi_2Te_{2.55}Se_{0.45}$ thermoelectric material with a Cu electrode at various temperatures from 250 to 325 °C for 60 min. Similar satisfactory interfaces without voids can be achieved in the diffusion-soldered $Bi_2Te_{2.55}Se_{0.45}$/Cu joints for such a long bonding time. Although the pre-coated Ag reaction layers underneath the Ni barrier layers have been exhausted after diffusion soldering at temperatures higher than 300 °C for times longer than 30 min, as shown in Figures 4c and 5b,c, sound interfaces without any voids appear between Ag_3Sn intermetallic compounds and Ni barrier layers.

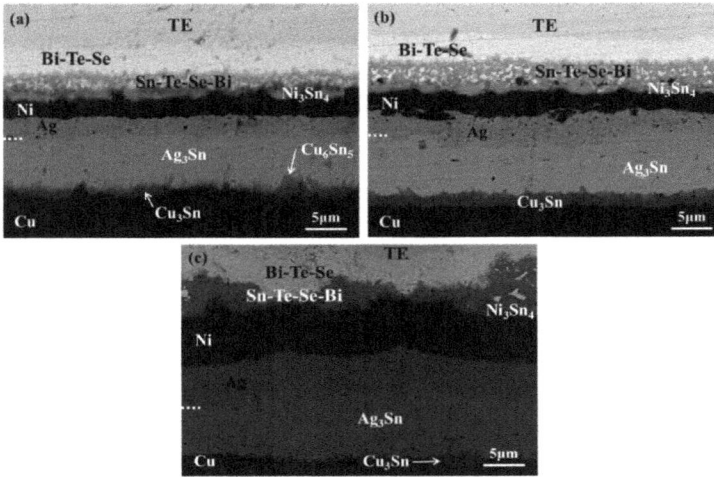

Figure 3. Morphology of intermetallic compounds formed after diffusion soldering between $Bi_2Te_{2.55}Se_{0.45}$ thermoelectric material and the Cu electrode with Sn interlayers at 250 °C for various times: (**a**) 5 min, (**b**) 10 min, (**c**) 30 min.

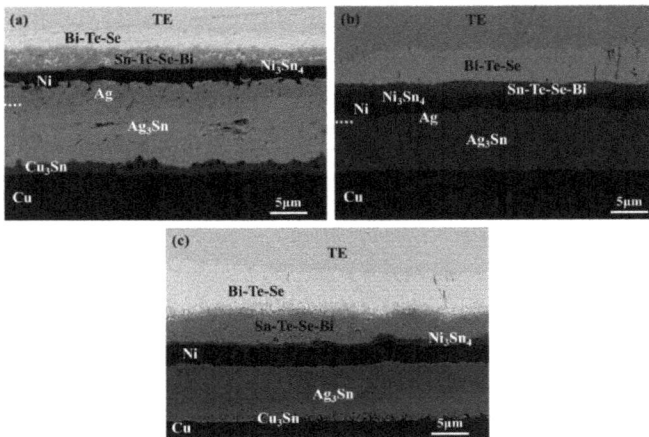

Figure 4. Morphology of intermetallic compounds formed after diffusion soldering between $Bi_2Te_{2.55}Se_{0.45}$ thermoelectric material and the Cu electrode with Sn interlayers at 300 °C for various times: (**a**) 5 min, (**b**) 10 min, (**c**) 30 min.

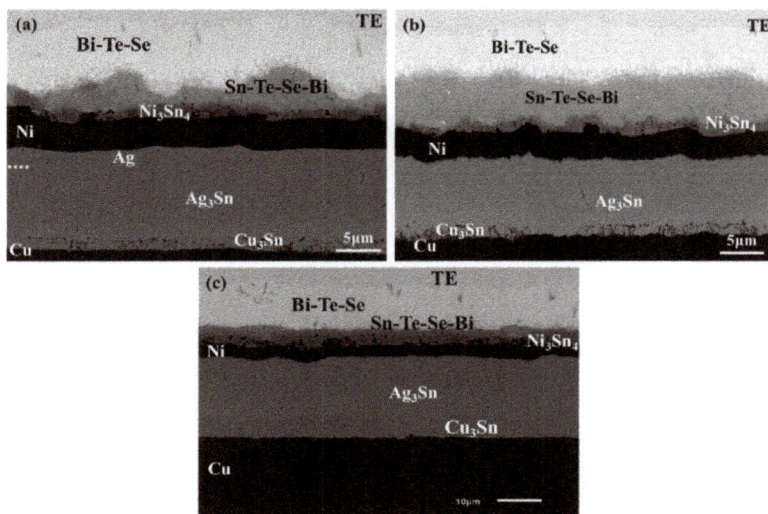

Figure 5. Morphology of intermetallic compounds formed after diffusion soldering between $Bi_2Te_{2.55}Se_{0.45}$ thermoelectric material and the Cu electrode with Sn interlayers at various temperatures for 60 min: (**a**) 250 °C, (**b**) 300 °C, (**c**) 325 °C.

The thicknesses (X) of these intermetallic layers that formed at various temperatures were measured and are plotted in Figure 6a–8a as a function of bonding time (t). For the kinetics analyses, LnX *versus* Lnt is plotted in Figure 6b–8b. The growth exponents (n) for Sn–Te–Se–Bi intermetallic layers calculated from the slopes of straight lines in Figure 6b ranged from 0.34 to 0.55. Similarly, the n values of and Ni_3Sn_4 growth calculated from the plots in Figure 7b ranged from 0.35 to 0.50. It is evidenced that the growth exponents (n) for both the Sn–Te–Se–Bi and the Ni_3Sn_4 intermetallics were close to that of a diffusion-controlled reaction ($n = 0.5$). However, the growth exponents (n) for the Ag_3Sn intermetallic layers, as shown in Figure 8b, ranged from 0.10 to 0.17, which are inconsistent with that of a diffusion mechanism. This discrepancy can be attributed to the release of partial Sn atoms in the Ag_3Sn intermetallic layers to react with the Cu electrodes to form Cu_3Sn intermetallic compounds after the exhaust of the Ag films on the Cu electrodes.

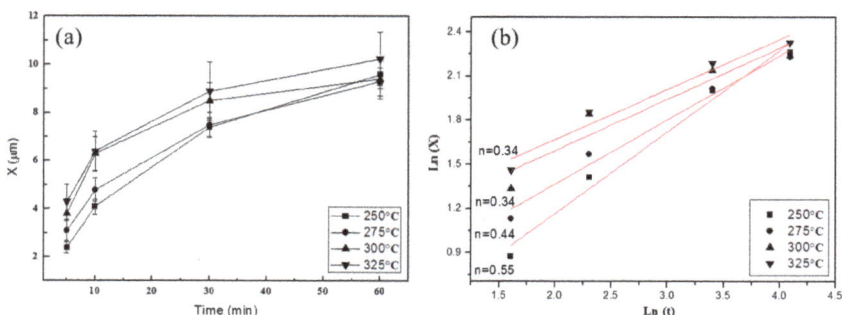

Figure 6. Thickness (X) of Sn–Te–Se–Bi phase formed during diffusion soldering of $Bi_2Te_{2.55}Se_{0.45}$ thermoelectric material with Cu electrode using Sn interlayers: (**a**) X *versus* t, (**b**) LnX *versus* Lnt.

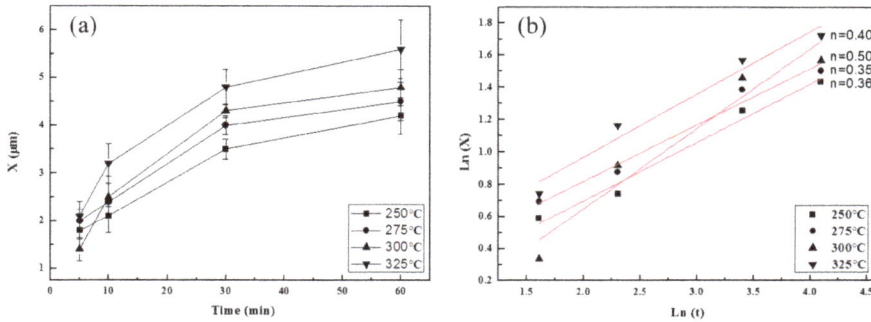

Figure 7. Thickness (X) of Ni_3Sn_4 intermetallic compound formed during diffusion soldering of $Bi_2Te_{2.55}Se_{0.45}$ thermoelectric material with Cu electrode using Sn interlayers: (**a**) X versus t, (**b**) LnX versus Lnt.

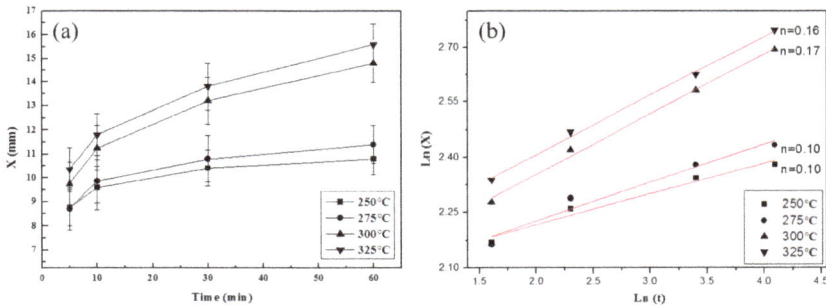

Figure 8. Thickness (X) of Ag_3Sn intermetallic compound formed during diffusion soldering of $Bi_2Te_{2.55}Se_{0.45}$ thermoelectric material with the Cu electrode using Sn interlayers: (**a**) X versus t, (**b**) LnX versus Lnt.

The shear strengths of the $Bi_2Te_{2.55}Se_{0.45}$/Cu joints, diffusion-soldered at 250 to 300 °C, ranged between 19.3 and 21.8 MPa, as shown in Figure 9. These strengths are much higher than the maximal strength of 10.7 MPa for the $Bi_{0.5}Sb_{1.5}Te_3$/Cu joints bonded in the study of Yang *et al.* under the same conditions [13]. The superior bonding of $Bi_2Te_{2.55}Se_{0.45}$/Cu relative to that of $Bi_{0.5}Sb_{1.5}Te_3$/Cu joints is attributed to the mechanical strength of the $Bi_2Te_{2.55}Se_{0.45}$ thermoelectric material, which is superior to that of $Bi_{0.5}Sb_{1.5}Te_3$ due to the embrittlement of the Sb element in BiTe material. In this case, the shear-tested $Bi_2Te_{2.55}Se_{0.45}$/Cu joints fractured through the interior of the $Bi_2Te_{2.55}Se_{0.45}$ thermoelectric material and Sn–Te–Se–Bi intermetallics layer, as shown in Figure 10a. However, lengthening the bonding time to 60 min at 275 and 300 °C decreased the shear strength to about 11 MPa. In addition, Figure 9 also reveals low strengths of about 7 MPa for the $Bi_2Te_{2.55}Se_{0.45}$/Cu joints bonded at 325 °C for various times. The micrographs in Figure 10b,c, for the shear-tested specimens bonded at 275 and 300 °C for 60 min, show that the $Bi_2Te_{2.55}Se_{0.45}$/Cu joints, having lower bonding strengths of about 11 MPa, also fractured through the interior of the $Bi_2Te_{2.55}Se_{0.45}$ thermoelectric material and the Sn–Te–Se–Bi intermetallics layer, similar to the case in Figure 10a. However, the amount of $Bi_2Te_{2.55}Se_{0.45}$ thermoelectric materials remaining in the fractography of these poorly bonded specimens was much less than that in Figure 10a, and many cracks appeared in the Sn–Te–Se–Bi intermetallic phase. Furthermore, diffusion soldering at 325 °C resulted in the fracture of shear-tested $Bi_2Te_{2.55}Se_{0.45}$/Cu joint through the interior of the $Bi_2Te_{2.55}Se_{0.45}$ thermoelectric material and Sn–Te–Se–Bi/Ni interface, as shown in Figure 10d. The insufficient bonding effect of

$Bi_2Te_{2.55}Se_{0.45}$/Cu joints at 325 °C can be correlated to the over-growth of the Sn–Te–Se–Bi intermetallics and preferential cracking at Sn–Te–Se–Bi/Ni interface during the shear tests. In addition, the cooling of $Bi_2Te_{2.55}Se_{0.45}$/Cu joints from 325 °C to room temperature can cause large thermal stress and exacerbate the decay of bonding strength. Similarly, the thickening of Sn–Te–Se–Bi intermetallics and high thermal stress can also occur in the $Bi_2Te_{2.55}Se_{0.45}$/Cu joints after diffusion soldering at 275 and 300 °C for 60 min, leading to the drastic drop of shear strengths in Figure 9. For the bonding at 250 °C, such detrimental effects are negligible, and the shear strengths of $Bi_2Te_{2.55}Se_{0.45}$/Cu joints can remain unchanged as shown in Figure 9.

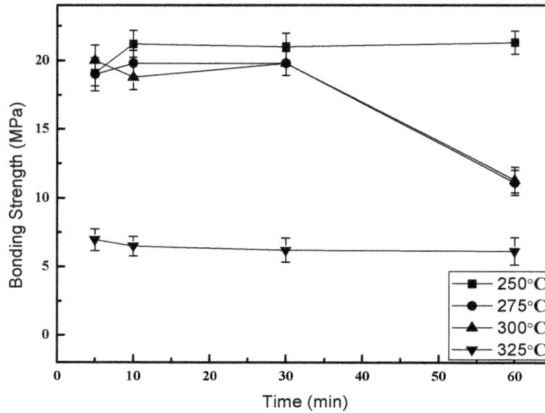

Figure 9. Shear strengths of $Bi_2Te_{2.55}Se_{0.45}$/Cu joints bonded with the diffusion soldering process using a Sn interlayer.

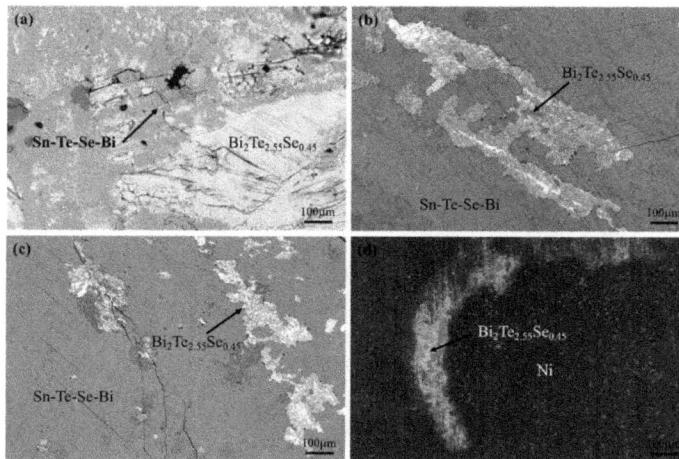

Figure 10. Fractured surfaces of shear-tested $Bi_2Te_{2.55}Se_{0.45}$/Cu joints, diffusion-soldered at various temperatures for 60 min using Sn interlayers: (**a**) 250 °C, (**b**) 275 °C, (**c**) 300 °C, (**d**) 325 °C.

Furthermore, the reliability of the diffusion-soldered $Bi_2Te_{2.55}Se_{0.45}$/Cu joints in correlation to the thermal stress formed in the multilayer structure of intermetallic compounds is considered. The coefficients of thermal expansion (CTE) of Ni_3Sn_4, Cu_3Sn, Ni and Cu as reported by Frear *et al.* [19]

are 13.7, 19.0, 12.9, and 17.1, respectively. The CTE value of the Sn–Te–Se–Bi intermetallic compound could be lower than that of pure Sn (23.0). The size of $Bi_2Te_{2.55}Se_{0.45}$ thermoelectric elements in this study is $3 \times 3 \times 3$ mm. It is known that the optimized thermoelectric temperature of a $Bi_2Te_{2.55}Se_{0.45}$ module is 120 to 150 °C. In this case, the thermal stress caused by the CTE mismatch between various materials in such small $Bi_2Te_{2.55}Se_{0.45}$/Cu joints results in a slight risk of failure during the operation of this thermoelectric module.

4. Conclusions

$Bi_2Te_{2.55}Se_{0.45}$ thermoelectric modules can be manufactured by the diffusion soldering method using a 4 μm Sn thin film interlayer inserted between the Ag/Ni-coated $Bi_2Te_{2.55}Se_{0.45}$ and Ag-coated Cu electrode. For this purpose, pre-coating a 1 μm Sn thin film on the thermoelectric element and heating at 250 °C for 3 min enhances the cohesion between the $Bi_2Te_{2.55}Se_{0.45}$ and Ni barrier. After diffusion soldering, a multi-layer of Bi–Te–Se, Sn–Te–Se–Bi, and Ni_3Sn_4 intermetallic phases can be found at the $Bi_2Te_{2.55}Se_{0.45}$/Ni interface. In addition, the employment of Ag interlayers in this Ni/Ag/Sn/Ag/Cu bonding system eliminates the voids that frequently appear at the Ni/Sn and Cu/Sn interfaces. Shear tests for the $Bi_2Te_{2.55}Se_{0.45}$/Cu joints, diffusion-soldered at 250 to 300 °C for 5 to 30 min, resulted in satisfactory bonding strengths ranging from 19.3 to 21.8 MPa. However, diffusion soldering at 275 and 300 °C for 60 min caused the over-growth and cracking of the Sn–Te–Se–Bi intermetallic layer, which resulted in lower shear strengths of about 11 MPa. The bonding strengths of $Bi_2Te_{2.55}Se_{0.45}$/Cu joints further decayed to values of about 7 MPa after diffusion soldering at 325 °C for various times.

Acknowledgments: This study was sponsored by the Small Business Innovation Research (SBIR) program of the Ministry of Economic Affairs (MOEA), Taiwan, under Grant No.1Z1041847, and by the industrial and academic cooperation program of Wire Technology Co. and the Ministry of Science and Technology, Taiwan, under Grant No. MOST 103-2622-E-002-012-CC2.

Author Contributions: C.H. Chuang conceived and designed the SLID bonding processes for thermoelectric modules, analyzed the interfacial reactions and wrote the paper. Y.C. Lin and C.W. Lin prepared the thermoelectric material, performed the bonding experiments and measured the bonding strengths.

Conflicts of Interest: The authors declare no conflict of interest.

References

1. Zhang, H.; Jing, H.Y.; Han, Y.D.; Xu, L.Y. Interfacial reaction between n-and p-type thermoelectric materials and SAC305 solders. *J. Alloy. Compd.* **2013**, *576*, 424–431. [CrossRef]
2. Chien, P.Y.; Yeh, C.H.; Hsu, H.H.; Wu, A.T. Polarity Effect in a $Sn_3Ag_{0.5}Cu$/Bismuth Telluride Thermoelectric System. *J. Electron. Mater.* **2014**, *43*, 284–289. [CrossRef]
3. Ritzer, T.M.; Lau, P.G.; Bogard, A.D. A Critical Evaluation of Today's Thermoelectric Modules. In Proceedings of the 16th International Conference on Thermoelectrics, Dresden, Germany, 26–29 August 1997; pp. 619–623.
4. Zybala, R.; Wojciechowski, K.T.; Schmidt, M. Junctions and Diffusion Barriers for High Temperature Thermoelectric Modules. *Mater. Ceram./Ceram. Mater.* **2010**, *62*, 481–485.
5. Jacobson, D.M.; Humpston, G. Diffusion Soldering. *Solder. Surf. Mt. Technol.* **1992**, *10*, 27–32. [CrossRef]
6. Chuang, T.H.; Lin, H.J.; Tsao, C.W. Intermetallic compounds formed during diffusion soldering of $Au/Cu/Al_2O_3$ and Cu/Ti/Si with Sn/In interlayer. *J. Electron. Mater.* **2006**, *35*, 1566–1570. [CrossRef]
7. Liang, M.W.; Hsieh, T.E.; Chang, S.Y.; Chuang, T.H. Thin-film reactions during diffusion soldering of Cu/Ti/Si and $Au/Cu/Al_2O_3$ with Sn interlayers. *J. Electron. Mater.* **2003**, *32*, 952–956. [CrossRef]
8. Welch, W.C.; Chae, J.; Najafi, K. Transfer of metal MEMS packages using a wafer-level solder transfer technique. *IEEE Trans. Adv. Packag.* **2005**, *28*, 643–649. [CrossRef]
9. Made, R.I.; Gan, C.L.; Yan, L.L.; Yu, A.; Yoon, S.W.; Lau, J.H.; Lee, C.K. Study of Low-Temperature Thermocompression Bonding in Ag-In Solder for Packaging Applications. *J. Electron. Mater.* **2009**, *38*, 365–371. [CrossRef]
10. Li, J.F.; Agyakwa, P.A.; Johnson, C.W. Kinetics of Ag_3Sn growth in Ag-Sn-Ag system during transient liquid phase soldering process. *Acta Mater.* **2010**, *58*, 3429–3443. [CrossRef]

11. Yan, L.L.; Lee, C.K.; Yu, D.Q.; Yu, A.B.; Choi, W.K.; Lau, J.H.; Yoon, S.U. A Hermetic Seal Using Composite Thin-Film In/Sn Solder as an Intermediate Layer and Its Interdiffusion Reaction with Cu. *J. Electron. Mater.* **2009**, *38*, 200–207. [CrossRef]

12. Bader, S.; Gust, W.; Hieber, H. Rapid formation of intermetallic compounds interdiffusion in the Cu-Sn and Ni-Sn systems. *Acta Metall. Mater.* **1995**, *43*, 329–337.

13. Yang, C.L.; Lai, H.J.; Hwang, J.D.; Chuang, T.H. Diffusion Soldering of $Bi_{0.5}Sb_{1.5}Te_3$ Thermoelectric Material with Cu Electrode. *J. Mater. Eng. Perform.* **2013**, *22*, 2029–2037. [CrossRef]

14. Yang, C.L.; Lai, H.J.; Hwang, J.D.; Chuang, T.H. Diffusion Soldering of Pb-Doped GeTe Thermoelectric Modules with Cu Electrodes Using a Thin-Film Sn Interlayer. *J. Electron. Mater.* **2012**, *42*, 359–365. [CrossRef]

15. Chuang, T.H.; Lin, H.J.; Chuang, C.H.; Yeh, W.T.; Hwang, J.D.; Chu, H.S. Solid Liquid Interdiffusion Bonding of (Pb, Sn)Te Thermoelectric Modules with Cu Electrodes Using a Thin-Film Sn Interlayer. *J. Electron. Mater.* **2014**, *43*, 4610–4618. [CrossRef]

16. Chuang, T.H.; Yeh, W.T.; Chuang, C.H.; Hwang, J.D. Improvement of bonding strength of a (Pb, Sn)Te–Cu contact manufactured in a low temperature SLID-bonding process. *J. Alloy. Compd.* **2014**, *613*, 46–54. [CrossRef]

17. Chiu, C.N.; Wang, C.H.; Chen, S.W. Interfacial Reactions in the Sn–Bi/Te Couples. *J. Electron. Mater.* **2007**, *37*, 40–44. [CrossRef]

18. Chang, J.Y.; Chang, T.C.; Chuang, T.H.; Lee, C.Y. Dual-Phase Intermetallic Interconnections Structure and Method of Fabricating the Same. U.S. Patent 8,742,600 B2, 3 June 2014.

19. Frear, D.R.; Burchett, S.N.; Morgan, H.S.; Lau, J.H. *The Mechanics of Solder Alloy Interconnects*; Van Nostrand Reinhold: New York, NY, USA, 1994; p. 60.

Article

Joining of TiAl to Steel by Diffusion Bonding with Ni/Ti Reactive Multilayers

Sónia Simões [1], Ana S. Ramos [2], Filomena Viana [1], Maria Teresa Vieira [2] and Manuel F. Vieira [1],*

[1] CEMUC, Department of Metallurgical and Materials Engineering, University of Porto, R. Dr. Roberto Frias, Porto 4200-465, Portugal; ssimoes@fe.up.pt (S.S.); fviana@fe.up.pt (F.V.)

[2] CEMUC, Department of Mechanical Engineering, University of Coimbra, R. Luís Reis Santos, Coimbra 3030-788, Portugal; sofia.ramos@dem.uc.pt (A.S.R.); teresa.vieira@dem.uc.pt (M.T.V.)

* Correspondence: mvieira@fe.up.pt; Tel.: +351-225081424; Fax: +351-220414900

Academic Editor: Hugo F. Lopez
Received: 2 March 2016; Accepted: 20 April 2016; Published: 25 April 2016

Abstract: Dissimilar diffusion bonds of TiAl alloy to AISI 310 stainless steel using Ni/Ti reactive multilayers were studied in this investigation. The Ni and Ti alternating layers were deposited by d.c. magnetron sputtering onto the base materials, with a bilayer thickness of 30 and 60 nm. Joining experiments were performed at 700 and 800 °C for 60 min under pressures of 50 and 10 MPa. The effectiveness of using Ni/Ti multilayers to improve the bonding process was assessed by microstructural characterization of the interface and by mechanical tests. Diffusion bonded joints were characterized by scanning electron microscopy (SEM), energy dispersive X-ray spectroscopy (EDS), electron backscatter diffraction (EBSD), transmission electron microscopy (TEM) and selected area electron diffraction (SAED), high resolution TEM (HRTEM) and Fast Fourier transform (FFT). The bonding interfaces are thin (approximately 5 μm thick) with a layered microstructure. For all joints, the interface is mainly composed of equiaxed grains of NiTi and $NiTi_2$. The thickness and number of layers depends on the joining conditions and bilayer thickness of the multilayers. Mechanical characterization of the joints was performed by nanoindentation and shear tests. Young´s modulus distribution maps highlight the phase differences across the joint´s interface. The highest shear strength value is obtained for the joint produced at 800 °C for 60 min under a pressure of 10 MPa using Ni/Ti multilayers with 30 nm of bilayer thickness.

Keywords: diffusion bonding; TiAl; stainless steel; multilayers

1. Introduction

The development of TiAl intermetallics has attracted the interest of several researchers due to its unique properties, which are suitable for a wide field of applications. However, a limiting factor for the integration of TiAl into structures is the lack of reliable and efficient joining techniques, especially for dissimilar joints. Dissimilar joining of these alloys is very difficult, due to their high reactivity and the tendency to form brittle intermetallic phases [1,2]. The development of joining technologies has become an essential requirement for application of these alloys in replacing traditional metallic alloys in aerospace and automotive applications [3–6]. However, there are only a few reports regarding the techniques used to join TiAl alloys with other materials [6–12].

The joining of TiAl and steel is relevant for some applications, in particular for missile and tank engine turbo components. The use of fusion welding techniques to join TiAl to steel promotes the formation of brittle phases, TiC and Ti–Fe intermetallics, and high residual stress produced during solidification, leading to weak joints [4,5].

Diffusion bonding is one of the most promising processes for successfully joining TiAl to steel with high mechanical properties, avoiding the problems associated with fusion welding processes.

Dissimilar joints can be produced by direct diffusion bonding of TiAl to steel at 850–1000 °C for 1 to 60 min under a pressure ranging from 5 to 40 MPa [9]. Close to the TiAl base material, the interfaces are composed of $Ti_3Al + FeAl + FeAl_2$ intermetallic compounds. The specimens bonded at low temperature exhibit very low tensile strength (18 MPa); a maximum tensile strength of 170–185 MPa was measured for the optimum bond conditions (930–960 °C for 5–6 min under a pressure of 20–25 MPa). These demanding bonding conditions are not suitable for the industrial implementation of the process and can lead to plastic deformation of the base materials.

Reactive multilayer thin films are an alternative for reducing the temperature and/or the pressure needed for diffusion bonding, since these multilayers can simultaneously improve the diffusivity, due to their nanocrystalline nature and high density of defects, and can also act as a local heat source, as the result of the heat released by the exothermic reaction of the multilayers to form intermetallic compounds [13–23]. Previous studies [23] have shown that Ni/Al multilayers are effective in the diffusion bonding of TiAl to steel. The use of these multilayers in this process allows lower joining conditions. However, these joints present unbonded areas, which impair the shear strength. The process could be rendered more effective by improving the contact of mating surfaces, which can be achieved by the use of different multilayer systems.

In this context, the main objective of this study is the diffusion bonding of TiAl alloy to AISI 310 austenitic stainless steel using Ni/Ti reactive multilayers. Although this multilayer system is less exothermic than others, such as Ni/Al, it has been successfully used in dissimilar joining of NiTi alloy to Ti_6Al_4V [24]. The present study centers on the microstructural characterization of the interface and on the mechanical characterization of the joints. Interface characterization was performed by scanning electron microscopy (SEM), energy dispersive X-ray spectroscopy (EDS), electron backscatter diffraction (EBSD), transmission electron microscopy (TEM) and selected area electron diffraction (SAED), high resolution TEM (HRTEM) and Fast Fourier transform (FFT); mechanical properties were evaluated by nanoindentation and shear strength tests.

2. Materials and Methods

2.1. Materials

The base materials used in this investigation were TiAl (Ti–45Al–5Nb at. %) and AISI 310 austenitic stainless steel (25Cr–20Ni–0.25C wt. %).

Ni/Ti multilayers were used as an interlayer for diffusion bonding between these two base materials. Ni/Ti multilayer thin films with ~2.5 μm total thickness were deposited onto the base materials (substrates) by d.c. magnetron sputtering from titanium (99.99% pure) and nickel (93 wt. %Ni, 7 wt. %V) targets. In the presence of vanadium, the Ni target becomes nonmagnetic, resulting in more stable depositions. Before entering the deposition chamber, the surfaces of the base materials were ground using SiC abrasive papers, followed by polishing down to 1 μm diamond suspension. The final polishing was carried out using colloidal silica. The samples were cleaned in ultrasound bath in acetone and ethanol. After attaining a base pressure of ~5 × 10^{-4} Pa, the depositions were carried out at an argon pressure of 0.25 Pa. The targets' power densities were adjusted in order to obtain near equiatomic average chemical composition. To avoid heating during deposition, the base materials were mounted in a thick rotating copper block, which acts as a heat sink. With this procedure the deposition temperature remains close to 100 °C. Multilayer thin films with 30 and 60 nm bilayer thickness (modulation period) were prepared by varying the substrates' rotation speed with the Ti and Ni targets operating simultaneously. During one rotation the substrates pass once in front of the Ti target and once in front of the Ni target, giving rise to a bilayer (Ti layer + Ni layer). After several rotations a stack of bilayers is obtained multilayered. When the substrates' rotation speed increases, the time that the substrates are in front of each target decreases, resulting in lower individual layer thicknesses.

2.2. Diffusion Bonding Experiments

Diffusion bonding of dissimilar (TiAl to steel) joints was produced at 700 and 800 °C under pressures of 10 and 50 MPa with a bonding time of 60 min, in a vertical furnace with a vacuum level better than 10^{-2} Pa. The diffusion bonding apparatus has been described elsewhere [13,17].

2.3. Joints Characterization

2.3.1. Microstructural Characterization

In order to perform the microstructural and chemical characterization of the interface, cross-sections of the joints were prepared using standard metallographic techniques. The cross-section specimens for TEM were prepared using a focused ion beam (FIB) (FEI FIB200, FEI company, Hillsboro, OR, USA), by means of the lift-out technique, at 5–30 keV. The bond interfaces were characterized by SEM (FEI company, Hillsboro, OR, USA) and analyzed by EDS (Oxford Intruments, Oxfordshire, UK), TEM (JEOL Ltd., Tokyo, Japan) and HRTEM (JEOL Ltd., Tokyo, Japan). A high resolution FEI QUANTA 400 FEG SEM (FEI company, Hillsboro, OR, USA) with EDAX Genesis X4M (Oxford Intruments, Oxfordshire, UK) was used for the SEM and EDS analyses. HRTEM images were obtained in a JEOL 2010F (JEOL Ltd., Tokyo, Japan) and a FEI Tecnai G2 (FEI company, Hillsboto, OR, USA), operating at 200 keV.

The EDS measurements were made at an accelerating voltage of 15 keV by the standardless quantification method. The results obtained by this method provide a fast quantification with an automatic background subtraction, matrix correction and normalization to 100% for all the elements in the peak identification list.

Crystallographic information from the joint interfaces was obtained by EBSD (FEI QUANTA 400 FEG SEM). The diffraction of backscattered electrons forms a Kikuchi pattern characteristic of the underlying crystal phase. For the indexation of the patterns, the ICDD PDF2 (2006) database was used. EBSD is a surface technique, since only the topmost 50 nm of the specimen contributes to the diffraction pattern; for Ti and Ni the minimum lateral spread of the interaction volume ranges from 50 to 100 nm [25], with an accelerating voltage of 15 keV. Seeing that a careful preparation of the surface is the key factor in achieving good quality EBSD patterns, the specimens were submitted to a final chemo-mechanical polishing stage, using colloidal silica, to remove the damage and deformation of the surface of the cross-section specimens.

Selected area electron diffraction (SAED) in TEM and Fast Fourier transform (FFT) in HRTEM were also applied in order to study the crystallographic structure of the phases present at the interface.

2.3.2. Mechanical Characterization

The mechanical behavior was evaluated by nanoindentation using a Micro Materials-Nano hardness apparatus equipped with a Berkovich diamond indenter. Hardness and Young's modulus were determined by the Oliver and Pharr analysis method [26]. Before the depth-sensing indentation experiments, the tip area function was calibrated using fused silica as a standard. The loading/ unloading experiments were run up to a maximum load of 4 mN with a dwell bilayer thickness of 30 s at 0.4 mN during unloading, for thermal drift correction. Indentation matrices with a minimum of 90 measurements were selected in order to test the joint's interface and both base materials. The scheme of the nanoindentation tests across the joints is represented in Figure 1. Prior to the indentation tests the joints were polished using standard metallographic procedures until a mirror-like surface was achieved.

The shear tests were performed at room temperature at a rate of 0.2 mm/min. The apparatus and specimens of the shear tests have been described elsewhere [23]. For each joint, three specimens were tested. SEM combined with EDS was used to identify the phases present at the fracture surfaces of the shear specimens.

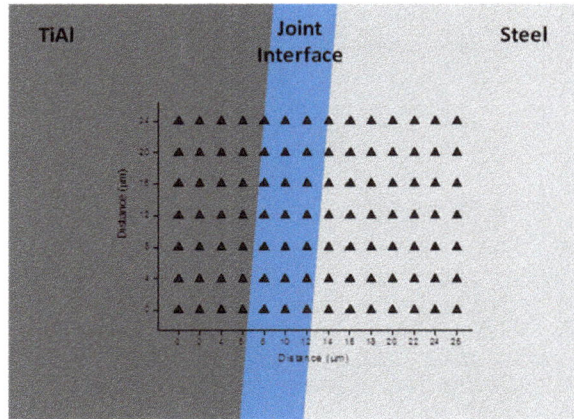

Figure 1. Scheme of the nanoindentation tests position (▲) across the joints.

3. Results and Discussion

The effectiveness of using Ni/Ti multilayers to improve dissimilar bonds between TiAl and steel was assessed through microstructural and mechanical characterizations. The influence of different processing conditions (700 or 800 °C under a pressure of 10 or 50 MPa) and different bilayer thickness (30 and 60 nm) was evaluated.

3.1. Microstructural Characterization

The microstructural characterization reveals that interfaces with apparent soundness were produced during the solid-state diffusion bonding of TiAl and steel using Ni/Ti multilayers. The bond between these metallic materials using Ni/Ti multilayers with 30 and 60 nm of bilayer thickness was obtained successfully at 700 °C under a pressure of 50 MPa and at 800 °C under a pressure of 10 MPa, for 60 min. However, some unbonded areas are observed at the edge of the samples bonded at 700 °C and at 800 °C using multilayers with 60 nm of bilayer thickness.

Figure 2 shows SEM images of the interface of the diffusion bonds between TiAl alloys and steel processed with Ni/Ti multilayers. Whatever the bilayer thickness, to achieve a sound joint at 700 °C it was necessary to apply a 50 MPa pressure, while at 800 °C the pressures could be reduced down to 10 MPa. The bond interfaces are thin (thickness close to 5 μm) and exhibit six layers: a thicker layer (4 μm) at the center divided by a very thin dark line and, adjacent to each of the base materials, two diffusion layers with an overall thickness of not more than 1 μm. The thickness of the interface layers, especially those close to the base materials, increase with increasing bonding temperature and multilayer bilayer thickness. These microstructures are different from those obtained using Ni/Al multilayers [23], which exhibited a thick reaction layer near the steel substrate. From the EDS profile of Figure 2d it can be seen that the interfaces mainly consist of Ni and Ti, as expected. The diffusion of elements from the base material to adjacent layers is observed for all bonds. Fe and Cr are detected in zones closest to the stainless steel, while Al is detected in the thin layers closest to TiAl.

The identification of the phases was carried out through the combination of the EDS, EBSD, SAED and FFT results. Due to the small interaction volume of EBSD analysis, this technique is essential for the crystallographic characterization of the phases of thinner layers and for identifying the formation of nanometric phases at the interface. For phase identification in areas smaller than the EBSD resolution, TEM/SAED and HRTEM/FFT were used.

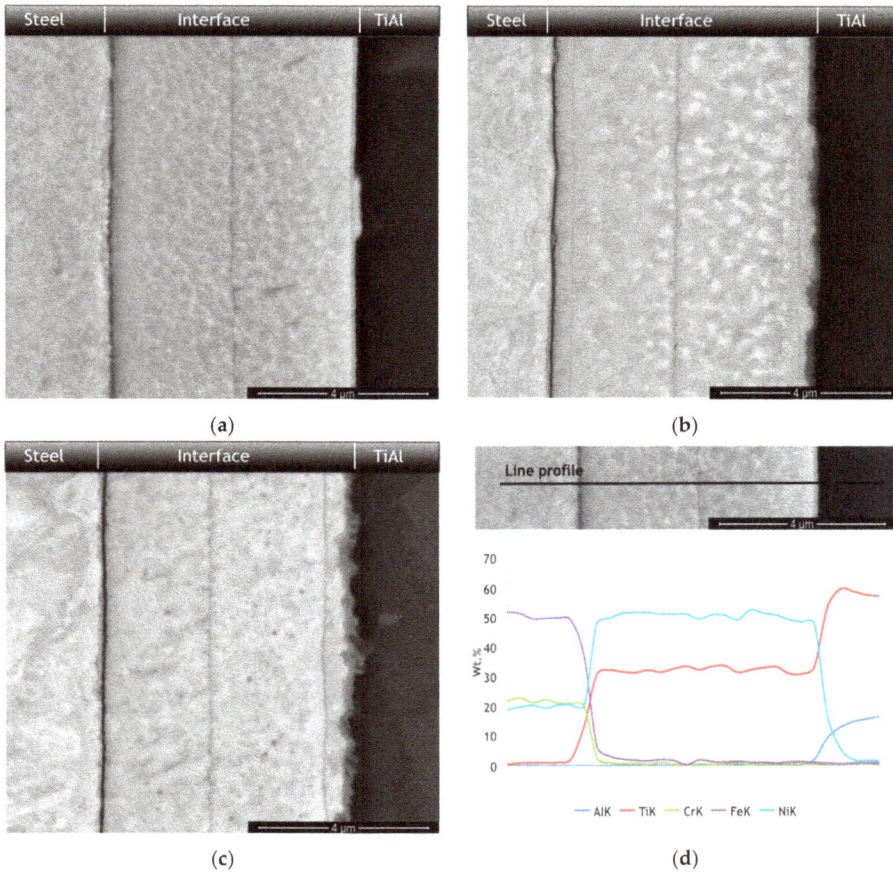

Figure 2. SEM images of the TiAl/steel joints produced at: (**a**) 700 °C under a pressure of 50 MPa using Ni/Ti multilayers with 30 nm of bilayer thickness; (**b**) 800 °C under a pressure of 10 MPa using Ni/Ti multilayers with 30 nm of bilayer thickness; (**c**) 800 °C under a pressure of 10 MPa using Ni/Ti multilayers with 60 nm of bilayer thickness and (**d**) EDS profile across the interface presented in (**a**).

Figure 3 shows the EBSD patterns obtained for the thin diffusion layers close to the base materials and for the thick central zone. The greatest area of the interface comprises equiaxed grains of NiTi and $NiTi_2$ (Kikuchi patterns 3 and 4 of Figure 3b). The indexation of the EBSD patterns of the layers close to the base materials identified the presence of AlNiTi and $AlNi_2Ti$ close to the TiAl (Kikuchi patterns 1 and 2 of Figure 3b) and α-Fe and $NiTi_2$ near the steel (Kikuchi patterns 5 and 6 of Figure 3b). The main differences between the interfaces produced using Ni/Ti and Ni/Al reactive multilayers are the phases that constitute the central area, NiTi + $NiTi_2$ for joints produced with Ni/Ti, and NiAl for those produced with Ni/Al. Phases richer in Ni, such as $AlNi_2Ti$, rather than phases richer in Al, such as Al_2NiTi, were also observed. The phases α-Fe and AlNiTi were also identified in the interfaces produced using Ni/Al multilayers [23].

(a)

(b)

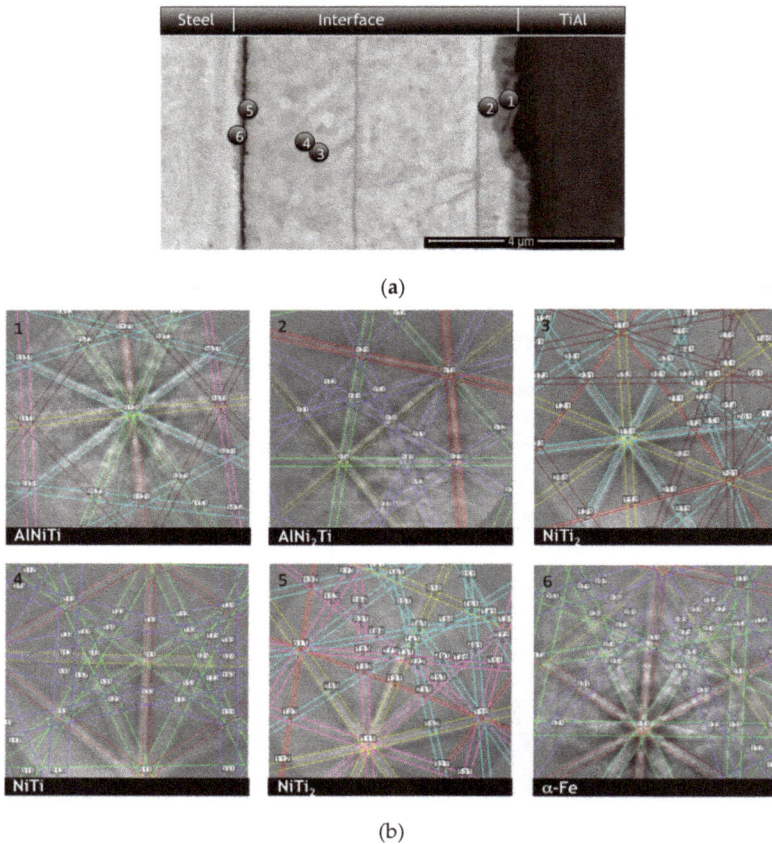

Figure 3. (**a**) SEM images of the TiAl/steel joints produced at 800 °C under a pressure of 10 MPa and (**b**) EBSD Kikuchi pattern indexation of the zones marked in (**a**).

Figure 4 shows the TEM and HRTEM images of the interface central zone. HRTEM images and FFT analysis revealed that this region is composed of a mixture of equiaxed grains of NiTi with nanometric $NiTi_2$ grains divided by a very thin layer with $NiTi_2$ aligned nanograins.

The thin layers close to the steel were also characterized by TEM and HRTEM combined with SAED and FFT, respectively, as shown in Figure 5. The TEM image (Figure 5a) of these layers shows large grains adjacent to the steel identified by SAED as α-Fe (Figure 5b) and a line of nanometric grains identified by FFT of HRTEM images as Fe_2Ti (Figure 5c) and $NiTi_2$ (Figure 5d).

In summary, the interface is composed mainly of NiTi and $NiTi_2$ phases, formed during the multilayer reaction. In addition, thin layers of α-Fe and Fe_2Ti + $NiTi_2$ were identified close to the steel while AlNiTi and $AlNi_2Ti$ were detected close to the TiAl base material, resulting from the interdiffusion of multilayer and base materials.

Figure 4. (a) TEM image of the interface central zone of a TiAl/steel joint produced at 800 °C using Ni/Ti multilayers with 30 nm of bilayer thickness; (b) and (c) HRTEM images and FFT indexation as NiTi and NiTi$_2$ grains at the center of the joint interface; and (d) HRTEM images and FFT indexation as NiTi$_2$ of a nanometric grain at the central line.

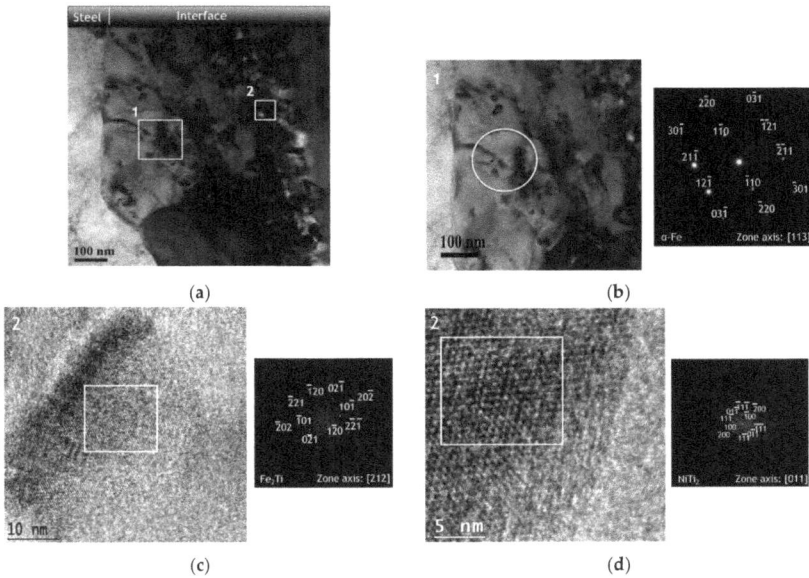

Figure 5. (a) TEM image of the zone close to the steel; (b) TEM images and SAED indexation as α-Fe of a grain at the region marked as 1 in (a); and (c) HRTEM images and FFT indexation as Fe$_2$Ti of a grain at the region marked as 2 in (a).

3.2. Mechanical Characterization

The mechanical characterization of the diffusion bonds was performed by nanoindentation and shear tests.

Figure 6 shows the hardness and reduced Young's modulus (*Er*) distribution maps for the joints produced at 800 °C using the multilayers with 30 nm bilayer thickness. The interface of the joints shows hardness values similar to those of the base materials with the exception of the thin layers close to the base materials, which are harder, in particular on the steel side. These layers exhibit a high hardness value of about 10–12 GPa. These values may be associated with the formation of AlNiTi, AlNi$_2$Ti and Fe$_2$Ti intermetallics together with nanometric microstructures. In the diffusion bonding of these materials with Ni/Al multilayers, similar results were observed. The reduced Young's modulus maps clearly reproduce the phase variations across the base materials and diffusion bond, with the joint interface and the diffusion layers perfectly identifiable. According to the nanoindentation results the interface has a lower Young's modulus (*Er* ≈ 85–120 GPa) in comparison with the base materials (*Er*$_\text{TiAl}$ ≈ 185–210 GPa and *Er*$_\text{steel}$ ≈ 210–260 GPa), which was expected due to the different phases that compose the interface. The variation of the *Er* values obtained by nanoindentation is in accordance with the Young's modulus values found in the available literature ($E_\text{stainless steel}$ > E_TiAl > E_NiTi > E_NiTi2) [27,28].

The mechanical characterization of the diffusion bonds was also performed by shear tests. tab:metals-06-00096-t001 shows the values of shear strength of the bonds produced. The macroscopic images of the fracture surfaces are present in Figure 7. These images show that the fracture always begins at the interface.

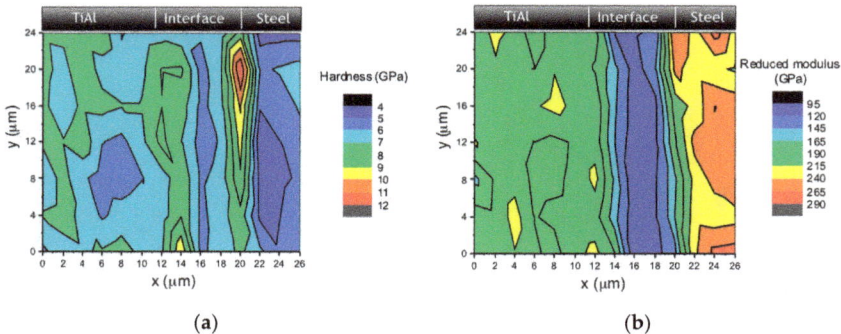

Figure 6. (**a**) Hardness and (**b**) reduced Young's modulus distribution maps of the joint produced at 800 °C using a Ni/Ti multilayer with 30 nm of bilayer thickness.

Table 1. Shear strength values of TiAl/steel joints produced using Ni/Ti multilayers.

Bilayer Thickness (nm)	Bonding Conditions (Temp./Time/Pressure)	Shear Strength Values (MPa)			Average Shear Strength (MPa)
30	700 °C/60 min/50 MPa	44	68	50	54
30	800 °C/60 min/10 MPa	216	225	231	225
60	800 °C/60 min/10 MPa	66	99	83	83

The highest shear strength value of 225 MPa was obtained for the diffusion bonding of TiAl and steel using Ni/Ti multilayers with 30 nm of bilayer thickness at 800 °C for 60 min under a pressure of 10 MPa.

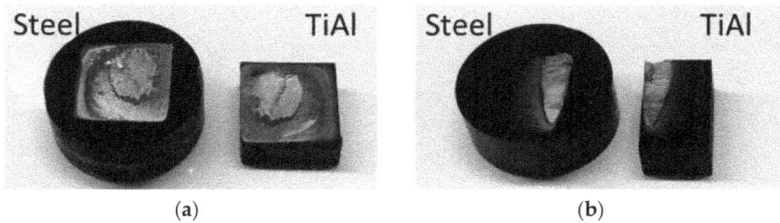

(a)   (b)

Figure 7. Macroscopic images of fracture surfaces of shear test specimens of the joints produced at 800 °C using Ni/Ti multilayers with: (**a**) 30 nm and (**b**) 60 nm of bilayer thickness.

The microstructural features and hardness distribution across the interfaces are similar for all tested conditions and cannot explain the lower strength of the joints produced at 700 and 800 °C using multilayers with 60 nm of bilayer thickness. The higher shear strength of the joints produced at 800 °C with 30 nm bilayer thickness may be explained by a smaller unbonded region, which is clearly observed in Figure 7.

From the shear strength results it can be concluded that the use of Ni/Ti multilayers with 30 nm of bilayer thickness provides advantages in solid-state diffusion bonding of steel to TiAl in comparison with the more reactive Ni/Al multilayers that present an average shear strength of 144 MPa [23]. This may be associated with the mechanical properties of the phases formed at the interface; with the use of Ni/Al multilayers, the interface comprises essentially NiAl, which is harder and has a higher elastic modulus than the NiTi and NiTi$_2$ that form when the Ni/Ti multilayers are used.

Diffusion bonding of TiAl to steel using Ni/Ti multilayers can be performed under lower temperature and produces joints with higher shear strength values than the ones without the interlayer and with Ni/Al multilayers.

4. Conclusions

Diffusion bonding of TiAl alloys to stainless steel can be enhanced by the use of Ni/Ti multilayers as interlayers. The use of these multilayers enables the production of sound joints at 800 °C for 60 min under a pressure of 10 MPa and multilayers with 30 nm of bilayer thickness with an average shear strength of 225 MPa. The joints are produced at a lower temperature than those required to diffusion bond TiAl to steel without multilayers and have higher strength than those obtained with Ni/Al multilayers. The interfaces are mainly composed of NiTi and NiTi$_2$ grains showing thin layers of α-Fe and Fe$_2$Ti + NiTi$_2$ close to the steel and AlNiTi and AlNi$_2$Ti close to the TiAl alloy.

Acknowledgments: This study was supported by the Portuguese Foundation for Science and Technology through the SFRH/BDP/109788/2015 grant and UID/EMS/00285/2013 project. The authors are grateful to CEMUP-Centro de Materiais da Universidade do Porto for expert assistance with SEM.

Author Contributions: S. Simões implemented and conducted the diffusion bonding experiments and the shear tests; A.S. Ramos and M.T. Vieira produced and characterized the multilayers and performed the nanoindentation tests; S. Simões, F. Viana and M.F. Vieira characterized the interfaces and analyzed and discussed the results; all the authors participated in the design of the experiments and cooperated in writing this paper.

Conflicts of Interest: The authors declare no conflict of interest.

References

1. Shiue, R.K.; Wu, S.K.; Chem, S.Y. Infrared brazing of TiAl using Al-based braze alloys. *Intermetallics* **2003**, *11*, 616–671. [CrossRef]
2. Çam, G.; İpekoglu, G.; Bohm, K.H.; Koçak, M. Investigation into the microstructure and mechanical properties of diffusion bonded TiAl alloys. *J. Mater. Sci.* **2006**, *41*, 5273–5282. [CrossRef]

3. Kim, Y.W. Ordered intermetallic alloys, Part III: Gamma-titanium aluminides. *JOM* **1994**, *46*, 30–39. [CrossRef]

4. Dimiduk, D.M. Gamma titanium aluminide alloys—An assessment within the competition of aerospace structural materials. *Mater. Sci. Eng. A* **1999**, *263*, 281–288. [CrossRef]

5. Loria, E.A. Quo vadis gamma titanium aluminide. *Intermetallics* **2001**, *9*, 997–1001. [CrossRef]

6. Cao, J.; Qi, J.; Song, X.; Feng, J. Welding and joining of titanium aluminides. *Materials* **2014**, *7*, 4930–4962. [CrossRef]

7. Noda, T.; Shimizu, T.; Okabe, M.; Iikubo, T. Joining of TiAl and steels by induction brazing. *Mater. Sci. Eng. A* **1997**, *239–240*, 613–618. [CrossRef]

8. Dong, H.; Yang, Z.; Yang, G.; Dong, C. Vacuum brazing of TiAl alloy to 40Cr steel with $Ti_{60}Ni_{22}Cu_{10}Zr_8$ alloy foil as filler metal. *Mater. Sci. Eng. A* **2013**, *561*, 252–258. [CrossRef]

9. He, P.; Feng, J.C.; Zhang, B.G.; Qian, Y.Y. Microstructure and strength of diffusion-bonded joints of TiAl base alloy to steel. *Mater. Charact.* **2002**, *48*, 401–406. [CrossRef]

10. He, P.; Feng, J.C.; Zhang, B.G.; Qian, Y.Y. A new technology for diffusion bonding intermetallic TiAl to steel with composite barrier layers. *Mater. Charact.* **2003**, *50*, 87–92. [CrossRef]

11. Han, W.; Zhang, J. Diffusion bonding between TiAl based alloys and steels. *J. Mater. Sci. Technol.* **2001**, *17*, 191–192.

12. He, P.; Yue, X.; Zhang, J.H. Hot pressing diffusion bonding of a titanium alloy to a stainless steel with an aluminum alloy interlayer. *Mater. Sci. Eng. A* **2008**, *486*, 171–176. [CrossRef]

13. Duckham, A.; Spey, S.J.; Wang, J.; Reiss, M.E.; Weihs, T.P.; Besnoin, E.; Knio, O.M. Reactive nanostructured foil used as a heat source for joining titanium. *J. Appl. Phys.* **2004**, *96*, 2336–2342. [CrossRef]

14. Ramos, A.S.; Vieira, M.T.; Duarte, L.; Vieira, M.F.; Viana, F.; Calinas, R. Nanometric multilayers: A new approach for joining TiAl. *Intermetallics* **2006**, *14*, 1157–1162. [CrossRef]

15. Duarte, L.I.; Ramos, A.S.; Vieira, M.F.; Viana, F.; Vieira, M.T.; Koçak, M. Solid-state diffusion bonding of gamma-TiAl alloys using Ti/Al thin films as interlayers. *Intermetallics* **2006**, *14*, 1151–1156. [CrossRef]

16. Ramos, A.S.; Vieira, M.T.; Simões, S.; Viana, F.; Vieira, M.F. Joining of superalloys to intermetallics using nanolayers. *Adv. Mater. Res.* **2009**, *59*, 225–229. [CrossRef]

17. Simões, S.; Viana, F.; Ventzke, V.; Koçak, M.; Ramos, A.S.; Vieira, M.T.; Vieira, M.F. Diffusion bonding of TiAl using Ni/Al multilayers. *J. Mater. Sci.* **2010**, *45*, 4351–4357. [CrossRef]

18. Simões, S.; Viana, F.; Koçak, M.; Ramos, A.S.; Vieira, M.T.; Vieira, M.F. Diffusion bonding of TiAl using reactive Ni/Al nanolayers and Ti and Ni foils. *Mater. Chem. Phys.* **2011**, *128*, 202–207. [CrossRef]

19. Simões, S.; Viana, F.; Koçak, M.; Ramos, A.S.; Vieira, M.T.; Vieira, M.F. Microstructure of reaction zone formed during diffusion bonding of TiAl with Ni/Al multilayer. *J. Mater. Eng. Perform.* **2012**, *21*, 678–682. [CrossRef]

20. Namazu, T.; Takemoto, H.; Fujita, H.; Nagai, Y.; Inoue, S. Self-propagating explosive reactions in nanostructured Al/Ni multilayer films as a localized heat process technique for MEMS. In Proceedings of 19th IEEE International Conference on Micro Electro Mechanical Systems, Istanbul, Turkey, 22–26 January 2006; pp. 286–289.

21. Zhang, J.; Wu, F.S.; Zou, J.; An, B.; Liu, H. Al/Ni multilayer used as a local heat source for mounting microelectronic components. In Proceedings of the International Conference on Electronic Packaging Technology & High Density Packaging, ICEPT-HDP 09, Beijing, China, 10–13 August 2009; pp. 838–842.

22. Simões, S.; Viana, F.; Ramos, A.S.; Vieira, M.T.; Vieira, M.F. Anisothermal solid-state reaction of Ni/Al nanometric multilayer. *Intermetallics* **2011**, *19*, 350–356. [CrossRef]

23. Simões, S.; Viana, F.; Ramos, A.S.; Vieira, M.T.; Vieira, M.F. Reaction-assisted diffusion bonding of TiAl alloy to steel. *Mater. Chem. Phys.* **2016**, *161*, 73–82. [CrossRef]

24. Simões, S.; Viana, F.; Ramos, A.S.; Vieira, M.T.; Vieira, M.F. Reaction Zone Formed during Diffusion Bonding of TiNi to Ti_6Al_4V using Ni/Ti Nanolayers. *J. Mater. Sci.* **2013**, *48*, 7718–7727.

25. Kim, J.S.; LaGrange, T.; Reed, B.W.; Taheri, M.L.; Armstrong, M.R.; King, W.E.; Browning, N.D.; Campbell, G.H. Imaging of transient structures using nanosecond *in situ* TEM. *Science* **2008**, *321*, 1472–1475. [CrossRef] [PubMed]

26. Oliver, W.C.; Pharr, G.M. An improved technique for determining hardness and elastic modulus using load and displacements sensing indentation experiments. *J. Mater. Res.* **1992**, *7*, 1564–1583. [CrossRef]

27. Kipp, D.O. Material Data Sheets. *MatWeb, LLC*, 2010; Online version. Available online: http://www.matweb.com (accessed on 25 April 2016).
28. Toprek, D.; Belosevic-Cavor, J.; Koteski, V. *Ab initio* studies of the structural, elastic, electronic and thermal properties of NiTi$_2$ intermetallic. *J. Phys. Chem. Solids* **2015**, *85*, 197–205. [CrossRef]

metals

MDPI

Article

Enhanced Mechanical Properties of MgZnCa Bulk Metallic Glass Composites with Ti-Particle Dispersion

Pei Chun Wong [1], Tsung Hsiung Lee [2], Pei Hua Tsai [2], Cheng Kung Cheng [1,*], Chuan Li [1], Jason Shian-Ching Jang [2,3,*] and J. C. Huang [4]

[1]   Department of Biomedical Engineering, National Yang-Ming University, Taipei 11221, Taiwan;
      s0925135546@gmail.com (P.C.W.); cli10@ym.edu.tw (C.L.)
[2]   Institute of Materials Science and Engineering, National Central University, Taoyuan 32001, Taiwan;
      pshunterbabu@gmail.com (T.H.L.); peggyphtsai@gmail.com (P.H.T.)
[3]   Department of Mechanical Engineering, National Central University, Taoyuan 32001, Taiwan
[4]   Department of Materials and Optoelectronic Science, National Sun Yat-Sen University,
      Kaohsiung 80424, Taiwan; jacobc@mail.nsysu.edu.tw
*   Correspondence: ckcheng@ym.edu.tw (C.K.C.); jscjang@ncu.edu.tw (J.S.C.J.); Tel.: +886-2-2826-7020 (C.K.C.);
     +886-3-426-7379 (J.S.C.J.); Fax: +886-2-2822-8557 (C.K.C.); +886-3-425-4501 (J.S.C.J.)

Academic Editor: Ana Sofia Ramos
Received: 16 February 2016; Accepted: 10 May 2016; Published: 17 May 2016

Abstract: Rod samples of $Mg_{60}Zn_{35}Ca_5$ bulk metallic glass composites (BMGCs) dispersed with Ti particles have been successfully fabricated via injection casting. The glass forming ability (GFA) and the mechanical properties of these Mg-based BMGCs have been systematically investigated as a function of the volume fraction (V_f) of Ti particles. The results showed that the compressive ductility increased with V_f. The mechanical performance of these BMGCs, with up to 5.4% compressive failure strain and 1187 MPa fracture strength at room temperature, can be obtained for the Mg-based BMGCs with 50 vol % Ti particles, suggesting that these dispersed Ti particles can absorb the energy of the crack propagations and can induce branches of the primary shear band into multiple secondary shear bands. It follows that further propagation of the shear band is blocked, enhancing the overall plasticity.

Keywords: composite; Mg-Zn-Ca; bulk metallic glass; self-degraded ability; mechanical properties

1. Introduction

Organ protection, muscle connection, and muscle action promotion are the most important functions of bone [1]. Once the bone is fractured due to impact or other external forces during, for example, sports activities or accidents, it needs to be reconstructed to recover its normal function. Therefore, bone screws, bone plates, and bone substitute materials are usually applied to fixate the fractured bones via medical surgery. In general, austenitic stainless steel, Co–Cr alloys, and Ti alloys have been commercially used in orthopedic implants, taking advantage of their good mechanical properties, biocompatibility, and corrosion resistance. However, most metallic orthopedic implant metals (except Ti alloys) are potentially cytotoxic when the implant parts are retained in the body for long periods of time. Therefore, medical doctors may suggest that the patient undergo a second operation to take out the old prosthesis and to replace a new one. Accordingly, to avoid secondary surgery and reduce surgical risk, biodegradable materials have become a trend for biomaterials development. Metallic biodegradable materials have attracted great attention because of their promising mechanical properties, good biocompatibility, and low degradation rate. The metallic biodegradable materials, including W-based, Ca-based, Fe-based, Zn-based, and Mg-based alloys, have been wildly studied for their physical properties, chemical properties, and their biocompatibility in the last decade.

Among these metallic biodegradable materials, Mg-based alloys have attracted more attention as biomedical materials because of their good biocompatibility and appropriate biodegradability. In parallel, the strength and Young's modulus of Mg-based alloys are closer to those of bone tissue, reducing the stress shielding effect. Though the Mg-based materials have great potential for applications on many biomedical instruments and orthopedic implants, the corrosion and degradation rates of most Mg-based crystalline alloys are still too fast. Conversely, MgZnCa bulk metallic glass composites (BMGCs) were found to present much lower corrosion rates than the other Mg-based crystalline alloys. Unfortunately, monolithic MgZnCa bulk metallic glasses are quite brittle, limiting their further applications [2–6]. There have hitherto been a number of studies reporting on effective methods that enhance the strength as well as the plasticity of Mg-based BMGCs through the dispersion of ductile metallic particles such as those of Mo, Fe, Nb, and Ti [7–17]. These ductile metallic particles can absorb the shear strain energy of shear bands, confine the propagation of shear bands, and significantly improve their plasticity. In this study, Ti particles were selected to add to the MgZnCa metallic glasses to improve their plasticity because Ti metal is well-known for having non-toxicity, good biocompatibility, and an immiscible reaction with the parent Mg metal.

2. Experimental

$Mg_{60}Zn_{35}Ca_5$-based BMGCs with different volume fractions of Ti particles (V_f = 20, 30, 40, and 50 vol %) were prepared by induction melting under an argon atmosphere. At first, high purity Mg, Zn, Ca (>99.9%), and pure Ti particles (three distributions in diameters, namely, 20–75, 75–105 and 105–130 μm) were melted together by induction melting under the argon atmosphere. During melting, the melt was churned mechanically to assure a final ingot containing a homogeneous mixture of Ti particles and melt. The $Mg_{60}Zn_{35}Ca_5$–Ti composite ingot was re-melted in a quartz tube and injected into a water-cooled Cu mold by argon pressure to form Mg-based BMGC rods, measuring 20 mm in length and 2 mm in diameter.

The amorphous state of $Mg_{60}Zn_{35}Ca_5$-BMGCs was examined by X-ray diffraction (XRD, Shimadzu XRD6000, Shimadzu Corporation, Kyoto, Japan) with monochromatic Cu–Kα radiation. The chemical compositions of samples were verified by energy dispersive spectroscopy (EDS, FEI, Quanta 3D FEG, FEI Corporation, Hillsboro, OR, USA) to confirm their compositions as the original design. The thermal properties were analyzed by differential scanning calorimetry (DSC, Mettler DSC1, Mettler-Toledo International Inc, Greifensee, Switzerland) under an argon atmosphere at a heating rate of 40 K/min. The DSC results were then used to calculate the glass forming ability (GFA). The uniaxial compression tests were carried out by a universal test system (Hung Ta, HT9102, Taipei, Taiwan) at a strain rate of 1×10^{-4} s^{-1}. The specimens were cut from the as-cast rods into compression test specimens measuring 4 mm in length and 2 mm in diameter. The latter specimens were polished for both ends to confirm the surface flatness. The tests of hardness and fracture toughness (K_C, stress concentration factor) for all Mg-based BMGCs were carried out via the indentation method [18] with a microhardness tester (Mitutoyo, HM-221, Kanagawa, Japan) with a load of 1–2 kgf. Both the surface morphology and fractography of compressed specimens were examined by scanning electron microscopy (SEM. FEI Quanta 3D FEG).

3. Result and Discussion

The XRD patterns of the $Mg_{60}Zn_{35}Ca_5$ base alloy, $Mg_{60}Zn_{35}Ca_5$-based BMGCs with 50 vol % Ti particles, and pure Ti particles are shown in Figure 1. Both XRD patterns of the base alloy with and without Ti particles show an amorphous nature with a broadened diffuse hump, except for several peaks from the crystalline Ti particles co-existing in the XRD pattern of the composite sample. Some of the crystalline peaks were quite unclear because of another phase precipitated out around the Ti particles, as shown in Figure 2.

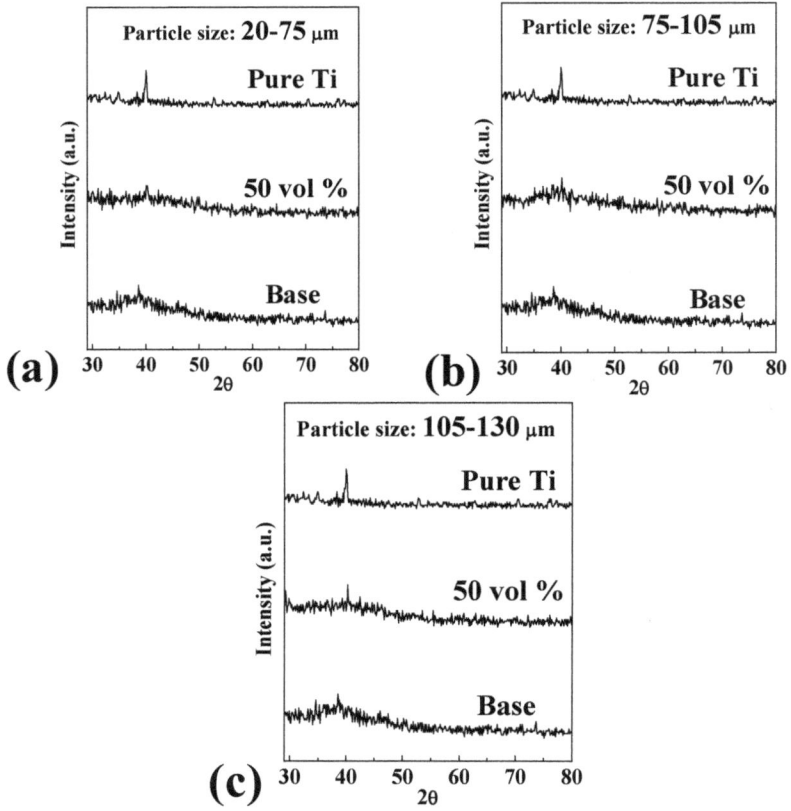

Figure 1. X-ray diffraction (XRD) pattern of the pure Ti particle, $Mg_{60}Zn_{35}Ca_5$ base alloy, and $Mg_{60}Zn_{35}Ca_5$-based bulk metallic glass composites (BMGCs) containing 50 vol % Ti particles with different particle sizes of (**a**) 20–75 μm; (**b**) 75–105 μm; (**c**) 105–130 μm.

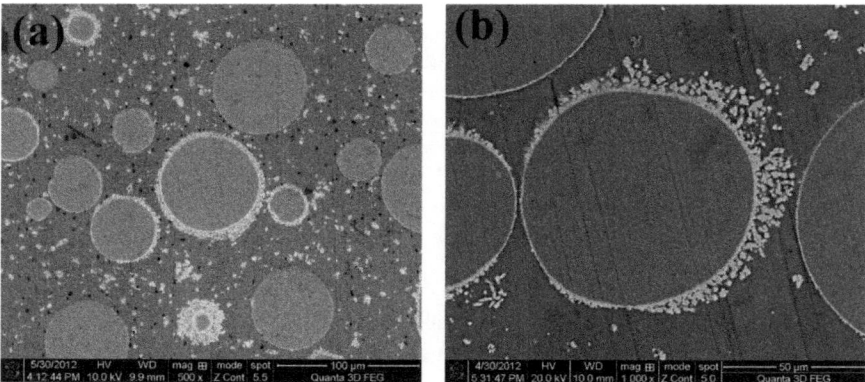

Figure 2. *Cont.*

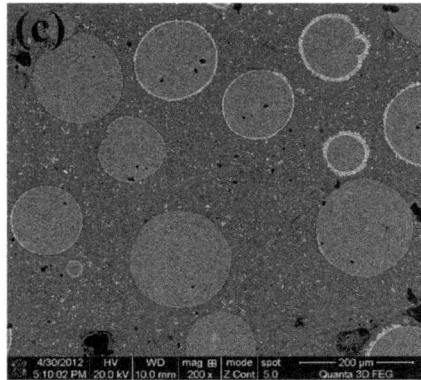

Figure 2. Back-scattering electron images of $Mg_{60}Zn_{35}Ca_5$-based BMGC samples containing different particle size of 50 vol % Ti particles: (**a**) 20–75 μm; (**b**) 75–105 μm; and (**c**) 105–130 μm.

The representative DSC scans of the $Mg_{60}Zn_{35}Ca_5$-based BMGCs with different volume fractions of 75–105 μm Ti particles are shown in Figure 3. Regardless of the inserted Ti particle size, both the glass transition temperature (T_g) and crystallization temperature (T_x) of $Mg_{60}Zn_{35}Ca_5$-based BMGCs exhibit a decreasing trend with an increasing volume fraction of Ti particles. This result indicates that both the glass transition and crystallization were trigged earlier by the presence of a large amount of crystalline Ti particles. The mechanism of such changes might be attributed to the numerous increases of heterogeneous nucleation sites from the interfaces between the Mg-based amorphous matrix and Ti particles. In addition, the calculated GFA indices were also found to decrease with an increasing volume fraction of Ti particles as shown in Table 1, supporting the mentioned mechanism of a lowering reaction barrier for crystallization.

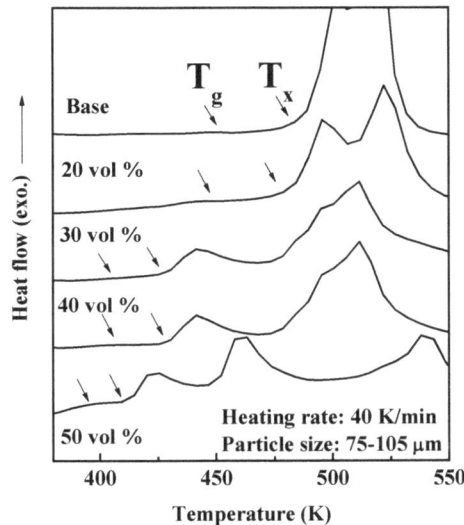

Figure 3. Representative differential scanning calorimetry (DSC) scans of $Mg_{60}Zn_{35}Ca_5$-based BMGC samples containing different volume fractions of 75–105-μm-sized Ti particles.

Table 1. Glass forming ability of $Mg_{60}Zn_{35}Ca_5$base alloy and $Mg_{60}Zn_{35}Ca_5$-based BMGCs.

Particle Size (D)	vol %	T_g (K)	T_x (K)	T_l (K)	T_{rg}	γ	γ_m	ΔT_x
Base		446	488	700	0.638	0.426	0.758	42
20–75 μm	20	400	430	675	0.593	0.400	0.681	30
	30	400	429	671	0.597	0.400	0.683	29
	40	403	432	674	0.598	0.401	0.684	29
	50	395	418	676	0.585	0.391	0.654	23
75–105 μm	20	394	433	679	0.580	0.404	0.698	40
	30	402	430	673	0.599	0.400	0.681	28
	40	404	429	672	0.601	0.399	0.677	25
	50	393	414	673	0.584	0.388	0.646	21
105–130 μm	20	408	435	671	0.609	0.403	0.688	27
	30	405	431	681	0.595	0.396	0.670	25
	40	400	425	680	0.588	0.394	0.663	25
	50	400	425	682	0.587	0.392	0.658	24

$$T_{rg} = \frac{T_g}{T_l}; \gamma = \frac{T_x}{T_g+T_l}; \gamma_m = \frac{2T_x-T_g}{T_l}.$$

Some representative back-scattering SEM images of the Mg-based BMGCs containing different sizes and a fixed volume fraction of Ti particles are shown in Figure 2. At a fixed volume fraction of Ti particles, the average inter-particle space of Ti particles clearly shows a decreasing trend with a decreasing size of Ti particles in the $Mg_{60}Zn_{35}Ca_5$-based BMGCs. The numerical values of inter-particle space of Ti particles for various combinations of particle size and volume fraction are summarized in Table 2. In addition, an extra phase was found to precipitate along the interface of the Ti particle/amorphous matrix around some Ti particles. These precipitates around the Ti particles were identified to be the TiZn compound based on its chemical composition (Ti: 52 atom %/Zn 48 atom %) analyzed by energy dispersive spectroscopy.

Table 2. Inter-particle spacing of $Mg_{60}Zn_{35}Ca_5$-based BMGCs.

Particle Size (D)	vol %	Inter-Particle Spacing (Edge to Edge, μm)	
		Calculated	Measured
20–75 μm (Average 48 μm)	20	59	53 ± 11
	30	40	42 ± 7
	40	28	36 ± 8
	50	18	19 ± 3
75–105 μm (Average 90 μm)	20	110	91 ± 21
	30	74	74 ± 14
	40	52	68 ± 12
	50	37	34 ± 4
105–130 μm (Average 120 μm)	20	148	135 ± 26
	30	99	117 ± 21
	40	68	83 ± 26
	50	50	65 ± 11

Figure 4 shows the compression stress–strain curves for the $Mg_{60}Zn_{35}Ca_5$-based BMGC samples (V_f = 20, 30, 40 and 50 vol %) tested at a strain rate of 1×10^{-4} s^{-1} at room temperature. The results show that the *ex-situ*-added Ti particles exhibit a positive effect on strengthening the $Mg_{60}Zn_{35}Ca_5$-based BMG. For the base Mg-based BMG, it only shows a compression strength of 655 MPa with a brittle fracture mode. On the contrary, the fracture strength level of $Mg_{60}Zn_{35}Ca_5$-based BMGCs was significantly increased by increasing the volume fraction of *ex situ*-added Ti particles. The fracture strength can reach 1190 MPa for the Mg-based BMGC with 50 vol % 20–75 μm Ti particles.

However, most BMGCs only show low failure strains, less than 3%. The brittle behavior of all Mg-based BMGC samples are presumed due to the brittle TiZn intermetallic compound that forms around the interface of the Ti particle/amorphous matrix and embrittle the Mg-based BMGC samples. Therefore, it is suggested that the Ti dispersoids cannot entirely restrict the propagation of the shear band due to its weak interface adhesion to the matrix. This can be evidenced from Figure 6, in which Ti particles were sheared apart from the matrix.

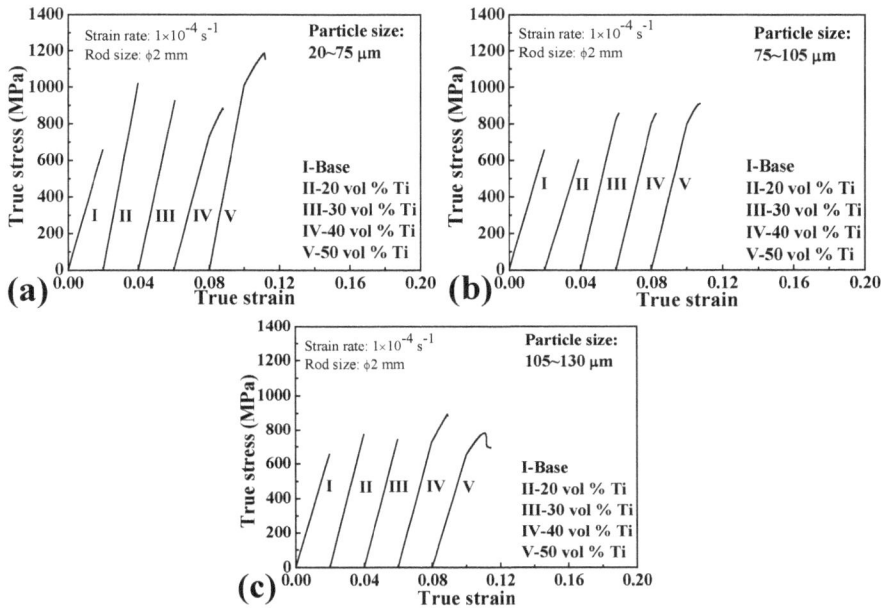

Figure 4. True stress–strain curve of $Mg_{60}Zn_{35}Ca_5$ BMGC rods containing different volume fractions and particles sizes of the Ti particles: (**a**) 20–75 μm; (**b**) 75–105 μm; (**c**) 105–130 μm.

The *ex situ* method may be a good possible way to enhance the plasticity of Mg-based BMGCs. In previous studies, the energy of the shear band can be absorbed by the dispersed particles (such as Fe, Mo, Ti, or Nb particles) in the Mg-based amorphous matrix alloys, and the shear band cannot easily propagate across the particles, significantly enhancing the plasticity of Mg-based BMGCs. Therefore, the combination of the large volume fraction and small particle size would create small inter-particle spacing of the dispersion phase to confine the propagation of shear bands and avoid the materials failure [15,18]. However, this theory does not seem to be applicable in every situation. In this study, the Mg-based BMGC samples with a smaller dispersoid size (20–75 μm) would present a lower failure strain than the Mg-based BMGC samples with a larger dispersoid size (75–105 μm). The reason is postulated to be a result of the brittle TiZn intermetallic compound precipitated out along the interface of the Ti particle/amorphous matrix, deteriorating the adhesion between the Ti particle and the amorphous matrix. Under the same volume fractions of the particles, the smaller particle size would create more interface areas of the Ti particle/amorphous matrix. Therefore, the interface of the Ti particle/amorphous matrix becomes a critical position of the crack initiation, leading to lower plastic deformation.

The dependency of the stress concentration factor (*Kc*) [19] and compression strain as a function of the interspacing of Ti particles for the Mg-based BMGCs is presented in Figure 5. Both the fracture toughness and the compression failure strain show a similar trend; they increase with a decreasing interspacing of Ti particles, the mean free path of the shear band. It is apparent that the mean free path plays an important role in enhancing the toughness of $Mg_{60}Zn_{35}Ca_5$-based BMGCs.

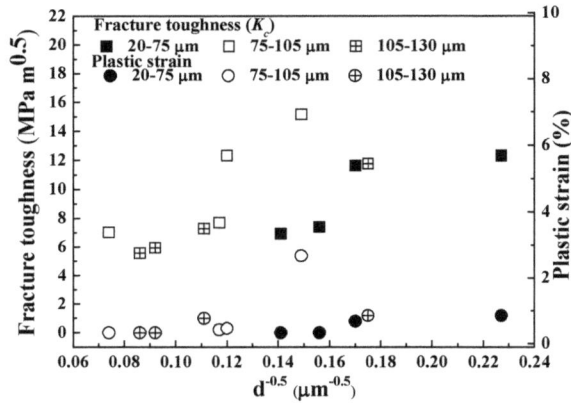

Figure 5. K_c (stress concentration factor) and compressive plastic strain as a function of the interspacing of various Ti particle additions of the $Mg_{60}Zn_{35}Ca_5$ BMGC rods with 2 mm in diameter.

The fracture surface of these $Mg_{60}Zn_{35}Ca_5$-based BMGC samples after the compression test exhibits two kinds of morphology. One is the rough fracture surface with a vein pattern and the other is the wavy surface with plastically sheared Ti particles, as shown in Figure 6. Figure 6a shows the relatively homogeneous distributed Ti particles on the fracture surface of Mg-based BMGC samples. Some Ti particles were peeled off from the matrix, implying a weak adhesion of some interfaces of the Ti particle/amorphous matrix. In addition, the melting trace and vein pattern can be seen clearly in Figure 6b. The plastically deformed Ti particles on the fracture surface indicate that the energy of the shear band has been effectively absorbed by these ductile Ti particles. This suggests that the dispersed Ti particles can limit the propagation of shear banding.

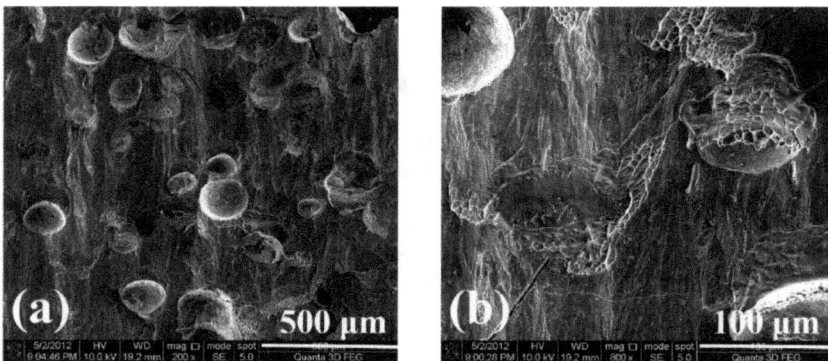

Figure 6. Scanning electron microscopy (SEM) micrographs of (**a**) fracture surface of specimen after compression test for the $Mg_{60}Zn_{35}Ca_5$ BMGC containing 50 vol % Ti particles with sizes of 75–105 μm; (**b**) Enlarged image of Figure 6 (**a**), some areas display vein pattern mixed with melting trace around Ti particles.

4. Conclusions

In summary, Mg-based BMG composites with different particle sizes and the volume fraction of spherical Ti particles were successfully developed in this study. The results show that *ex-situ*-added Ti particles exhibit a positive effect on strengthening $Mg_{60}Zn_{35}Ca_5$-based BMGCs. In addition, both the fracture toughness and the compression failure strain show a similar trend, increasing with a decreasing mean free path of the shear band (the interspacing of Ti particles). However, the performance of these Mg-based BMGCs is strongly dependent on the adhesion ability between the interface of the Ti particle/amorphous matrix. Overall, the dispersion of Ti particles in $Mg_{60}Zn_{35}Ca_5$ BMG alloys is considered to be a possible way in enhancing their plasticity as well as yield strength.

Acknowledgments: The authors would like to grateful acknowledge the sponsorship from Ministry of Science and Technology, Taiwan, under the Contract No. NSC101-2221-E-008-043-MY3, MOST103-2120-M-110-004, and MOST103-2221-E-008-028-MY3.

Author Contributions: Pei Chun Wong and Jason Shian-Ching Jang conceived and designed the experiments; Pei Chun Wong, Tsung Hsiung Lee and Pei Hua Tsai performed the experiments; Pei Chun Wong, Tsung Hsiung Lee and Pei Hua Tsai analyzed the data; Pei Chun Wong, Cheng Kung Cheng, Chuan Li, Jason Shian-Ching Jang and J. C. Huang contributed reagents, materials and analysis tools; Pei Chun Wong, Pei Hua Tsai and Jason Shian-Ching Jang wrote the paper.

Conflicts of Interest: The authors declare no conflict of interest.

References

1. Mordin, M.; Frankel, V.H. *Basic Biomechanics of the Musculoskeletal System*, 4th ed.; Lippincott Williams & Willkin: Philadelphia, PA, USA, 2012.

2. Zberg, B.; Uggowitzer, P.J.; Loffler, J.F. MgZnCa glasses without clinically observable hydrogen evolution for biodegradable implants. *Nat. Mater.* **2009**, *8*, 887–891. [CrossRef] [PubMed]

3. Gu, X.N.; Zheng, Y.F.; Zhong, S.P.; Xi, T.F.; Wang, J.Q.; Wang, W.H. Corrosion of, and cellular response to Mg-Zn-Ca bulk metallic glasses. *Biomaterials* **2010**, *31*, 1093–1103. [CrossRef] [PubMed]

4. Li, Q.F.; Weng, H.R.; Suo, Z.Y.; Ren, Y.L.; Yuan, X.G.; Qiu, K.Q. Microstructure and mechanical properties of bulk Mg-Zn-Ca amorphous alloy and amorphous matrix composites. *Mater. Sci. Eng. A* **2008**, *487*, 301–308. [CrossRef]

5. Zhao, Y.Y.; Ma, E.; Xu, J. Reliability of compressive fracture strength of Mg-Zn-Ca bilk metallic glass: Flaw sensitivity and Weibull statistics. *Scr. Mater.* **2009**, *58*, 496–499. [CrossRef]

6. Zberg, B.; Arata, E.R.; Uggowitzer, P.J.; Loffler, J.F. Tensile properties of glassy MgZnCa wires and reliability analysis using Weibull statistics. *Acta Mater.* **2009**, *57*, 3223–3231. [CrossRef]

7. Jang, J.S.C.; Li, T.H.; Jian, S.R.; Huang, J.C.; Nieh, T.G. Effects of characteristics of Mo dispersions on the plasticity of Mg-based bulk metallic glass composites. *Intermetallics* **2011**, *19*, 738–743. [CrossRef]

8. Hsieh, P.J.; Yang, L.C.; Su, H.C.; Lu, C.C.; Jang, J.S.C. Improvement of mechanical properties in MgCuYNdAg bulk metallic glasses with adding Mo particles. *J. Alloy. Compd.* **2010**, *504*, 98–101. [CrossRef]

9. Jang, J.S.C.; Chang, Y.S.; Li, T.H.; Hsieh, P.J.; Huang, J.C.; Tsao, C.Y. Plasticity enhancement of Mg58Cu28.5Gd11Ag2.5 based bulk metallic glass composites dispersion strengthened by Ti particles. *J. Alloy. Compd.* **2010**, *504*, 102–105. [CrossRef]

10. Kinaka, M.; Kato, H.; Hasegawa, M.; Inoue, A. High specific strength Mg-based metallic glass matrix composite highly ductilized by Ti dispersoid. *Mater. Sci. Eng. A* **2008**, *494*, 299–303. [CrossRef]

11. Shanthi, M.; Gupta, M.; Jarfors, A.E.W.; Tan, M.J. Synthesis characterization and mechanical properties of nano alumina particulate reinforced magnesium based bulk metallic glass composites. *Mater. Sci. Eng. A* **2011**, *528*, 6045–6050. [CrossRef]

12. Xu, Y.K.; Ma, H.; Xu, J.; Ma, E. Mg-based bulk metallic glass composites with plasticity and gigapascal strength. *Acta Mater.* **2005**, *53*, 1857–1866. [CrossRef]

13. Jang, J.S.C.; Jian, S.R.; Li, T.H.; Huang, J.C.; Tsao, C.Y.; Liu, C.T. Structural and mechanical characterizations of ductile Fe particles-reinforced Mg-based bulk metallic glass composites. *J. Alloy. Compd.* **2009**, *485*, 290–294. [CrossRef]

14. Jang, J.S.C.; Ciou, J.Y.; Li, T.H.; Huang, J.C.; Nieh, T.G. Dispersion toughening of Mg-based bulk metallic glass reinforced with porous Mo particles. *Intermetallics* **2010**, *18*, 451–458. [CrossRef]

15. Jang, J.S.C.; Li, J.B.; Lee, S.L.; Chang, Y.S.; Jian, S.R.; Huang, J.C.; Nieh, T.G. Prominent plasticity of Mg-based bulk metallic glass composites by *ex-situ* spherical Ti particles. *Intermetallics* **2012**, *30*, 25–29. [CrossRef]

16. Wang, J.F.; Huang, S.; Wei, Y.Y.; Guo, S.F.; Pan, F.S. Enhanced mechanical properties and corrosion resistance of a Mg-Zn-Ca bulk metallic glass composite by Fe particle addition. *Mater. Lett.* **2013**, *91*, 311–314. [CrossRef]

17. Wang, J.F.; Huang, S.; Li, Y.; Wei, Y.Y.; Guo, S.F.; Pan, F.S. Ultrahigh strength MgZnCa eutectic alloy/Fe particle composites with excellent plasticity. *Mater. Lett.* **2014**, *137*, 139–142. [CrossRef]

18. Xi, X.K.; Zhao, D.Q.; Pan, M.X.; Wang, W.H.; Wu, Y.; Lewandowski, J.J. Fracture of Brittle Metallic Glasses: Brittleness or Plasticity. *Phys. Rev. Lett.* **2005**, *94*. [CrossRef] [PubMed]

19. Meyers, M.A.; Chawla, K.K. *Mechanical Behavior of Materials*; Cambridge University Press: Cambridge, UK, 2009.

metals

MDPI

Article

The Effects of Al and Ti Additions on the Structural Stability, Mechanical and Electronic Properties of D8$_m$-Structured Ta$_5$Si$_3$

Linlin Liu [1], Jian Cheng [1], Jiang Xu [1,2,*], Paul Munroe [3] and Zong-Han Xie [2,4,*]

[1] Department of Material Science and Engineering, Nanjing University of Aeronautics and Astronautics, 29 Yudao Street, Nanjing 210016, China; liulin060410311@126.com (L.L.); 18855533109@126.com (J.C.)
[2] School of Mechanical & Electrical Engineering, Wuhan Institute of Technology, 693 Xiongchu Avenue, Wuhan 430073, China
[3] School of Materials Science and Engineering, University of New South Wales, Sydney NSW 2052, Australia; p.munroe@unsw.edu.au
[4] School of Mechanical Engineering, University of Adelaide, Adelaide SA 5005, Australia
*   Correspondence: xujiang73@nuaa.edu.cn (J.X.); zonghan.xie@adelaide.edu.au (Z.-H.X.); Tel.: +86-25-5211-2626 (J.X.); +61-883-133-980 (Z.-H.X.)

Academic Editor: Ana Sofia Ramos
Received: 6 January 2016; Accepted: 30 April 2016; Published: 26 May 2016

Abstract: In the present study, the influence of substitutional elements (Ti and Al) on the structural stability, mechanical properties, electronic properties and Debye temperature of Ta$_5$Si$_3$ with a D8$_m$ structure were investigated by first principle calculations. The Ta$_5$Si$_3$ alloyed with Ti and Al shows negative values of formation enthalpies, indicating that these compounds are energetically stable. Based on the values of formation enthalpies, Ti exhibits a preferential occupying the Ta^{4b} site and Al has a strong site preference for the Si^{8h} site. From the values of the bulk modulus (B), shear modulus (G) and Young's modulus (E), we determined that both Ti and Al additions decrease both the shear deformation resistance and the elastic stiffness of D8$_m$ structured Ta$_5$Si$_3$. Using the shear modulus/bulk modulus ratio (G/B), Poisson's ratio (υ) and Cauchy's pressure, the effect of Ti and Al additions on the ductility of D8$_m$-structured Ta$_5$Si$_3$ are explored. The results show that Ti and Al additions reduce the hardness, resulting in solid solution softening, but improve the ductility of D8$_m$-structured Ta$_5$Si$_3$. The electronic calculations reveal that Ti and Al additions change hybridization between Ta-Si and Si-Si atoms for the binary D8$_m$-structured Ta$_5$Si$_3$. The new Ta-Al bond is weaker than the Ta-Si covalent bonds, reducing the covalent property of bonding in D8$_m$-structured Ta$_5$Si$_3$, while the new strong Ti^{4b}-Ti^{4b} anti-bonding enhances the metallic behavior of the binary D8$_m$-structured Ta$_5$Si$_3$. The change in the nature of bonding can well explain the improved ductility of D8$_m$-structured Ta$_5$Si$_3$ doped by Ti and Al. Moreover, the Debye temperatures, Θ_D, of D8$_m$-structured Ta$_5$Si$_3$ alloying with Ti and Al are decreased as compared to the binary Ta$_5$Si$_3$.

Keywords: transition metal silicides; D8$_m$-Ta$_5$Si$_3$; first principle calculation; mechanical properties

1. Introduction

In order to meet the ever-increasing demand for high performing and durable structural components to be used in harsh environments, much attention has been focused upon transition metal silicides [1,2]. As the largest family of intermetallic compounds, transition metal silicides are of great interest to a wide variety of applications due to their unique properties, such as ultra-high melting temperatures, excellent electrical properties, superior thermal stability, high oxidation resistance, good creep tolerance, and excellent mechanical strength at elevated temperature [3–5]. Compared with other transition metal silicides, such as Mo and Ti silicides, that have been extensively investigated over the past few decades, Ta-Si compounds with higher melting point than the former two silicides systems have received limited attention. As the most refractory compound in the Ta-Si binary system [6], Ta_5Si_3 has a melting point of 2550 °C, and thus has a great potential for structural applications at ultra-high temperatures. Moreover, although the high cost and density of Ta make its direct clinical application in a bulk form difficult, tantalum-based compounds are particularly suitable for use as implant coatings to improve the performance of metal substrates, due to their good corrosion resistance, exceptional biocompatibility, osseointegration properties, hemocompatibility, and high radiopacity [7]. The Ta_5Si_3 was reported to have three different polymorphs: Mn_5Si_3, W_5Si_3 and Cr_5B_3 type [8]. A $D8_8$-structured Ta_5Si_3 (Mn_5Si_3 prototype) with the $P6_3/mcm$ symmetry is a metastable phase, a $D8_1$-structured Ta_5Si_3 (Cr_5B_3 prototype) with the $I4/mcm$ symmetry is the low-temperature phase and a $D8_m$-structured Ta_5Si_3 (W_5Si_3 prototype) with the $I4/mcm$ symmetry is the high-temperature phase. Tao *et al.* [9] predicted the ductile behavior of the three prototype structures for Ta_5Si_3 and they found that the Cr_5B_3-prototype phase is slightly prone to brittleness, and the W_5Si_3-prototype phase is more prone to ductility. Therefore, in comparison to the other two prototype structures, $D8_m$-structured Ta_5Si_3 exhibits more potential as a candidate material for structural applications under an aggressive environment and has been chosen as the subject of our present work. However, as with many other intermetallic materials, owing to its strong covalent-dominated atomic bonds and intrinsic difficulties of dislocation movement, its low toughness poses a serious obstacle to its commercial application. To address this problem, several effective strategies have been developed, including densification [10,11], grain refinement [12], substitutional alloying [13], and the incorporation of a second reinforcing phase to form a composite [14]. In contrast to the ionic or covalent bonding that is prevalent in ceramics, the partial metallic character of bonding endows metal silicides with a degree of alloying ability, making it possible to improve toughness through an alloying approach. An alloying approach is achieved either by doping the metal silicides with interstitial atoms [15], or by the incorporation of substitutional atoms into the lattice of metal silicides crystal [13].

Compared with additions of interstitial atoms, incorporation of substitutional atoms into Ta_5Si_3 is more complex, and the site substitution of the added alloying elements in Ta_5Si_3 plays an important role in influencing the electronic structure and hence the mechanical properties of the material. Unfortunately, to our knowledge, no first-principles calculations have been employed to elucidate the effects of substitutional elements on the electronic structure and mechanical properties of $D8_m$-structured Ta_5Si_3. In this paper, Ti and Al were selected to investigate the site occupancy behavior of the two elements in $D8_m$-structured Ta_5Si_3 and, based upon that, the elastic moduli including bulk modulus, shear modulus, and Young's modulus were calculated for the binary and ternary Ta_5Si_3 with $D8_m$-structure. This, in turn, was used to gain insight into the effect of the two substitutional elements on the electronic structure and mechanical properties of $D8_m$-structured Ta_5Si_3.

2. Calculation Details

The first-principal calculations in this work were performed based on density functional theory (DFT) by using the Cambridge Sequential Total Energy Package code (CASTEP) [16]. The electron exchange and correlations were treated within the generalized gradient approximation using the Perdew, Burke and Ernzerhof function (GGA-PBE) [17]. The interactions between the electron and ionic cores were described by the projector augmented wave (PAW) [18] method. The ultrasoft pseudo-potentials set $5p^65d^36s^2$ for Ta, $3s^23p^2$ for Si, $3s^23p^1$ for Al and $3p^63d^24s^2$ for Ti as valence electrons, respectively, for each element. A 350 eV cutoff energy determined by the converged test for the total-energy remained constant for both the binary and ternary Ta_5Si_3 compounds. The special points sampling integration over the Brillouin zone were carried out using the Monkhorst-Pack *k*-points (the reciprocal space) mesh grid [19]. Confirming convergence to a precision of 1 meV/atom, a $4 \times 4 \times 4$ Monkhorst-Pack grid of *k*-points was adopted. Geometry optimization was processed to create a steady bulk structure with the lowest energy close to the natural structure by adjusting the lattice parameters of the models. In the optimization process, the energy change, maximum force, maximum stress, and maximum displacement tolerances were set as 5.0×10^{-6} eV/atom, 0.01 eV/Å, 0.02 GPa and 5.0×10^4 Å, respectively. All calculations were performed under non-spin-polarized condition.

A stoichiometric $D8_m$-structured Ta_5Si_3 unit cell consisting of 20 Ta atoms and 12 Si atoms is used for the computations. The atomic configuration of $D8_m$-structured Ta_5Si_3 is shown in Figure 1. The stacking sequence of layers along the c-axis for this structure is the same as that for $D8_m$-structured Mo_5Si_3 and can be viewed as repeated ABAC ... stacking (see Figure 1b). The A-layer has a more open atomic environment (*i.e.*, lower atomic packing density), which links together with the more close-packed B- and C-layers through the transition metal-silicon bonds [20]. Both Ta and Si have two symmetrically non-equivalent positions in the lattice. The elements that are localized on the more close-packed planes are termed Ta^{16k} and Si^{8h}, and those on the less close-packed planes are termed Ta^{4b} and Si^{4a}. For the determination of which of the doped sites is the preferential site for the alloying elements under consideration (Ti and Al). $Ta_{20}Si_{12}$ is substituted by either one Ti or one Al atom (that is, the additions concentration is 3.125 at.% (atomic concentration)), resulting in four possible substitution structures: $Ta_{19}M^{16k}Si_{12}$, $Ta_{19}M^{4b}Si_{12}$, $Ta_{20}Si_{11}M^{8h}$ and $Ta_{20}Si_{11}M^{4a}$ (M = Ti or Al). The enthalpies of formation of the four substitutional structures were calculated according to the following equation:

$$E_{form} = \frac{1}{N}\left(E(Ta_\alpha Si_\beta M) - \alpha E(Ta) - \beta E(Si) - E(M)\right) \tag{1}$$

where M denotes the alloying element Ti or Al, N represents the total atom number of $Ta_\alpha Si_\beta M$ (M is Ti or Al), and α, β means the atom number of Ta, Si. $E(Ta_\alpha Si_\beta M)$ is the value for the enthalpy of formation of $Ta_\alpha Si_\beta M$. $E(Ta)$, $E(Si)$ and $E(M)$ are the ground states enthalpy of per atom for Ta, Si, Al and Ti, respectively.

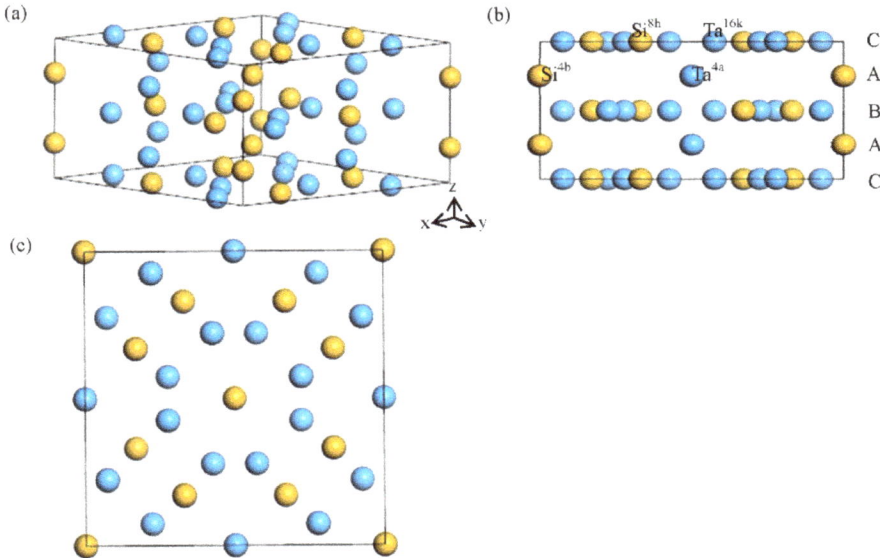

Figure 1. The crystal structure of $D8_m$ Ta_5Si_3: (**a**) 3D model; (**b**) side view; (**c**) top view. Ta atoms (**blue** balls) and Si atoms (**yellow** balls) have two symmetrically non-equivalent positions. The elements on the close-packed planes (represented by the B-layer and C-layer) are termed Ta^{16k} and Si^{8h}, and those on the less close-packed planes (represented by the A-layer) are termed Ta^{4a} and Si^{4b}.

The elastic constants can gauge the mechanical stress required to produce a given deformation. Both stress $[\sigma_{\alpha\beta}]$ and strain $[\varepsilon_{\alpha\beta}]$ are symmetric tensors of rank two, which can be described by matrix σ_i ($i = 1, 2, \ldots, 6$) and ε_j ($j = 1, 2, \ldots, 6$). As a result, the elastic constants can be represented by a 6×6 symmetric matrix $[C_{ij}]$ under the conditions of small stresses and small strains:

$$\sigma_i = \sum_j C_{ij}\varepsilon_j \tag{2}$$

Meanwhile, under the conditions of small stresses and small strains on a crystal, Hooke's law is applicable and the elastic energy ΔE is a quadratic function of the strains:

$$\Delta E = V \sum_{i,j=1}^{6} \frac{1}{2} C_{ij} e_i e_j \tag{3}$$

where V is the total volume of the unit cell, e_i are the components of the strain matrix [21]

$$\varepsilon' = \begin{pmatrix} e_1 & e_6 & e_5 \\ e_6 & e_2 & e_4 \\ e_5 & e_4 & e_3 \end{pmatrix} \tag{4}$$

For the symmetry of a tetragonal crystal ($a = b \neq c$, $\alpha = \beta = \gamma = 90°$), there are six independent elastic constants: C_{11}, C_{12}, C_{33}, C_{13}, C_{44}, and C_{66}. In order to determine these independent elastic constants, a set of six independent total-energy calculations are performed [22]. The lattice is deformed uniformly in six different ways, and for each of them we calculated the total energy as a function of strain (up to 2%–3%). The new lattice axes a'_i are related to the original ones a'_j by $a'_i = (I + \varepsilon') \times a_j$, where I is the identity matrix and

$$\varepsilon' = \begin{pmatrix} e_1 & e_6/2 & e_5/2 \\ e_6/2 & e_2 & e_4/2 \\ e_5/2 & e_4/2 & e_3 \end{pmatrix} \tag{5}$$

Thus, if the elastic energy of a crystal is known, the elastic constants at equilibrium volumes can be calculated using Equations (2) and (3).

For the cubic symmetry ($a = b = c$, $\alpha = \beta = \gamma = 90°$), its geometry renders some elastic constant components equal, resulting in the remaining three independent elastic constants; namely, C_{11}, C_{12} and C_{44}. The details of the calculations can be found elsewhere [9] and are not detailed here.

The effective elastic moduli of polycrystalline aggregates are usually calculated by two approximations according to Voigt [23] and Reuss [24], where uniform strain and stress are assumed throughout the polycrystal. Hill [25] has shown that the Voigt and Reuss averages are limited and suggested that the actual effective modulus can be approximated by the arithmetic mean of the two bounds, referred to as the Voigt-Reuss-Hill (VRH) value.

The Voigt bounds for the cubic system are:

$$B_V = 1/3(C_{11} + 2C_{12}) \tag{6}$$

$$G_V = 1/5(C_{11} - C_{12} + 3C_{44}) \tag{7}$$

and the Reuss bounds are:

$$B_R = 1/3(C_{11} + 2C_{12}) \tag{8}$$

$$G_R = \frac{5(C_{11} - C_{12})C_{44}}{4C_{44} + 3(C_{11} - C_{12})} \tag{9}$$

The Voigt bounds for tetragonal systems are:

$$B_V = 1/9[2(C_{11} + 2C_{12}) + C_{33} + 4C_{14})] \tag{10}$$

$$G_V = 1/30(M + 3C_{11} - C_{12} + 12C_{44} + 6C_{66}) \tag{11}$$

and the Reuss bounds are:

$$B_R = C^2/M \tag{12}$$

$$G_R = 15\{(18B_V/C^2) + [6/(C_{11} - C_{12}) + (6/C_{44}) + (3/C_{66})]\}^{-1} \tag{13}$$

$$M = C_{11} + C_{12} + 2C_{33} - 4C_{13} \tag{14}$$

$$C^2 = (C_{11} + C_{12})C_{13} - 2C_{13}{}^2 \tag{15}$$

For both cubic and tetragonal crystals, the VRH mean values are finally computed by:

$$B = 1/2(B_V + B_R) \tag{16}$$

$$G = 1/2(G_V + G_R) \tag{17}$$

The Young's modulus and the Poisson's ratio can be calculated based on the values of both bulk and shear modulus by the relations:

$$E = 9BG/(3B + G) \tag{18}$$

$$\upsilon = (3B - 2G)/2(3B + G) \tag{19}$$

3. Results and Discussion

3.1. Lattice Parameter

The structures of both pure elements and compounds are optimized by relaxing the simulation cell and the corresponding calculated lattice constants are obtained at a minimum energy. The phases of b.c.c tantalum, diamond-structured silicon, h.c.p titanium and f.c.c Al are used as the reference ground states. The calculated lattice constants, together with available experimental data and theoretical values, are listed in Table 1. The lattice constants of the pure elements are in good agreement with the values reported in the literature [26–30], giving confidence to the reliability of the data for this investigation. Figure 2 shows the change of unit cell volume *versus* the substitution sites occupied by one Ti or one Al atom. As shown in Figure 2, it can be seen that the change of unit cell volume is only over a range from −0.75% to 1.75%, indicating that the lattice distortion is very small when one of the four possible substitution sites is replaced by either Ti or Al. The substitution of Ti for Si in D8$_m$-structured Ta$_5$Si$_3$ slightly increases unit cell volume because the Ti atom has a larger atomic radius than the Si. After the substitution of Ti for Ta, the unit cell volume of D8$_m$-structured Ta$_5$Si$_3$ hardly changes due to a small difference of the atomic radius between Ti and Ta. The lattice expansion for Ti atom occupying the Ta^{4b} site is significantly smaller than that for Ti atom occupying the other three sites (*i.e.*, Ta^{16k}, Si^{4a} and Si^{8h}), implying a lower lattice distortion. Meanwhile, from Table 1, it is clear that only the Ti substitution for Ta^{4b} site, the lattice constants a and b are of equal value, which can maintain the tetragonal structure of D8$_m$ Ta$_5$Si$_3$. For Al substitutional alloying, it can be observed that Al substitution for Ta and Si cause lattice contraction and lattice expansion, respectively. As can be seen from Figure 2, Al atom occupying the Ta^{16k} site yields the lowest lattice distortion. However, from Table 1, only Al substitution for Si^{8h} site can retain the tetragonal structure of D8$_m$ Ta$_5$Si$_3$.

Table 1. Calculated lattice constants and enthalpies of formation for $D8_m$-structured $Ta_{20}Si_{12}$ and alloyed $D8_m$ $Ta_{20}Si_{12}$.

Phase	Type	Substitution Sites	Lattice Parameters			Enthalpy of Formation (eV/Atom)
			a (Å)	b (Å)	c (Å)	
Ta	W(A2)	-	3.309 3.306 [a]	-	-	-
Ti	-	-	2.981 2.954 b	-	-	-
$D8_m$ $Ta_{20}Si_{12}$	W_5Si_3	-	10.08 10.01 [c] 9.862 d	-	5.183 5.106 [c] 5.05 [d]	−0.619 −0.544 [c]
$Ta_{19}TiSi_{12}$	-	Ta^{16k}	10.09	10.09	5.199	−0.634
-	W_5Si_3	Ta^{4b}	10.084	-	5.197	−0.639
$Ta_{20}Si_{11}Ti$	-	Si^{4a}	10.12	10.12	5.221	−0.554
-	-	Si^{8h}	10.15	10.11	5.210	−0.557
$Ta_{19}AlSi_{12}$	-	Ta^{16k}	10.10	10.06	5.175	−0.591
-	-	Ta^{4b}	10.09	10.10	5.137	−0.602
$Ta_{20}Si_{11}Al$	-	Si^{4a}	10.08	10.15	5.199	−0.605
-	W_5Si_3	Si^{8h}	10.12	-	5.180	−0.605
Si	Si(A4)	-	5.462 5.433 e	-	-	-
Al	-	-	4.071 4.052 f	-	-	-

[a] Experiment [26]; [b] Experiment [27]; [c] Calculation [9]; [d] Experiment [28]; [e] Experiment [29]; [f] Experiment [30].

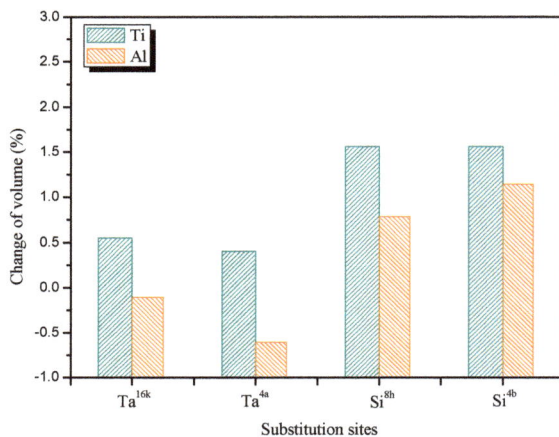

Figure 2. The change of unit cell volume for Ti and Al additions occupying the four possible substitution sites for $D8_m$-structured $Ta_{20}Si_{12}$.

3.2. Site Occupancy of Ternary Elements

It is generally assumed that the preference sites of the solute atoms are decided by their atomic radius and electronegativity. In this regard, Ti shows a preference for occupying the Ta sublattice and Al prefers to occupy the Si sublattice to maintain the minimum energy state. To confirm this assumption, the energetics of the preferential sites of M(Ti or Al) in the $D8_m$ $Ta_{20}Si_{12}$ structure can be studied based on the substitutional sites energy difference, which is defined as the difference between two different formation enthalpies [31]:

$$E_{site} = E_{site\text{-}Ta} - E_{site\text{-}Si} \qquad (20)$$

where $E_{site-Ta}$ represents the formation enthalpies of $Ta_{19}MSi_{12}$ including $Ta_{19}M^{16k}Si_{12}$ and $Ta_{19}M^{4b}Si_{12}$; $E_{site-Si}$ denotes the formation enthalpies of $Ta_{20}Si_{11}M$ including $Ta_{19}Si_{11}M^{4a}$ and $Ta_{19}Si_{11}M^{8h}$ structures. If $E_{site} < 0$, occupying the Ta sublattice is energetically more favorable, and if $E_{site} > 0$, M is prone to occupying the Si sites.

Table 1 shows the enthalpy of formation for $D8_m$ $Ta_{20}Si_{12}$ which shows good agreement between experimental data [28] and theoretical results [9]. The calculations show that all of the binary and ternary $D8_m$ $Ta_{20}Si_{12}$ compounds have negative values of enthalpy of formation, meaning that all these compounds are energetically stable. Furthermore, it is clear that the E_{site} is less than zero for the Ti substitution and the enthalpy of formation of $Ta_{19}Ti^{4b}Si_{12}$ is smaller than that for $Ta_{19}Ti^{16k}Si_{12}$, implying that Ti exhibits the tendency to occupy Ta^{4b} site. In the case of one Al atom substitution, the E_{site} is larger than zero and $Ta_{20}Si_{11}Al^{8h}$ has a relatively low enthalpy of formation, suggesting that Al has a strong site preference for the Si^{8h} site.

3.3. Elastic and Mechanical Properties

The elastic properties of a solid are essential for understanding the macroscopic mechanical properties of the solid and for its potential technological applications. In order to shed some light on the influence of these two alloying elements on the mechanical properties of $D8_m$-structured Ta_5Si_3, when only one atom in $D8_m$-structured Ta_5Si_3 is substituted by ternary elements, the elastic constants of the pure elements and binary and ternary $D8_m$ Ta_5Si_3 have been calculated and are listed in Table 2. The calculated results for b.c.c Ta and diamond-structured Si are in good consistence with the experimental data [32,33] and the elastic properties for the alloying element Al are comparable to the values reported by Kang et al. [34], which is indicative of the reliability and accuracy of the theoretically predicted results. As is well known, the elastic stability is a necessary condition for a stable crystal. For the tetragonal system, the mechanical stability criteria are decided by the following restrictions of its elastic constants, $C_{11} > 0$, $C_{33} > 0$, $C_{44} > 0$, $C_{66} > 0$, $C_{11} - C_{12} > 0$, $(C_{11} + C_{33} - 2C_{13}) > 0$, and $2(C_{11} + C_{12}) + C_{33} + 4C_{13} > 0$. The calculated elastic constants of all the binary and ternary $D8_m$-structured Ta_5Si_3 compounds satisfy the above criteria, indicating that these compounds are mechanically stable. The elastic constants provide valuable information about the bonding character between adjacent atomic planes and the anisotropic character of the bonding. It can be seen from Table 2 that the values of C_{11} are larger than that of C_{33}, which indicates that their a-axis direction ([100] direction) and b-axis direction ([010] direction) are stiffer than c-axis direction ([001] direction). The values of $(C_{11} + C_{12})$ are greater than that of C_{33}, suggesting that the elastic modulus is higher in the (001) plane than along [001] direction. Regardless of the binary and ternary $D8_m$-structured Ta_5Si_3 compounds, the values of C_{66} is larger than C_{44}, indicating that the [100](001) shear is easier than the [100](010) shear. The elastic anisotropy of the transition metal silicides has an important impact on the mechanical properties of the systems, especially on dislocation structures and mechanisms [35]. The anisotropy factors for tetragonal phases can be estimated as $A = 2C_{66}/(C_{11} - C_{12})$. A is equal to one for an isotropic crystal. The values of A are 1.03, 1.09 and 1.13 for the $Ta_{20}Si_{12}$, $Ta_{20}Si_{11}Al^{8h}$ and $Ta_{19}Ti^{4b}Si_{12}$, indicating that the shear elastic properties of the (001) plane are nearly independent of the shear direction for all the binary and ternary $D8_m$-structured Ta_5Si_3 compounds. Similar properties have been found for the $D8_m$-structured phase of V_5Si_3 [36] and Mo_5Si_3 [37].

Based on the calculated elastic constants, the bulk modulus (*B*), shear modulus (*G*) and Young's modulus (*E*) for binary and ternary $D8_m$-structured Ta_5Si_3 are calculated based on the Voigt–Reuss–Hill approximation, and are shown in Figure 3. The calculated values for the bulk modulus (*B*), shear modulus (*G*) and Young's modulus (*E*) for $D8_m$-structured Ta_5Si_3 are also in good agreement with the previous calculations [9]. Therefore, with respect to the available data, the present calculations yield comparable values. The substitution of Ta by Ti simultaneously lowers the values of *B*, *G*, and *E*, while the substitution of Si by Al also lowers the values of *G* and *E*, but slightly enhances the value of B. Since bulk moduli are generally believed to be related to the cohesive energy of materials, the variation of bulk modulus is related to changes in cohesive strength between atoms [38]. The binary Ta_5Si_3 has the largest values of shear modulus and Young's modulus, confirming that it has a higher shear deformation resistance and elastic stiffness as compares to ternary Ta_5Si_3 compounds. The ductility or brittleness of a solid is very important, because ductile materials are more resistant to catastrophic failure under critical mechanical stress conditions. There is no universally accepted criterion to gauge ductility or brittleness of a given material purely based on elastic data derived from linear elastic theory, but a reasonable representation of it can be realized by ratio of bulk/shear modulus *B/G* proposed by Pugh [39]. According to Pugh's empirical rule, the critical number of this parameter separating the ductile and brittle nature of a material is 1.75. If *B/G* > 1.75, the material behaves in a ductile manner; otherwise, the material deforms in a brittle mode. As shown in Figure 4, the *B/G* ratios for all of binary and ternary Ta_5Si_3 compounds are slightly larger than this critical value, implying that all of binary and ternary Ta_5Si_3 compounds exhibit ductile behavior. Meanwhile, the additions of Ti and Al further increase in the calculated *B/G* ratios, suggesting that both Ti and Al are beneficial to the improvement of ductility of the Ta_5Si_3 compound. Based on the values of *B/G*, the addition of Al is more efficient than Ti addition to improve the ductility of Ta_5Si_3. Poisson's ratio (υ) is associated with the volume change during uniaxial deformation and has been used to quantify the stability of the crystal against shear. In addition, Poisson's ratio provides more information about the degree of directionality of the covalent bonding than any of the other elastic coefficients and its value can also be employed to evaluate the ductility or brittleness of materials [40]. It is generally assumed that when υ is less than 0.26, the material is brittle, and at higher values it is ductile. The variation of the values of Poisson's ratios for the binary and ternary Ta_5Si_3 compounds is the same as those for *B/G* ratios. The results indicate that the additions of Ti and Al reduce the directionality of atomic bonding of Ta_5Si_3 and the directionality in atomic bonding in Ta_5Si_3 alloyed with Al is weaker than that in Ta_5Si_3 alloyed with Ti. Cauchy pressure ($C_{12} - C_{44}$) describes the angular characteristics of interatomic bonds in metals and compounds, thus serving as an indication of ductility. According to Pettifor [41], if the value of Cauchy pressure is positive, the material is expected to be ductile in nature; otherwise, it is brittle in nature. It can be seen from Figure 5 that the values of Cauchy pressure for the binary and ternary Ta_5Si_3 compounds is positive, further confirming that they are expected to be ductile in nature.

As one of the most basic mechanical properties of a material, hardness of a given material is characterized by the ability to resist to both elastic and irreversible plastic deformations, when a force is applied [42]. Some researchers have tried to create a correlation between hardness and elasticity [43,44]. For the partially covalent transition metal-based materials, since the shear modulus scales with the stresses required to nucleate or move isolated dislocations and hence is proportional to the hardness, it is regarded as a good indicator of hardness [45]. Thus, in the term of the values of shear modulus, the hardness increases in the following sequence: $Ta_{20}Si_{11}Al^{8h} < Ta_{19}Ti^{4b}Si_{12} < Ta_{20}Si_{12}$. Chen *et al.* [46] assumed that unlike the moduli of *G* and *B*, which only measured the elastic response, the Pugh's modulus ratio (*B/G*) seemed to correlate much more reliably with hardness because it responded to both elasticity and plasticity, which were the most intrinsic features of hardness. They proposed a semiempirical formula to calculate Vickers hardness of polycrystalline materials:

$$H_v = 2(k^2 G)^{0.585} - 3 \qquad (21)$$

where k denotes the Pugh's modulus ratio, B/G. The Vickers hardness values calculated from Equation (21) are presented in Figure 5. The values of Vickers hardness for $Ta_{20}Si_{11}Al^{8h}$, $Ta_{19}Ti^{4b}Si_{12}$ and $Ta_{20}Si_{12}$ are estimated to 11.40, 11.48 and 12.27 GPa. Again, the calculated results predict that Ti and Al substitutions result in solid solution softening.

Table 2. Calculated elastic constants for the binary, ternary D8$_m$-structured $Ta_{20}Si_{12}$ and their constituent elements compared to the data obtained experimentally and theoretically.

Structural	Ta (BCC)	Si (Diamond)	Al (FCC)	Ti (Hexagonal)	$Ta_{20}Si_{12}$	$Ta_{19}Ti^{4b}Si_{12}$	$Ta_{20}Si_{11}Al^{8h}$
C_{11} (GPa)	260 266 [a]	160 168 [b]	102 97 [c]	196	386.80 410.19 [d]	365.51	378.39
C_{12} (GPa)	149 158 [a]	70 65 [b]	70 62 [c]	57	137.12 145.66 [d]	133.36	137.31
C_{13} (GPa)	-	-	-	69.59	106.55 125.45 [d]	112.33	116.75
C_{33} (GPa)	-	-	-	200.12	339.40 338.89 [d]	298.14	317.29
C_{44} (GPa)	85 87 [a]	75 80 [b]	22 21 [c]	45.98	86.68 92.94 [d]	84.55	85.22
C_{66} (GPa)	-	-	-	69.86	128.03 132.91 [d]	130.85	131.50

[a] Experiment [32]; [b] Experiment [33]; [c] Calculation [34]; [d] Calculation [9].

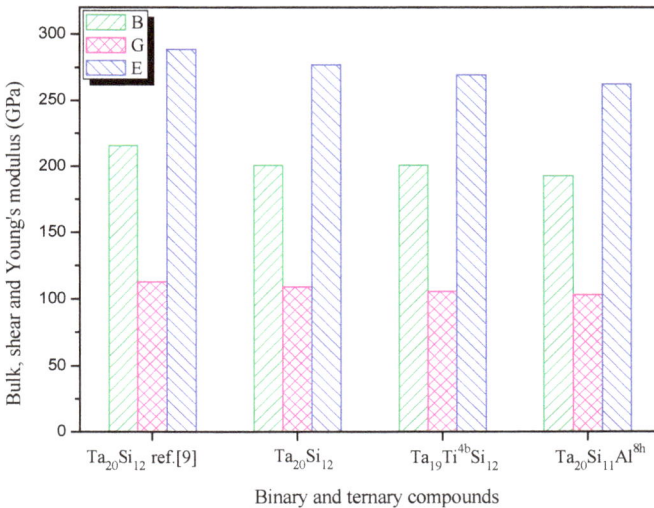

Figure 3. Calculated bulk modulus (*B*), shear modulus (*G*) and Young's modulus (*E*) for the binary and ternary D8$_m$-structured $Ta_{20}Si_{12}$.

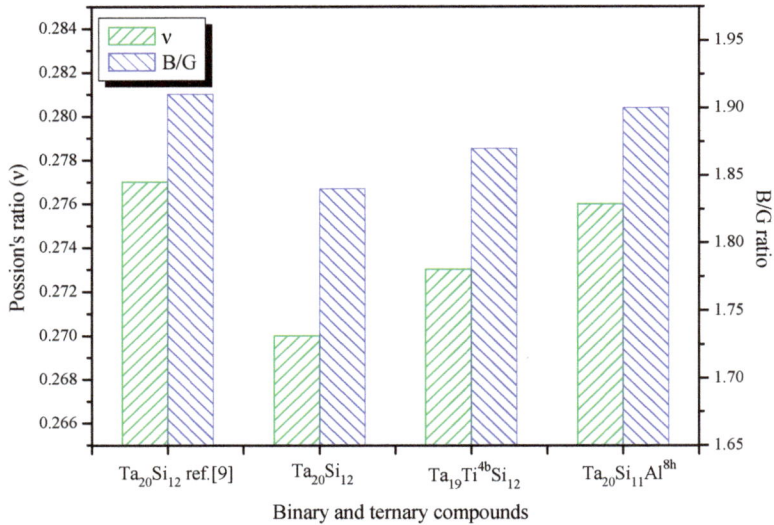

Figure 4. The Poisson's ratio, ν, and the Pugh modulus ratio, B/G, for $Ta_{20}Si_{12}$, $Ta_{20}Si_{11}Al^{8h}$ and $Ta_{19}Si_{12}Ti^{4b}$. Theoretical values reported in the literature [8] are shown in the figure for comparison.

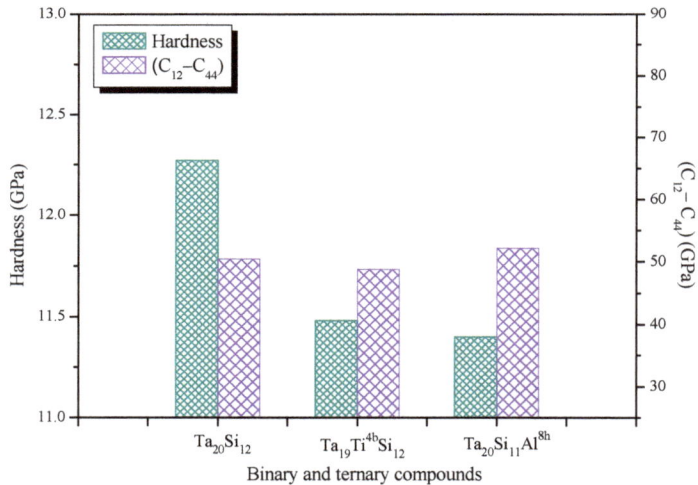

Figure 5. The Vickers hardness and Cauchy's pressure $(C_{12} - C_{44})$ for $Ta_{20}Si_{12}$, $Ta_{19}Ti^{4b}Si_{12}$ and $Ta_{20}Si_{11}Al^{8h}$.

3.4. Electronic Structure and Population Analysis

Mechanical properties of a material reflect its resistance to elastic and plastic deformation, which are controlled, in part, by the chemical bonding that is, in turn, determined by electronic structures. Hence, in order to gain insight into the mechanisms governing the mechanical properties of the binary and ternary D8$_m$-structured Ta$_5$Si$_3$, the electronic structures of these materials were investigated and compared. Figures 6 and 7 present the calculated total and partial density of states (DOS) for the binary and ternary D8$_m$-structured Ta$_5$Si$_3$, where the black vertical dashed of DOS represents the Fermi level.

The pseudogap (the low valley near the Fermi level) is likely to split the bonding and anti-bonding states. As shown in Figure 6, the introduction of either an Al atom or a Ti atom pushes the Fermi level towards a lower energy side of the pseudogap, denoting that some bonding states may be occupied by electrons from anti-bonding states. Compared with the binary Ta_5Si_3, the total DOS for ternary Ta_5Si_3 have two new peaks of bonding states near -6 and -2 eV and a new peak of bonding states near -2 eV, respectively. As can be seen from Figure 7a, the partial densities of states (PDOS) profile of binary Ta_5Si_3 mainly divides into three parts. The first part is from -12 to -6.7 eV, consisting mainly of Ta-5*d* states, Si-3*s* states, and part of Si-3*p* states. For the second part, the main bonding states located between -6.7 and -1.0 eV are dominated by Ta-5*d* states, Si-3*p* states and part of Ta-5*p*. The third part ranging from -1.0 to 2 eV is contributed mainly by Ta-5*d* states and part of Si-3*p* states. For the first part, the hybridization focuses on Ta-5*d* states and Si-3*s* states, forming Ta-Si bond. It is to be noted here that Si-3*s* states plays an important role in charge transfer. The second and third parts are different from the first part. In the second and third parts, the charge of Si atom transfers from 3*s* to 3*p* states and the Ta-5*d* states play an important role in electronic contribution. The hybridization between Ta-5*d* states and Si-3*p* states forms Ta-Si bond. In the last part, the anti-bonding states dominated by Ta-5*d* states forms Ta-Ta anti-bonding. For the PDOS profile of $Ta_{20}Si_{11}Al^{8h}$ (Figure 7b), the bonding states from -6 to -7 eV originate mainly from Ta-5*s* states and Al-3*s* sates, indicating strong electronic interactions for Ta-Al bonding. The Ta5*s*-Al3*s* bonding contribute to the new peak of bonding states near -6 eV in total DOS of $Ta_{20}Si_{11}Al^{8h}$ compound (Figure 7b). States from -6.2 to 2eV are dominated by Ta-5*d*, Si-3*p* and Al-3*p* states. Furthermore, the hybridized Al-3*p* and Ta-5*d* states occur in an energy range that is higher than that of the hybridized Si-3p and Ta-5d states, suggesting that the Al3*p*-Ta5*d* bonds are weaker than that of the Si3*p*-Ta5*d* bonds, reducing the covalent property of bonding in $D8_m$-structured Ta_5Si_3. The formation of the weaker Al-Ta covalent bonds explains why the ductility of the Ta_5Si_3 alloyed with Al is better than that of binary Ta_5Si_3. In the partial DOS profile of $Ta_{19}Ti^{4b}Si_{12}$ (Figure 7c), the Ti-3*d* states near -2 eV hybridize with Si-3*p* states, contributing to formation of the peak of bonding states in the total DOS of $Ta_{19}Ti^{4b}Si_{12}$(Figure 7b). The anti-bonding states are dominated by Ti-3*d* and Ta-5*d* states, showing evidence for the formation of strong Ti-Ta anti-bonding. Moreover, these Ti-Ta anti-bonding states occupy some bonding states between Ta and Si in the total DOS profile of $Ta_{19}Ti^{4b}Si_{12}$ as compares to binary $D8_m$ $Ta_{20}Si_{12}$, weakening the covalent property of bonding in $D8_m$ $Ta_{20}Si_{12}$. The strong Ti-Ta anti-bonding and the reduction of covalent property should be responsible for the improvement of ductility.

To further reveal the charge transfer and the chemical bonding properties of both non-doped and doped Ta_5Si_3, the difference in bonding charge density for the three compounds was investigated. The bonding charge density, also called the deformation charge density, is defined as the difference between the self-consistent charge density of the interacting atoms in the crystal and a reference charge density constructed from the superposition of the non-interacting atomic charge density at the lattice sites [47]. The distribution of the bonding charge densities in the (001) plane along the a- and b-axis for non-doped and doped Ta_5Si_3 is given in Figures 8 and 9 where the red region denotes a high charge density and the blue region denotes a relatively low charge density. It is noted in Figure 8 that the bonding charge density shows a depletion of the electronic density at the lattice sites together with an increase of the electronic density in the interstitial region. It is also noted that the bonding charge build-up in the Ta^{16k}-Si^{8h}-Ta^{16k} planar triangular bonding area is strong. This feature is consistent with the PDOS plots in Figure 7a, showing the importance of the hybridization between Ta^{16k}-5*d* states and Si^{8h}-3*p* states, forming the Ta^{16k}-Si^{8h} bonds.

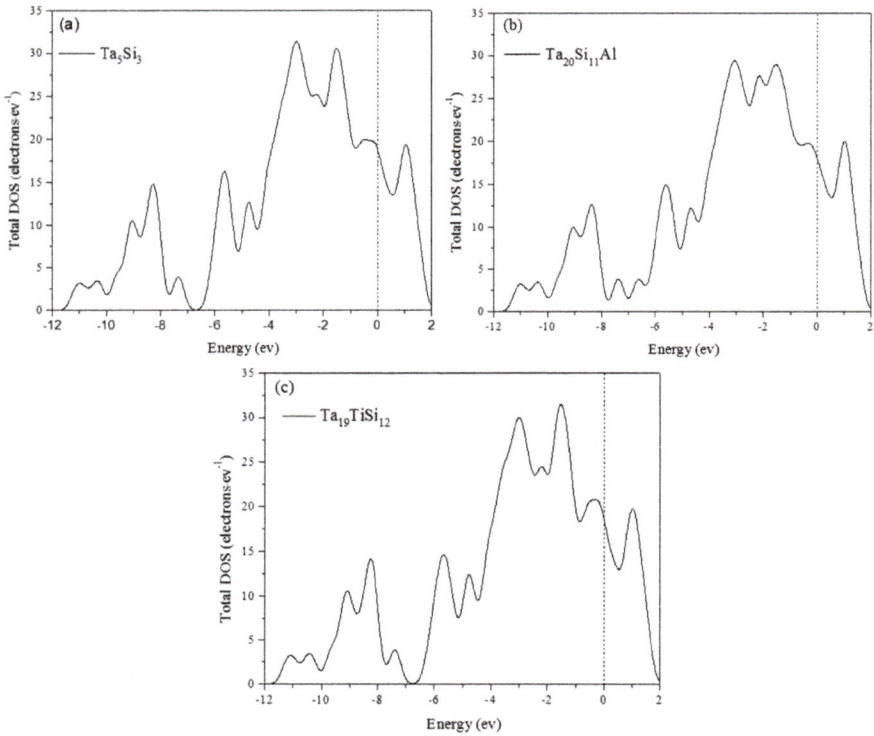

Figure 6. Calculated total density of states (DOS) for (**a**) $Ta_{20}Si_{12}$; (**b**) $Ta_{20}Si_{11}Al^{8h}$ and (**c**) $Ta_{19}Si_{12}Ti^{4b}$.

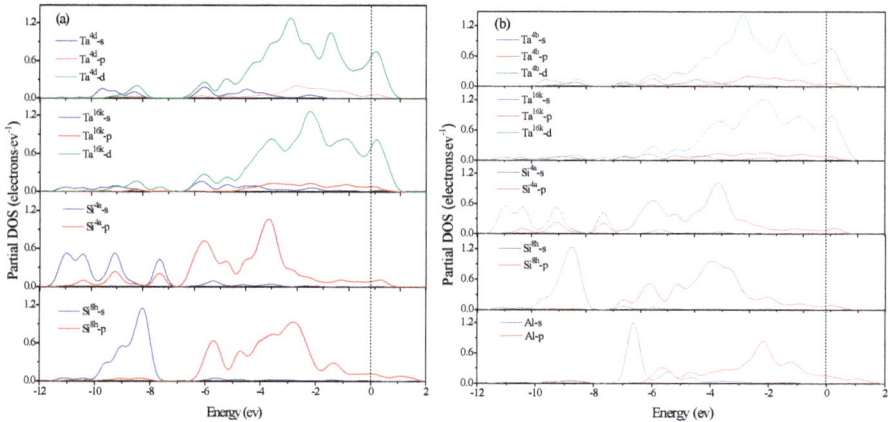

Figure 7. *Cont.*

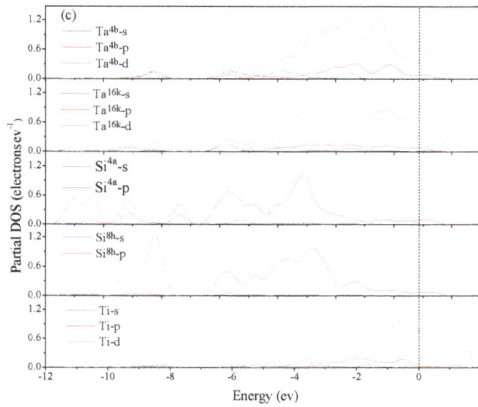

Figure 7. Calculated partial density of states (DOS) for (**a**) $Ta_{20}Si_{12}$; (**b**) $Ta_{20}Si_{11}Al^{8h}$ and (**c**) $Ta_{19}Ti^{4b}Si_{12}$.

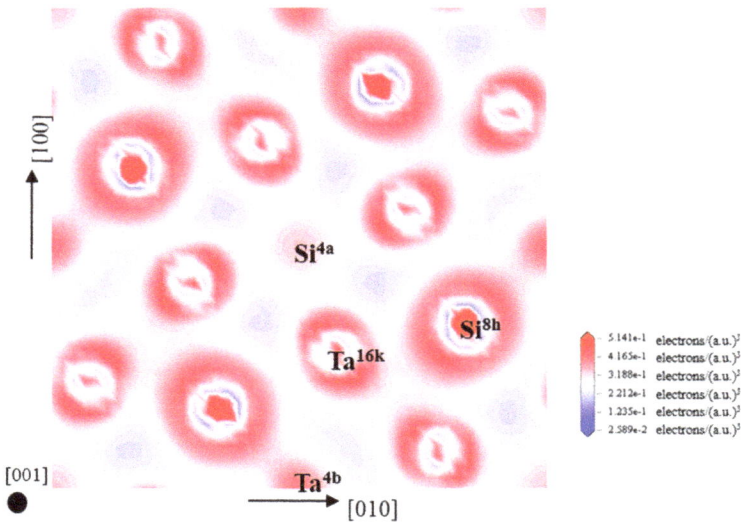

Figure 8. The difference density of contour plots for $Ta_{20}Si_{12}$ unit cell in the (001) plane. ● denotes the direction perpendicular paper from inside to outside. The Ta^{4b}, Ta^{16k}, Si^{4a} and Si^{8h} symbols designate the positions of, respectively, Ta in the 4b site, Ta in the 16k site, Si in the 4a, and Si in the 8h site.

After replacing a Si atom in the 8h site with an Al atom, a significant redistribution of the bonding charge in the interstitial region is seen in Figure 9a. The region around the substitution lattice site changes from red to blue, meaning that the charge density around the Al atom is much sparser than that around the Si atom. In addition, Ta^{16k}-Al^{8h}-Ta^{16k} planar triangular bonding area is weaker than the Ta^{16k}-Si^{8h}-Ta^{16k} planar triangular bonding area. This result is consistent with the view derived from the DOS plots that the strength of Al-Ta bonds is weaker than that of the Si-Ta bonds. When one Ta^{4a} site is occupied by a Ti atom, see Figure 9b, the region between substitution site and Si^{8h} site is almost near-white, showing the decline of the interacting atomic charge density for the Ti^{4b}-Si^{8h} planar bonding area as compares with Ta^{4b}-Si^{8h} planar bonding area.

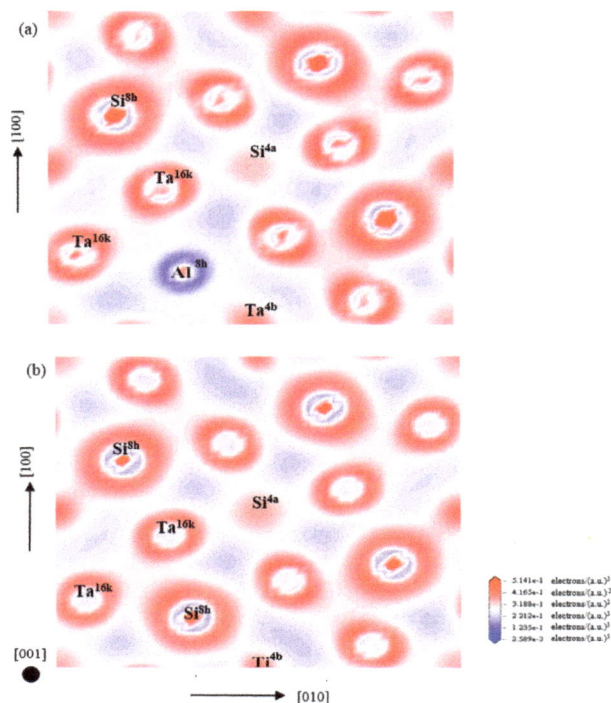

Figure 9. The difference density of contour plots for (**a**) $Ta_{20}Si_{11}Al^{8h}$ and (**b**) $Ta_{19}Ti^{4b}Si_{12}$ unit cell in (001) plane. ● denotes the direction perpendicular paper from inside to outside. The Ta^{4b}, Ta^{16k}, Si^{4a}, Si^{8h}, Al^{8h} and Ti^{4b} symbols designate the positions of, respectively, Ta in the 4b site, Ta in the 16k site, Si in the 4a, Si in the 8h site, Al in the 8h site and Ti in the 4b site.

Mulliken [48] analyzed correlations of overlap population with the degree of covalency of bonding and bond strength. The bonding (anti-bonding) states related to positive (negative) values of bond population, and the low (high) values imply that the chemical bond exhibits strong ionic (covalent) bonding [49].

As seen in Table 3, the covalent character for the binary and ternary Ta_5Si_3 are mainly attributed to the Ta^{16k}-Si^{8h} covalent bond with high overlap population and the highest bond number. The above discussion about DOS for $Ta_{19}Ti^{4b}Si_{12}$ indicates that the Ti substitution changes the anti-bonding states between the Ta atoms. This observation is confirmed by the following changes in the populations of bonds: (a) the number of metallic bonds Ta^{16k}-Ta^{16k} increases by four and its overlap population exhibits a significant decease; (b) a new strong anti-bonding Ti^{4b}-Ta^{4b} appears in the network of metallic bonds with the disappearance of weak Ta^{4b}-Ta^{4b} anti-bonding. The change (a) weakens the covalent character of $Ta_{19}Ti^{4b}Si$ and improves its metallic properties and (b) improves the metallic property of the 4b-4b bonds. Clearly, those Ta-Ta metallic bonds including Ta^{16k}-Ta^{16k} and Ta^{4b}-Ta^{4b} anti-bonding lower the deformation resistance, thereby reducing the elastic properties. Consequently, the change seen in both (a) and (b) intensifies the metallic property of bonds for the $Ta_{19}Ti^{4b}Si_{12}$ compound, resulting in a decrease of the elastic properties and an increase in ductility for $Ta_{19}Ti^{4b}Si_{12}$. In addition, the hybridization between Ti-3d and Si-3p forms four new Ti^{4b}-Si^{8h} bonds of which the overlap population are close to zero. This also means a decrease in covelency for $Ta_{19}Ti^{4b}Si_{12}$ compared with $Ta_{20}Si_{12}$. In the case of Si site substitution, the bonding length and the overlap populations of Si^{8h}-Ta^{16k} bonds becomes longer and smaller and the bonding length of the Ta^{16k}-Ta^{16k} metallic bonds becomes shorter, indicating that the Si^{8h}-Ta^{16k} bonds get weaker leading to charge transfer towards the Ta^{16k}-Ta^{16k} bonds. As such, the covalent character of 8h-16k bonds becomes weaker and Ta^{16k}-Ta^{16k} bonds become stronger, resulting in the improvement of ductility for $Ta_{20}Si_{11}Al^{8h}$.

Table 3. Mulliken overlap populations, bond lengths and numbers of bonds calculated for $Ta_{20}Si_{12}$, $Ta_{20}Si_{11}Al^{8h}$ and $Ta_{19}Ti^{4b}Si_{12}$ compounds.

Bond	$Ta_{20}Si_{12}$			$Ta_{20}Si_{11}Al^{8h}$			$Ta_{19}Ti^{4b}Si_{12}$		
	P	**L**	**N**	**P**	**L**	**N**	**P**	**L**	**N**
Ta^{4b}-Ta^{4b}	−0.08	2.592	2	0.01	2.597	2	0.01	2.599	1
Ta^{16k}-Ta^{16k}	0.46	2.918	8	0.45	2.903	8	0.34	2.965	12
Si^{4a}-Si^{4a}	0.41	2.592	2	0.40	2.598	2	0.40	2.599	2
Si^{4a}-Ta^{16k}	0.19	2.677	32	0.19	2.677	32	0.19	2.682	32
Ta^{4b}-Si^{8h}	0.31	2.677	16	0.31	2.69	14	0.35	2.696	12
Si^{8h}-Ta^{16k}	0.35	2.833	48	0.33	2.832	43	0.40	2.78	43
Al^{8h}-Ta^{4b}		-		0.30	2.735	2			
Al^{8h}-Ta^{16k}		-		0.41	2.930	6		-	
Ti^{4b}-Si^{8h}		-			-		0.13	2.678	4
Ti^{4b}-Ta^{4b}		-			-		−0.24	2.599	1

Data in the Table 3 are average values; P: Mulliken populations; L: Bond length (Å); N: Numbers of bonds.

3.5. Debye Temperature

As a fundamental parameter, the Debye characteristic temperature correlates with many physical properties of solids, such as specific heat, elastic constants and melting temperature [50]. Elastic and thermal properties are inseparable because they both originate in the lattice vibrational spectrum of a solid. Therefore, the Debye temperature can be determined from the elastic constants. Theoretically, the Debye temperature (Θ_D), at $T = 0$ K, can be calculated by using the average sound velocity (v_m) from the following relationship:

$$\Theta_D = \frac{h}{k_B} \left[\frac{3n}{4\pi} \left(\frac{N_A \rho}{M} \right) \right]^{\frac{1}{3}} \times v_m \tag{22}$$

here, h and k_B represent the Planck and Boltzmann constants; n is the number of atoms in per formula unit, N_A is Avogadro's number, M is the mass of per formula unit. The density, ρ, can be obtained based on a knowledge of the unit cell lattice parameters. The parameter v_m represents the average sound velocity which is defined as:

$$v_m = \left[\frac{1}{3} \left(\frac{1}{v_l^3} + \frac{2}{v_s^3} \right) \right]^{-\frac{1}{3}} \tag{23}$$

where v_l and v_s are the longitudinal and transverse sound velocities respectively, which can be obtained by bulk modulus (B) and shear modulus (G)

$$v_l = \left(\frac{B + \frac{4}{3}G}{\rho} \right) \text{ and } v_s = \left(\frac{G}{\rho} \right)^{\frac{1}{2}} \tag{24}$$

The calculated longitudinal, transverse and average sound velocities and Debye temperature of both non-alloyed and alloyed $D8_m$-structured Ta_5Si_3 and their constituent elements were calculated and are listed in Table 4. The results obtained agree well with previous studies [8,51,52]. The Debye temperatures of the three compounds are larger than that of tantalum. This is because the mass-density of Ta (16.58 g/cm³) is larger than that of these compounds (around 12.5 g/cm³), so that sound velocity in Ta is lower, that is, this leads to a lower Debye temperature. In contrast, Si and Al have relatively much lower densities (2.286 and 2.66 g/cm^3, respectively), resulting in larger Debye temperatures. Compared to the D8$_m$-structured Ta$_5$Si$_3$, the ternary Ta$_5$Si$_3$ compounds have lower Debye temperatures. Different Θ_D-values indicate different lattice dynamics or degrees of lattice stiffness [53]. Thus, the reasons for the observed decrease of Θ_D can be mainly attributed to structural softening and the relative low shear modulus (shown in Table 4), which results in a lower cut-off for the atomic vibrational frequency [54].

Table 4. Calculated density ρ (in g/cm^3), longitudinal, transverse, average sound velocities (in m/s) and Debye temperature Θ_D (in K) for the binary, ternary D8$_m$-structured Ta$_{20}$Si$_{12}$ and its constituent elements.

Phase	Ta	Si	Al	Ti	Ta$_{20}$Si$_{12}$	Ta$_{20}$Si$_{11}$Al^{8h}	Ta$_{19}$Ti^{4b}Si$_{12}$
ρ	16.58	2.286	2.66	4.56	12.5	12.5	12.1
v_l	5827	8906	6474	6398	5262	5227	5221
v_t	2940	5166	2980	3561	2954	2906	2918
v_m	3296	5733	3357	3965	3287	3236	3248
Θ	374	626	390	450	385	380	381
	253 [a]	635 [a]	428 [b]	374 [b]	388 [c]		

[a] Experiment [51]; [b] Calculation [52]; [c] Calculation [9].

4. Conclusions

In summary, the influence of substitutional alloying elements (Ti and Al) on the structural stability, mechanical properties, electronic properties and Debye temperature of Ta$_5$Si$_3$, exhibiting a D8$_m$ structure, were systematically investigated by first principle calculations. All of the binary and ternary Ta$_5$Si$_3$ compounds show negative enthalpy of formation values, indicating that these compounds are energetically stable. Based on the values of formation enthalpies, Ti exhibits a preferential occupying the Ta^{4b} site and Al has a strong site preference for the Si^{8h} site. The calculated results for the elastic constants show that doping with Ti simultaneously lowers the B, G, and E values of D8$_m$-structuted Ta$_5$Si$_3$. The variation of the shear modulus and the Young's modulus of this compound with an Al addition has a similar tendency to that of a Ti addition, although an Al addition slightly enhances the bulk modulus. The change of the ductility of the D8$_m$-structured Ta$_5$Si$_3$ is explained according to the shear modulus/bulk modulus ratio (G/B), the Poisson's ratio and the Cauchy's pressure. The results show that ductility was improved by doping with Ti and Al. Moreover, the hardness of the Ta$_5$Si$_3$ is reduced by both Ti and Al additions, resulting in solid solution softening. Electronic structure analysis indicates that alloying with Ti and Al introduces a more metallic nature to the bonding, resulting in the enhancement in ductility of D8$_m$ Ta$_5$Si$_3$. Moreover, the Debye temperatures, Θ_D, of D8$_m$-structured Ta$_5$Si$_3$ alloying with Ti and Al are decreased as compared to the binary Ta$_5$Si$_3$.

Acknowledgments: The authors acknowledge the financial support from the National Natural Science Foundation of China under Grant No. 51374130. This work is supported by Funding of Jiangsu Innovation Program for Graduate Education, the Fundamental Research Funds for the Central Universities (CXLX13-151). This study is also supported by the Australian Research Council Discovery Project (DP150102417).

Author Contributions: Linlin Liu, Jian Cheng and Jiang Xu conceived and designed the experiments; Linlin Liu performed the experiments; Linlin Liu and Jian Cheng analyzed the data; Paul Munroe and Zong-Han Xie contributed reagents, materials and analysis tools; Linlin Liu, Jian Cheng and Jiang Xu wrote the paper.

Conflicts of Interest: The authors declare no conflict of interest.

References

1. Nakano, T.; Azuma, M.; Umakoshi, Y. Tensile deformation and fracture behaviour in NbSi$_2$ and MoSi$_2$ single crystals. *Acta Mater.* **2002**, *14*, 3731–3742. [CrossRef]
2. Xu, J.; Leng, Y.; Li, H.; Zhang, H. Preparation and characterization of SiC/(Mo, W)Si$_2$ composites from powders resulting from a SHS in a chemical oven. *Int. J. Refract. Met. Hard Mater.* **2009**, *27*, 74–77. [CrossRef]

3. Datta, M.K.; Pabi, S.K.; Murty, B.S. Thermal stability of nanocrystalline Ni silicides synthesized by mechanical alloying. *Mater. Sci. Eng. A* **2000**, *1*, 219–225. [CrossRef]

4. Mason, K.N. Growth and characterization of transition metal silicides. *Prog. Cryst. Growth Charact. Mater.* **1981**, *2*, 269–307. [CrossRef]

5. Fujiwara, H.; Ueda, Y. Thermodynamic properties of molybdenum silicides by molten electrolyte EMF measurements. *J. Alloy. Compd.* **2007**, *441*, 168–173. [CrossRef]

6. Schlesinger, M.E. The Si-Ta (silicon-tantalum) system. *J. Phase Equilib.* **1994**, *15*, 90–95. [CrossRef]

7. Matsuno, H.; Yokoyama, A.; Watari, F.; Uo, M.; Kawasaki, T. Biocompatibility and osteogenesis of refractory metal implants, titanium, hafnium, niobium, tantalum and rhenium. *Biomaterials* **2001**, *22*, 1253–1262.

8. Tillard, M. The mixed intermetallic silicide $Nb_{5-x}Ta_xSi_3$ ($0 \leqslant x \leqslant 5$). Crystal and electronic structure. *J. Alloy. Compd.* **2014**, *584*, 385–392. [CrossRef]

9. Tao, X.; Jund, P.; Viennois, R.; Catherine, C.; Tedenacet, J.C. Physical properties of thallium–tellurium based thermoelectric compounds using first-principles simulations. *J. Phys. Chem. A* **2011**, *31*, 8761–8766.

10. Mitra, R. Microstructure and mechanical behavior of reaction hot-pressed titanium silicide and titanium silicide-based alloys and composites. *Metall. Mater. Trans. A* **1998**, *6*, 1629–1641. [CrossRef]

11. Rosenkranz, R.; Frommeyer, G.; Smarsly, W. Microstructures and properties of high melting point intermetallic Ti_5Si_3 and $TiSi_2$ compounds. *Mater. Sci. Eng. A* **1992**, *1*, 288–294. [CrossRef]

12. Korznikov, A.V.; Pakieła, Z.; Kurzydłowski, K.J. Influence of long-range ordering on mechanical properties of nanocrystalline Ni_3Al. *Scr. Mater.* **2001**, *45*, 309–315. [CrossRef]

13. Dasgupta, T.; Umarji, A.M. Thermal properties of $MoSi_2$ with minor aluminum substitutions. *Intermetallics* **2007**, *15*, 128–132. [CrossRef]

14. Gutmanas, E.Y.; Gotman, I. Reactive synthesis of ceramic matrix composites under pressure. *Ceram. Int.* **2000**, *7*, 699–707. [CrossRef]

15. Williams, J.J.; Ye, Y.Y.; Kramer, M.J.; Ho, K.M.; Hong, L.; Fu, C.L.; Malik, S.K. Theoretical calculations and experimental measurements of the structure of Ti_5Si_3 with interstitial additions. *Intermetallics* **2000**, *8*, 937–943. [CrossRef]

16. Du, W.; Zhang, L.; Ye, F.; Ni, X.; Lin, J. Intrinsic embrittlement of $MoSi_2$ and alloying effect on ductility: Studied by first-principles. *Phys. B Condens. Matter* **2010**, *7*, 1695–1700. [CrossRef]

17. Perdew, J.P.; Chevary, J.A.; Vosko, S.H.; Jackson, K.A.; Pederson, M.R.; Singh, D.J.; Fiolhais, C. Atoms, molecules, solids, and surfaces: Applications of the generalized gradient approximation for exchange and correlation. *Phys. Rev. B* **1992**, *11*. [CrossRef]

18. Vanderbilt, D. Soft self-consistent pseudopotentials in a generalized eigenvalue formalism. *Phys. Rev. B* **1990**, *11*. [CrossRef]

19. Pack, J.D.; Monkhorst, H.J. Special points for Brillouin-zone integrations—A reply. *Phys. Rev. B* **1977**, *4*.

20. Chu, F.; Thoma, D.J.; McClellan, K.J.; Peralta, P. Mo_5Si_3 single crystals: Physical properties and mechanical behavior. *Mater. Sci. Eng. A* **1999**, *261*, 44–52. [CrossRef]

21. Kocherzhinskij, Y.A.; Kulik, O.G.; Shishkin, E.A. Phase diagram of the tantalum-silicon system. *Dokl. Akad. Nauk SSSR* **1981**, *261*, 106–108.

22. Fu, C.L.; Wang, X.; Ye, Y.Y.; Ho, K.M. Phase stability, bonding mechanism, and elastic constants of Mo_5Si_3 by first-principles calculation. *Intermetallics* **1999**, *2*, 179–184. [CrossRef]

23. Voigt, W. A Determination of the elastic constants for beta-quartz lehrbuch de kristallphysik. *Terubner Leipzig.* **1928**, *40*, 2856–2860.

24. Reuss, A. Berechnung der Fließgrenze von Mischkristallen auf Grund der Plastizitätsbedingung für Einkristalle. *ZAMM J. Appl. Math. Mech.* **1929**, *9*, 49–58. [CrossRef]

25. Hill, R. The elastic behaviour of a crystalline aggregate. *Proc. Phys. Soc. A* **1952**, *65*, 349–354. [CrossRef]

26. Brauer, G.; Zapp, K.H. Die nitride des tantals. *Anorg. Allg. Chem.* **1954**, *277*, 129–139. [CrossRef]

27. Wang, F.E.; Buehler, W.J.; Pickart, S.J. Crystal structure and a unique "Martensitic" transition of TiNi. *J. Appl. Phys.* **1965**, *36*, 3232–3239. [CrossRef]

28. Nowotny, H.; Laube, E. The thermal expansion of high-melting phases. *Plansee. Pulver.* **1961**, *9*, 85–92.

29. Jette, E.R.; Foote, F. Precision determination of lattice constants. *J. Chem. Phys.* **1935**, *3*, 605–616. [CrossRef]

30. Sakakibara, N.; Takahashi, Y.; Okumura, K.; Hattori, K.H.; Yaita, T.; Suzuki, K.; Shimizu, H. Speciation of osmium in an iron meteorite and a platinum ore specimen based on X-ray absorption fine-structure spectroscopy. *Geochem. J.* **2005**, *39*, 383–389. [CrossRef]

31. Zhang, C.; Han, P.; Li, J.; Chi, M.; Yan, L.; Liu, Y.; Liu, X.; Xu, B. First-principles study of the mechanical properties of NiAl microalloyed by M (Y, Zr, Nb, Mo, Tc, Ru, Rh, Pd, Ag, Cd). *J. Phys. D Appl. Phys.* **2008**, *41*.
32. Featherston, F.H.; Neighbours, J.R. Elastic constants of tantalum, tungsten, and molybdenum. *Phys. Rev.* **1963**, *130*. [CrossRef]
33. McSkimin, H.J.; Andreatch, P., Jr. Elastic moduli of silicon *vs.* hydrostatic pressure at 25.0 °C and −195.8 °C. *J. Appl. Phys.* **1964**, *35*, 2161–2165. [CrossRef]
34. Kang, L.U.O.; Bing, Z.; Shang, F.U.; Jiang, Y.; Yi, D.Q. Stress/strain aging mechanisms in Al alloys from first principles. *Trans. Nonferr. Met. Soc. China* **2014**, *24*, 2130–2137.
35. Chu, F.; Lei, M.; Maloy, S.A.; Petrovic, J.J.; Mitchell, T.E. Elastic properties of C40 transition metal disilicides. *Acta Mater.* **1996**, *44*, 3035–3048. [CrossRef]
36. Tao, X.; Chen, H.; Tong, X.; Ouyang, Y.; Jund, P.; Tedenac, J.C. Structural, electronic and elastic properties of V_5Si_3 phases from first-principles calculations. *Comput. Mater. Sci.* **2012**, *53*, 169–174. [CrossRef]
37. Ström, E.; Eriksson, S.; Rundlöf, H.; Zhang, J. Effect of site occupation on thermal and mechanical properties of ternary alloyed Mo_5Si_3. *Acta Mater.* **2005**, *53*, 357–365. [CrossRef]
38. Pan, Y.; Lin, Y.; Xue, Q.; Ren, C.; Wang, H. Relationship between Si concentration and mechanical properties of Nb–Si compounds: A first-principles study. *Mater. Des.* **2016**, *89*, 676–683.
39. Pugh, S.F. XCII. Relations between the elastic moduli and the plastic properties of polycrystalline pure metals. *Lond. Edinb. Dublin Philos. Mag. J. Sci.* **1954**, *45*, 823–843. [CrossRef]
40. Chumakov, A.I.; Monaco, G.; Fontana, A.; Bosak, A.; Hermann, R.P.; Bessas, D.; Wehinger, B.; Crichton, W.A.; Krisch, M.; Baldi, G.; *et al.* Role of disorder in the thermodynamics and atomic dynamics of glasses. *Phys. Rev. Lett.* **2014**, *112*. [CrossRef] [PubMed]
41. Pettifor, D.G. Theoretical predictions of structure and related properties of intermetallics. *Mater. Sci. Technol.* **1992**, *8*, 345–349. [CrossRef]
42. Mukhanov, V.A.; Kurakevych, O.O.; Solozhenko, V.L. Thermodynamic aspects of materials' hardness: Prediction of novel superhard high-pressure phases. *High Press. Res.* **2008**, *28*, 531–537. [CrossRef]
43. Gilman, J.J. Hardness—A strength microprobe. In *The Science of Hardness Testing and Its Research Applications*; Westbrook, J.H., Conrad, H., Eds.; American Society of Metal: Metal Park, OH, USA, 1973.
44. Liu, A.Y.; Cohen, M.L. Structural properties and electronic structure of low-compressibility materials: β-Si_3N_4 and hypothetical β-C_3N_4. *Phys. Rev. B* **1990**, *41*, 10727–10734. [CrossRef]
45. Teter, D.M.; Bull, M.R.S. Hardness and fracture toughness of brittle materials. *Phys. Rev. B* **1998**, *23*.
46. Chen, X.Q.; Niu, H.; Franchini, C.; Li, D.; Li, Y. Hardness of T-carbon: Density functional theory calculations. *Phys. Rev. B* **2011**, *84*. [CrossRef]
47. Tao, X.; Jund, P.; Colinet, C.; Tédenac, J.C. First-principles study of the structural, electronic and elastic properties of W_5Si_3. *Intermetallics* **2010**, *18*, 688–693. [CrossRef]
48. Mulliken, R.S. Electronic population analysis on LCAO–MO molecular wave functions. I. *J. Chem. Phys.* **1955**, *23*, 1833–1840. [CrossRef]
49. Cao, Y.; Zhu, P.; Zhu, J.; Liu, Y. First-principles study of NiAl alloyed with Co. *Comput. Mater. Sci.* **2016**, *111*, 34–40. [CrossRef]
50. Wang, H.Z.; Zhan, Y.Z.; Pang, M.J. The structure, elastic, electronic properties and Debye temperature of $M2AlC$ (M = V, Nb and Ta) under pressure from first principles. *Comput. Mater. Sci.* **2012**, *54*, 16–22.
51. Shi, S.; Zhu, L.; Jia, L.; Zhang, H.; Sun, Z. Ab-initio study of alloying effects on structure stability and mechanical properties of α-Nb_5Si_3. *Comput. Mater. Sci.* **2015**, *108*, 121–127. [CrossRef]
52. Chen, Q.; Sundman, B. Calculation of Debye temperature for crystalline structures—A case study on Ti, Zr, and Hf. *Acta Mater.* **2001**, *49*, 947–961. [CrossRef]
53. Du, L.; Wang, L.; Zheng, B.; Du, H. Numerical simulation of phase separation in Fe–Cr–Mo ternary alloys. *J. Alloy. Compd.* **2016**, *663*, 243–248. [CrossRef]
54. Cuenya, B.R.; Frenkel, A.I.; Mostafa, S.; Behafarid, F.; Croy, J.R.; Ono, L.K.; Wang, Q. Anomalous lattice dynamics and thermal properties of supported size-and shape-selected Pt nanoparticles. *Phys. Rev. B* **2010**, *82*. [CrossRef]

metals

MDPI

Article

Influence of the Overlapping Factor and Welding Speed on T-Joint Welding of Ti6Al4V and Inconel 600 Using Low-Power Fiber Laser

Shamini Janasekaran [1], **Ai Wen Tan** [1,2], **Farazila Yusof** [1,2,*] **and Mohd Hamdi Abdul Shukor** [1,2]

[1] Department of Mechanical Engineering, Faculty of Engineering, University of Malaya, Kuala Lumpur 50603,
    Malaysia; shaminijp@gmail.com (S.J.); aiwen_2101@hotmail.com (A.W.T.); hamdi@um.edu.my (M.H.A.S.)
[2] Centre of Advanced Manufacturing and Material Processing (AMMP), Faculty of Engineering,
    University of Malaya, Kuala Lumpur 50603, Malaysia
*   Correspondence: farazila@um.edu.my; Tel.: +60-03-79677633; Fax: +60-03-79677669

Academic Editor: Ana Sofia Ramos
Received: 18 February 2016; Accepted: 12 April 2016; Published: 2 June 2016

Abstract: Double-sided laser beam welding of skin-stringer joints is an established method for many applications. However, in certain cases with limited accessibility, single-sided laser beam joining is considered. In the present study, single-sided welding of titanium alloy Ti6Al4V and nickel-based alloy Inconel 600 in a T-joint configuration was carried out using continuous-wave (CW), low-power Ytterbium (Yb)-fiber laser. The influence of the overlapping factor and welding speed of the laser beam on weld morphology and properties was investigated using scanning electron microscopy (SEM) and X-ray diffraction (XRD), respectively. XRD analysis revealed the presence of intermetallic layers containing NiTi and $NiTi_2$ at the skin-stringer joint. The strength of the joints was evaluated using pull testing, while the hardness of the joints was analyzed using Vickers hardness measurement at the base metal (BM), fusion zone (FZ) and heat-affected zone (HAZ). The results showed that the highest force needed to break the samples apart was approximately 150 N at a laser welding power of 250 W, welding speed of 40 mm/s and overlapping factor of 50%. During low-power single-sided laser welding, the properties of the T-joints were affected by the overlapping factor and laser welding speed.

Keywords: overlapping factor; Ti6Al4V; Inconel-600; fiber laser; T-joint; pull test; NiTi

1. Introduction

Titanium and nickel-based alloys are generally difficult-to-machine and hard-to-cut materials. Nickel-based alloys, such as Inconel 600, have broad operational temperature, which makes them suitable for high-temperature processes owing to their thermal fatigue resistance [1]. Ni-based super alloys, such as Inconel 718, Inconel 628 and Inconel 600, are also extensively employed in the aerospace industry because of their superior mechanical properties and excellent oxidation resistance at elevated temperatures. They are suitable to be manufactured as components in high-temperature regions of aero engines and gas turbines [2]. For extreme applications such as in Formula 1 race car exhaust systems, heat and pipe stress are high, meaning that Inconel is best suited due to its ability to withstand extreme heat levels [3]. Titanium alloys, specifically Ti6Al4V, have been widely applied in the aerospace, engine turbine, exhaust system and biomedicine fields due to their favorable mechanical properties including high strength-to-weight ratio, toughness, corrosion resistance, light weight, biocompatibility and high yield stress [4]. In the automotive field, manufacturing lighter racing car parts has prompted great interest in achieving rapid movement. In this context, titanium (Ti) alloy and nickel (Ni) super alloys offer the advantage of reduced race car weight through the custom design of headers and shafts [5].

The tungsten inert gas (TIG) welding technique utilizing non-consumable electrode tungsten has been primarily used for welding these alloys for exhaust system applications. Although thick workpiece with good quality are achievable during the TIG process, this technique is time consuming and requires tremendous skill to obtain a good finishing. Therefore, the metal inert gas (MIG) welding technique has been automated to replace TIG; only minimal training is thus necessary and a faster output can be attained. However, MIG cannot be used to fabricate thin flanges (less than 0.7 mm), such as in the case of exhaust tubing for race cars [6].

To overcome these disadvantages, many studies have been done on welding Ti and Ni alloys through advanced techniques such as laser beam welding (LBW). In recent years, the welding technology of dissimilar materials has drawn increasing attention in various industries because it is capable of offering complex functions with greater design flexibility and reduced material costs [7]. The increasing demand for dissimilar materials is to create new component functions in various industrial fields, but this is difficult due to the high reliability on strength [8]. The welding of dissimilar metals can be divided into two types. One is according to the differences between thermo-physical properties, including thermal conductivity (κ) and the temperature coefficient of surface tension ($d\gamma/dT$). Conductivity differences indirectly influence the weld composition and lead to asymmetric heat transport. The weld geometry pattern is produced by the differences in $d\gamma/dT$ and it influences the surface tension of the molten pool. Second, inhomogeneous molten flow leads to the phase formation of different crystals due to metallurgical differences [9]. Intermetallic layers are present for dissimilar metal welding due to the differences in thermo-physical properties of the materials. These phases can lead to brittleness and their existence can decrease the usage of the joint [10]. Using Nd:YAG laser for spot welding of these alloys creates cracks and incompletely mixed liquids [11]. Ti–Ni dissimilar butt welding using CO_2 laser showed asymmetric weld shapes and brittle intermetallic compounds. $NiTi_2$ and Ni_3Ti were formed with macroscopic cracks in the weld [12]. Carpinteri *et al.* investigated welded T-joints and stated that fatigue failure in structures is caused by stress concentrators, which are also the preferred points for crack initiation [13]. Corrosion properties also need to be studied in detail to maximize the lifespan of T-joints [14]. Considering dissimilar metal welding of Ti alloy and Ni alloy, some success has been achieved by using laser welding due to its rapid processing capability as well as the precise and localized heat input capability that reduces the heat-affected zone (HAZ), residual stress and residual distortion. Chen *et al.* succeeded in joining Ti6Al4V and Inconel 718 for aerospace applications with minimal cracks thanks to improved supply heat input and laser beam positioning [2]. In a study by Chartterjee *et al.*, dissimilar metal welding of pure Ti and Ni was performed using high-power CO_2 laser butt welding. The results showed that an asymmetric weld shape formed and brittle intermetallic compounds of $NiTi_2$ and Ni_3Ti were obtained in the macroscopic shape of the weld pool [12]. The crystallographic mismatch between Ni and Ti led to the formation of intermetallic layers such as $NiTi_2$. This layer contributes to brittle failure of joints and solidification cracking. Although attempts have been made to avoid or decrease the formation of intermetallic layers by adding brazing filler during dissimilar metal welding, joints with poor mechanical properties have been achieved and thus alternative methods need to be implemented [15,16].

Most of the existing literature cites high laser power during butt or lap welding of Ti-based alloys and nickel-based alloys with the aim of improving joint soundness by increasing the heat input. Nonetheless, the use of high laser power is not suitable for welding skin-stringer joints (T-joints) with thin sections, especially in the aerospace industry where reduced weight is required, due to the excessive penetration in the weld joint as a result of high heat input. In this sense, the welding of T-joints is usually carried out using single-sided laser welding to reduce the heat input and decrease possible thermal distortion. These parts normally undergo heavy-duty vibrations during the lifespan and must be fatigue resistant. The single-sided T-joint configuration is very helpful when accessibility to the joining seam is limited. The mechanical strength of the joint is influenced by the weld beads along the skin-stringer components [17]. T-joints are normally made on customized clamping fixtures in order to retain the position between the skin and stringer as well as preserve a precise shape. Nevertheless, little work has been done on low-power fiber laser welding of Ti6Al4V and Inconel 600, especially in the T-joint configuration. Therefore, in the present study, single-sided laser welding was performed on thin sheets of Ti_6Al_6V and Inconel 600 in a T-joint configuration using a low-power Ytterbium-fiber laser. Specifically, the effects of welding speed and the overlapping factor on microstructure and mechanical properties including microhardness and pull testing of the joints were systematically evaluated.

2. Experimental Setup

2.1. Materials and Sample Preparation

Inconel 600 sheets of $20 \times 20 \times 1$ mm were used as stringers and Ti6Al4V sheets of $20 \times 20 \times 0.5$ mm were used as skin in this study. The chemical composition of Inconel 600 and Ti6Al4V is presented in Table 1. Prior to laser welding, all the sheets were first mechanically polished using SiC abrasive paper (No. 600 grit size) to remove the cutting edge burr and then cleaned with acetone to remove any contaminants.

Table 1. Chemical composition of Inconel 600 and Ti6Al4V (wt. %) [18].

Element	Ti6Al4V	Inconel 600
Ni	-	72 min
Cr	-	14.0–17.0
Cu	-	0.5 max
S	-	0.015 max
Si	-	0.5 max
C	0–0.08	0.05–0.10
Fe	0–0.4	6.0–10.0
Ti	Balance	-
Al	5.5–6.75	-
V	3.5–4.5	-

2.2. Laser Welding

A CW Ytterbium (Yb)-fiber laser with a 1070 nm emitting wavelength and 300 W maximum power output (Starfiber 300) was used in this study. The focal distance was kept constant at 346 mm between the scanner and the workpiece. The single-sided laser welding of these dissimilar work pieces was conducted in a T-joint configuration at a tilt angle of $45°$. Figure 1 shows a schematic diagram of the experimental welding setup. The laser beam with 100 μm spot diameter hit the seam between the skin and the stringer, which was clamped in the T-joint configuration using a clamping fixture jig as shown in Figure 1. Argon was used as shielding gas to prevent oxidation of the weld surface during welding. Welding speed ranging between 40 and 50 mm/s and overlapping factor ranging between 30% and 50% were chosen as the parameters of interest in this study. The experiments were performed in two separate series to investigate their effects on weld quality. The detailed welding parameters are listed in Table 2.

Table 2. Welding parameters used in the study.

Sample	Power (W) P	Welding Speed (mm/s) v	Overlapping Factor (%) η
L1	250	40	30
L2	250	40	40
L3	250	40	50
L4	250	50	50

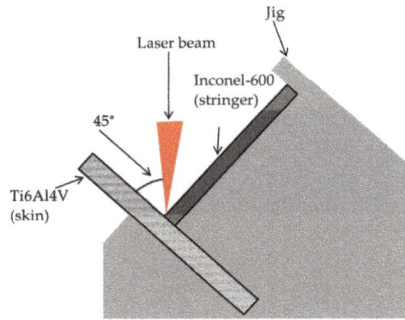

Figure 1. Schematic diagram of the welding experiment setup.

2.3. Microstructural Characterization

After welding, the samples were cut perpendicular to the welding direction for metallographic analysis. After grinding and polishing, the samples were etched using Kroll's reagent (92 mL distilled water, 6 mL nitric acid (HNO_3) and 2 mL hydrofluoric acid (HF)) for 20 s at room temperature. The microstructures were observed using an optical microscope (OM, Olympus, Tokyo, Japan) and SEM (Phenom Pro X, Crest System (M) Sdn. Bhd., Eindhoven, The Netherlands), and the chemical composition of the weld bead cross sections was estimated by energy dispersive X-ray (EDX, Phenom Pro X, Crest System (M) Sdn. Bhd.,Eindhoven, The Netherlands). The SEM employed was equipped with EDX. The formation of intermetallic weld phases was characterized by X-ray diffraction (XRD, PaNalytical Empyrean, DKHS Holdings (Malaysia) Bhd., Almelo, The Netherlands) using CuKα radiation at 2θ diffraction angles ranging from 30° to 80°.

2.4. Pull Test Measurement

The pull method was utilized in this study to evaluate the force needed to break the joints and was carried out according to a previous study [19]. To execute the pull test, a customized jig was used to clamp the workpiece, as shown in Figure 2. The pull test was then performed using a universal testing machine (INSTRON, Model: 3369, Necomb Sdn. Bhd., Singapore) at room temperature with crosshead speed of 5 mm/min and 10 kN load cell. The results were determined from the average value of three different samples.

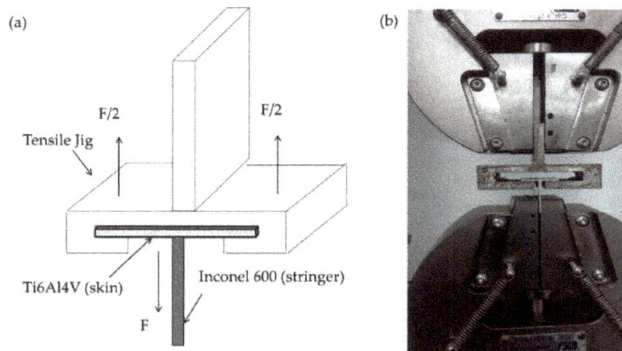

Figure 2. (**a**) Schematic diagram of tensile pull test for the T-joint configuration of Inconel 600 and Ti6Al4V; (**b**) Customized jig for tensile pull test.

2.5. Vickers Microhardness Measurement

The Vickers microhardness profile, including of the BM, FZ and HAZ, was evaluated using a pyramidal diamond indenter (HMV 2T E, SHIMAZDU, Kyoto, Japan) with a load of 200 g for a dwell time of 5 s. Figure 3 schematically illustrates the hardness distribution profile of the weld geometry. Nine indentations at an indentation interval of 100 μm were performed perpendicular to the stringer-skin weld seam to investigate the FZ and each HAZ variation between the two dissimilar base metals.

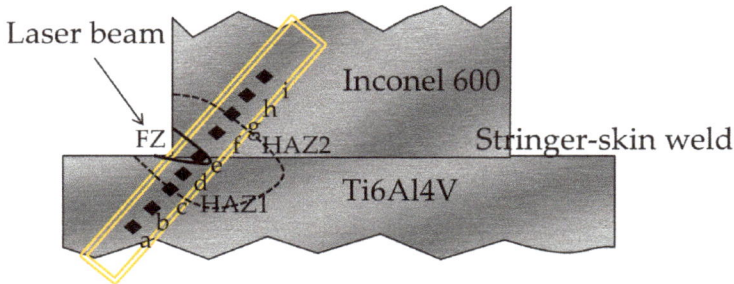

Figure 3. Schematic diagram of hardness distribution on the weld geometry.

3. Results and Discussion

3.1. Metallurgical Characterization

In the present study, single-sided laser beam welding was performed on Inconel 600 and Ti6Al4V in a T-joint configuration. Optical micrographs for the weld bead cross-sections of all samples are presented in Figure 4. An asymmetric welding seam was observed in every sample because only one side was welded. The welded area where the laser beam was introduced exhibited a concave shape, whereas a slightly convex shape was observed in the non-welded area. This observation is typical for single-sided laser welding and has been reported in other studies [20]. As shown in Figure 4, the weld pools displayed a certain asymmetry and a greater extent of melting was observed on Ti6Al4V skin. This is because the thermal diffusivity of Ti alloy is lower than Ni alloy, resulting in localized heating and greater melting of Ti6Al4V during welding [9]. Moreover, the color of the Ti6Al4V base metal remained consistent for all samples, whereas color variation was observed on the Inconel 600 base metal from one sample to another, as indicated in Figure 4. The reason for this is that Kroll's reagent is a recommended etchant for titanium alloys but not for nickel alloys. It has been reported that the nitric acid in Kroll's reagent acts as an oxidizing agent and reacts with the nickel in Inconel 600 to produce Ni^{3+}. Therefore, it can be inferred that the color variation of Inconel 600 is due to the intensity difference of Ni^{3+} production [18].

A sound joint was obtained when the welding speed was 40 mm/s and the overlapping factor was 50%. This is proven by the minor gap line of 0.202 mm indicated by sample L3 in Figure 4c. However, decreasing penetration depth and increasing gap line length were observed on the weld bead samples L1, L2 and L4, which possessed either lower overlapping factors or higher welding speeds. The gap lines were 0.637 mm, 0.527 mm and 0.422 mm long for samples L1, L2 and L4, respectively. Therefore, it can be concluded that the gap line length increased and the penetration depth decreased as the welding speed increased (comparing samples L3 and L4) while the overlapping factor decreased (comparing samples L1, L2 and L3).

As mentioned above, optimum welding was achieved for sample L3 because the heat input under this welding condition was sufficient to produce a sound T-joint with satisfactory penetration between the skin and stringer. It is reported that the amount of heat input determines the degree of dilution and cooling rate in the weld. Insufficient heat input during the welding process resulted in a faster cooling rate, thus causing limited melting between the skin and stringer and thereby leading to insufficient penetration. Heat input is critical in defining the joint's geometry and can be manipulated by varying the welding parameters [21]. Among these, laser welding speed and the laser beam overlapping factor are commonly known as the most notable variables affecting the heat input. In this study, the heat input was calculated according to the following equation:

$$\text{Heat Input} = P/[\pi \left(\frac{d-\eta}{2}\right)^2 \times v] \tag{1}$$

where P = laser power, v = welding speed, d = beam diameter and η = overlapping factor.

Based on the equation above, it is proven that increasing the overlapping factor and decreasing the laser welding speed will increase the heat input on the workpiece. According to Figure 5, the affected areas are exposed to more accumulated heat input when the overlapping factor increases [22,23].

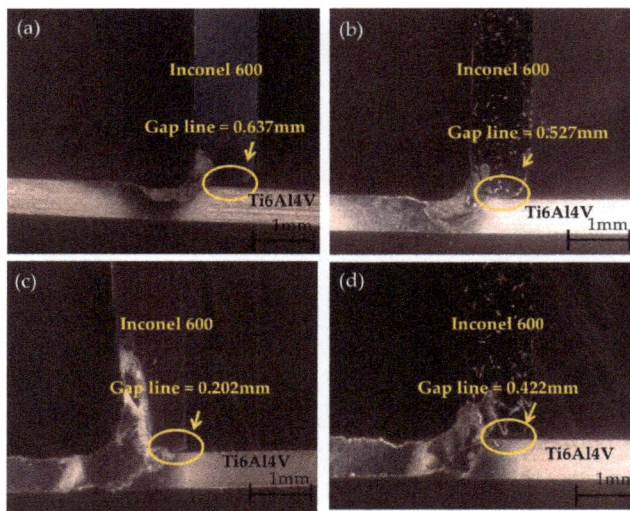

Figure 4. Optical micrographs of samples (**a**) L1; (**b**) L2; (**c**) L3 and (**d**) L4.

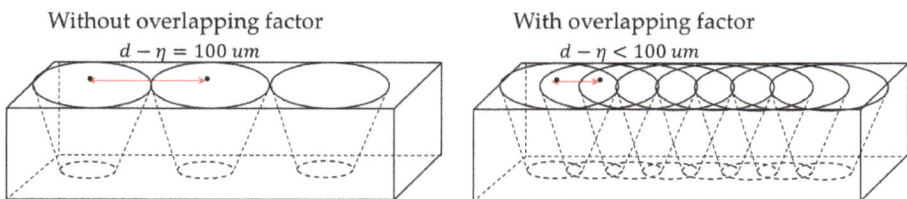

Figure 5. Schematic diagram showing the effect of the overlapping factor on the laser-irradiated area.

3.2. Microstructural Observation and Elemental Composition Analysis of the Weld Bead

Figure 6 shows the microstructure of the interface between the FZ and HAZ of Ti6Al4V (denoted as HAZ1) and Inconel 600 (denoted as HAZ2) cross-sections of sample L3. The grains become notably finer in HAZ compared to those in the base metals, indicating a large thermal gradient at the FZ and HAZ interface during the laser welding process. Such a phenomenon can be attributed to rapid solidification, whereby the grain size in the microstructure is rather small in the laser-affected areas and surroundings [24]. At FZ, an equiaxed pattern was observed as grain sizes were smaller due to rapid melting and solidification. The fusion zone had a similar identification of columnar beta grains of Ti that occurred during two-dimensional heat flow condition [25].

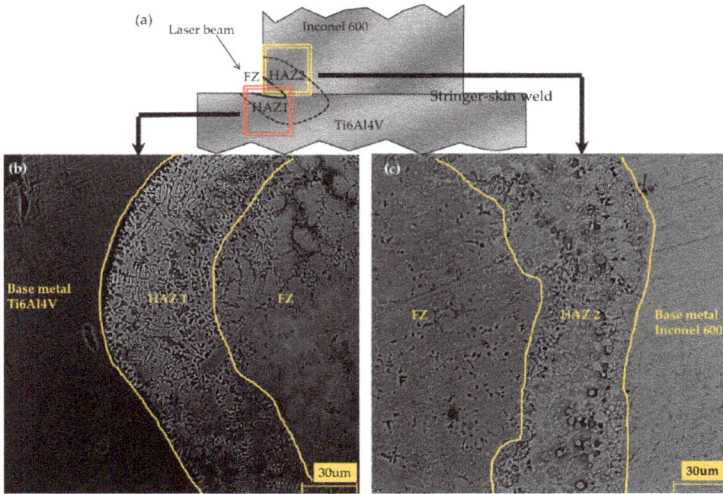

Figure 6. (**a**) Schematic diagram of FZ, HAZ1, HAZ2 and base of sample, SEM micrographs of microstructure sample; (**b**) L3 for FZ, HAZ1 and base metal, Ti6Al4V (**c**) L3 for FZ, HAZ2 and base metal, Inconel 600.

Figure 7 shows the presence of NiTi and NiTi$_2$ microstructures, proving the formation of these phases between Ni base alloy and Ti base alloy during laser welding. This observation is further corroborated by the EDX analysis displayed in Table 3, where the atomic percentage of Ni and Ti for each sample was determined. According to the Ni–Ti phase diagram, the NiTi phase was produced in primary crystallization from the melt, whereas the NiTi$_2$ phase was formed by peritectic reaction between the NiTi phase and the cooling after NiTi columnar dendrite melting [26]. Points A, C, D and E had nearly the same percentage of Ni and Ti, proving NiTi was present, while point B and F showed that the ratio of Ti to Ni was 2:1, indicating the presence of NiTi$_2$. As displayed in Figure 7, NiTi$_2$ was only detected near the weld pool for samples L1 and L4 due to the lower heat input that allowed the phase formation. Meanwhile, NiTi was detected in all samples, as shown in Figure 7. The results obtained are similar to the study done by Chatterjee *et al.* who investigated the microstructural development of NiTi and NiTi$_2$ during dissimilar metal welding of Ti and Ni [12]. The phases were detected at the weld pool beside the HAZ of Inconel 600. The formation of these intermetallic phases is reportedly attributed to the crystallographic mismatch between Ni and Ti.

Table 3. The element atomic percentage in L1, L2, L3 and L4.

Point	Ni (at. %)	Ti (at. %)	V (at %)	Al (at. %)	Cr (at. %)	Fe (at. %)	Total (%)	Phase
A	39.5	40.5	3.4	5.4	7.7	3.5	100.0	NiTi
B	28.4	56.4	3.3	5.3	3.6	3.0	100.0	NiTi2
C	42.1	38.9	3.2	4.0	8.3	3.5	100.0	NiTi
D	40.5	40.9	3.5	3.9	8.1	3.1	100.0	NiTi
E	39.4	40.6	3.3	4.4	7.6	4.7	100.0	NiTi
F	29.1	55.4	3.5	6.0	3.3	2.7	100.0	NiTi2

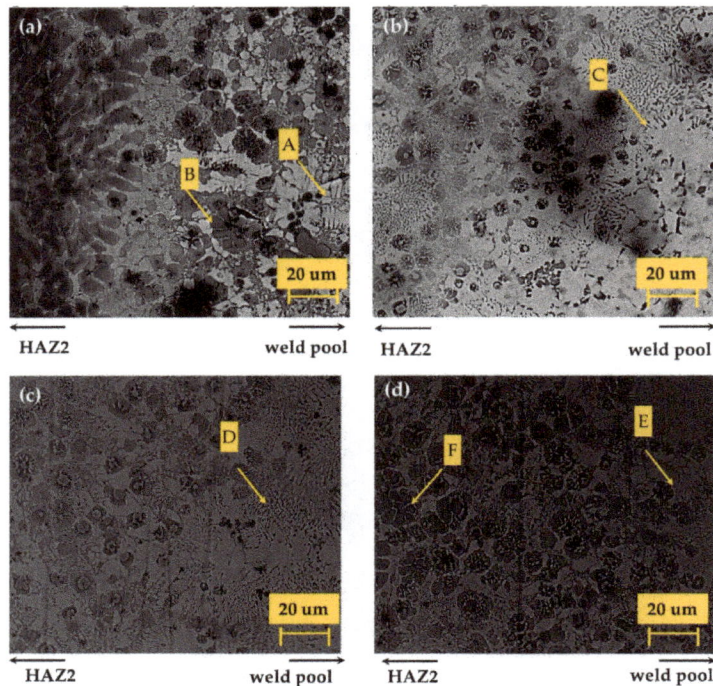

Figure 7. SEM micrographs of NiTi and NiTi$_2$ intermetallic phases formed in the FZ at samples (**a**) L1; (**b**) L2; (**c**) L3 and (**d**) L4.

The Ti–Ni system had different conductivities of $\kappa_{Ni} \approx 4\kappa_{Ti}$, resulting in a Ti-rich melt pool combined with metallurgical dissimilarity. Due to the differences in thermal diffusivity and conductivity, Ti from Ti6Al4V flowed towards the weld pool and reacted with Ni from Inconel. This resulted in inhomogeneous molten flow (Ti-rich melt pool) and asymmetric heat transfer during the welding process, contributing to the formation of these intermetallic phases due to metallurgical dissimilarity as shown in the phase diagram of Ti–Ni [9]. NiTi is reportedly being more erosion-resistant than stainless steels [27,28]. Therefore, the presence of NiTi as an intermetallic compound in the weld zone is an advantage in improving the joining properties. During dissimilar laser welding of titanium alloy and nickel alloy, base metal grains grew into the welding pool. A steep composition gradient was observed at the interface. NiTi and NiTi$_2$ dendrites grew to form a band at the HAZ. The NiTi liquidus line had a steeper slope compared to NiTi$_2$, suggesting the latter is likely to form heterogeneously. According to the phase diagram, both NiTi and NiTi$_2$ constituted the bulk of the weld microstructure. The microstructures may have been caused by the precipitation of NiTi$_2$ from the rich Ti β2 phase and entrapment due to growing NiTi dendrites. This particle can nucleate before NiTi in certain circumstances [29].

The XRD analysis indicated that NiTi and NiTi$_2$ formed in the FZ and HAZ. The NiTi$_2$ phase is not recommended in large quantities, because it can lead to a strong tendency of hot cracking, which is prone to cause weaker joints [9,15,16,29]. From Figure 8 it can be seen that the high-temperature B2 phase (austenite) from NiTi was found in samples L2 and L3, which further changed to low-temperature B19' phase (martensite) when the overlapping factor was 30% for sample L1, and the welding speed increased to 50 mm/s for sample L4. The amount of NiTi also decreased when the welding speed increased (L4) and the overlapping factor reduced (L1). Based on the XRD analysis, NiTi$_2$ formed in all samples. However, the peak $NiTi_2$ was detected only in L1 and L4. This may be related to the peritectic reaction that did not occur in samples L2 and L3. At lower heat input where shorter heating and cooling time are needed for laser welding, the penetration of Ni from Inconel 600 was insufficient, resulting in the XRD peak of $NiTi_2$ and martensitic phase of NiTi. However, with higher heat input and slightly longer welding time, the penetration of Ni from Inconel 600 was sufficient, thus causing the XRD peak of the NiTi austenitic phase to become more apparent [27]. This was proven by the increased number of NiTi counts with the increasing overlapping factor, whereas the NiTi count reduced when laser welding speed increased. The changes in XRD counts show the differences clearly. Similar phases were detected in related studies done by other researchers, proving that the X-ray beam hit the welded area [9,27]. However, limited peaks were found during XRD, indicating a tentative phase indexation of these phases. The overlapping factor and laser welding speed clearly affected the microstructural properties of the weld joint.

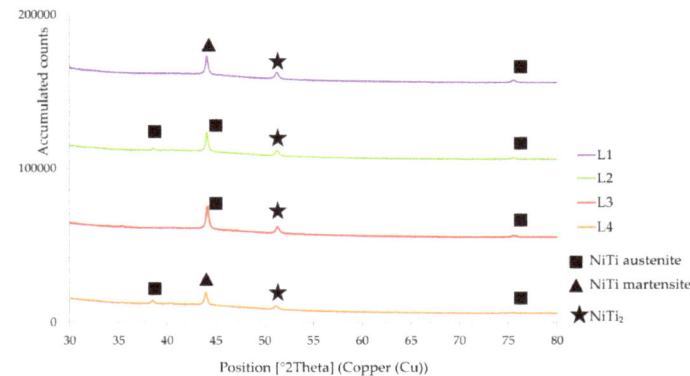

Figure 8. XRD patterns of samples L1, L2, L3 and L4.

3.3. Characterization of Mechanical Properties

The effects of the overlapping factor and welding speed on the breaking force (the force needed to fracture samples) of the weld beads are presented in Figure 9. By comparing samples L1, L2 and L3, it was noted that the breaking force increased by increasing the overlapping factor. However, the opposite trend was observed when studying the effect of welding speed, where the breaking force decreased with increasing welding speed as seen for samples L3 and L4. In the present study, the maximum force of 150 N was achieved in sample L3 when the overlapping factor was 50% and the welding speed was 40 mm/s at a given constant power of 250 W. The increase was mainly because the laser welding at higher overlapping factor or lower welding speed induced greater heat input and thus a wider fusion zone with deeper penetration that required higher breaking force. The results are in accordance with the results obtained from metallurgical characterization, indicating that heat input was the main factor affecting the geometry and breaking force of the weld joint [27].

Figure 10 depicts the effects of the overlapping factor and welding speed on the hardness distribution of the welded bead. The hardness was studied across the entire profile of the weld bead (including base metal, heat-affected zone and fusion zone). The hardness of Inconel 600 base material was around 140–160 HV, whereas the hardness of Ti6Al4V base material was around 260–300 HV. As the overlapping factor increased, the FZ and HAZ hardness increased. The hardest sample was achieved at a welding speed of 40 mm/s and overlapping factor of 50%. The hardness at HAZ1 was higher than HAZ2, as the HAZ1 was closer to the Ti6Al4V base material that had higher hardness. As shown in Figure 10, sample L3 displayed the highest hardness among its counterparts with 554 HV at FZ, 715 HV at HAZ1 and 570 HV at HAZ2. This is attributed to the presence of NiTi, as the Vickers hardness of NiTi alloy reportedly ranges between 380 and 440 HV [30]. Both skin and stringers in this study were annealed by undergoing heat treatment. When welding in this condition, there is a possibility that micro fissures will form in the HAZ region due to the coarser grains formed. The measured FZ region hardness was higher than the base metals. Meanwhile, the hardness in the HAZ region was even higher than the FZ region. The hardness of the FZ region was observed to be lower than the HAZ region, as shown in Figure 10. This was a result of the higher amount of molten metal in FZ due to the lack of precipitation strengthening caused by particles dispersing from the precipitate. A hardness transition between FZ and the base metal occurred and the HAZ region measured the highest hardness among the three regions. Other comparable studies have been done, where similar areas were indicated for hardness measurement [31].

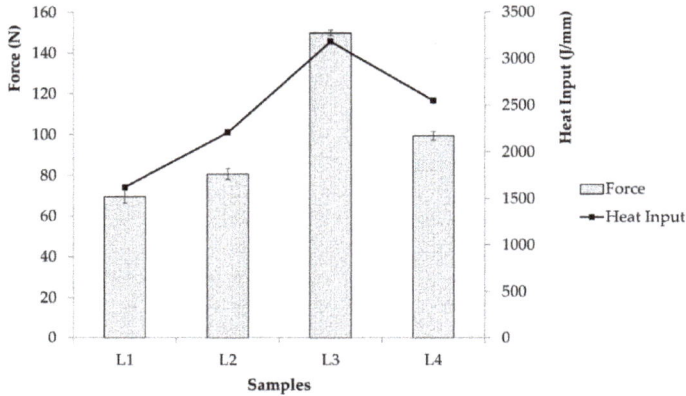

Figure 9. Force needed to fracture the samples and heat input for samples L1, L2, L3 and L4.

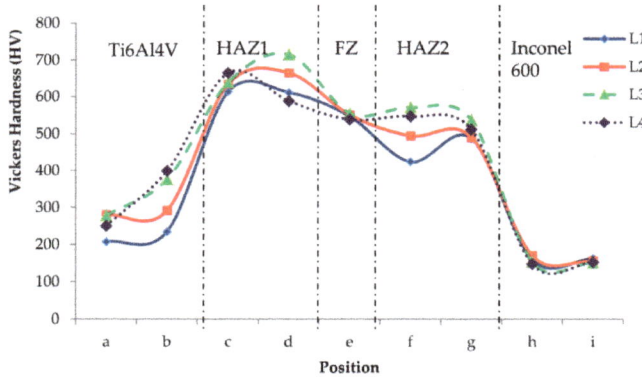

Figure 10. Vickers microhardness profiles of all samples.

4. Conclusions

According to the experimental results reported in this study, it can be concluded that:

1. Single-sided laser welding was successfully performed on thin sheets of Ti6Al4V and Inconel 600 in a T-joint configuration using low-power Ytterbium-fiber laser with the influence of laser welding speed and the overlapping factor.

2. The highest breaking force and hardness were obtained when the welding speed was 40 mm/s and the overlapping factor was 50% at laser power of 250 W, proving a sound joint was obtained because sufficient heat input was produced.

3. NiTi and NiTi$_2$ intermetallic layers formed in the FZ and HAZ regions due to crystallographic mismatch between Ni and Ti as well as the differences in thermo-physical properties of the materials.

4. The extent of skin-stringer penetration was related to the heat input, which was manipulated by varying the laser welding speed and overlapping factor.

Acknowledgments: The authors greatly acknowledge University of Malaya for providing the necessary facilities and resources for this research. This research is funded by the Postgraduate Research Fund (PPP) with Grant No: PG001-2013A and University Malaya Research Grant (UMRG) with Grant No.: RP010A-13AET. Authors would like to acknowledge Mr Zaharudin Md Salleh from Geology Department and Madam Hartini Baharum from Engineering Faculty for their support in laboratory works.

Author Contributions: Shamini Janasekaran conducted and analyzed the experiments, and wrote the article. Ai Wen Tan contributed by analyzing data and revising the writing. Farazila Yusof and Mohd Hamdi Abdul Shukor supervised the work and further analyzed the data.

Conflicts of Interest: The authors declare no conflict of interest.

Abbreviations

The following abbreviations were used in this manuscript:

SEM	Scanning electron microscopy
XRD	X-ray diffraction
LBW	Laser beam welding
BM	Base metal
CW	Continuous wave
FZ	Fusion zone
HAZ	Heat-affected zone
TIG	Tungsten inert gas
MIG	Metal inert gas
Ti	Titanium
Ni	Nickel
Yb	Ytterbium
HNO$_3$	Nitric acid
HF	Hydrofluoric acid
OM	Optical microscope
EDX	Energy dispersive X-ray

References

1. Shokrani, A.; Dhokia, V.; Newman, S.T. Environmentally conscious machining of difficult-to-machine materials with regard to cutting fluids. *Int. J. Mach. Tool. Manuf.* **2012**, *57*, 83–101. [CrossRef]

2. Chen, H.C.; Pinkerton, A.J.; Li, L. Fibre laser welding of dissimilar alloys of Ti–6Al–4V and inconel 718 for aerospace applications. *Int. J. Adv. Manuf. Technol.* **2011**, *52*, 977–987. [CrossRef]

3. Mavrigian, M. *Performance Exhaust Systems: How to Design, Fabricate, and Install*; CarTech Inc.: North Branch MN, USA, 2014; p. 144.

4. Shen, Y.; Liu, Y.; Sun, W.; Dong, H.; Zhang, Y.; Wang, X.; Zheng, C.; Ji, R. High-speed dry compound machining of Ti6Al4V. *J. Mater. Process. Technol.* **2015**, *224*, 200–207. [CrossRef]

5. Warwick Manufacturing Group. Off to the races. *Met. Powder Rep.* **2012**, *67*, 22–24.

6. Welding exhaust systems—part 1. Available online: http://www.burnsstainless.com/weldingarticle1.aspx (accessed on 12 April 2016).

7. Guo, J.F.; Chen, H.C.; Sun, C.N.; Bi, G.; Sun, Z.; Wei, J. Friction stir welding of dissimilar materials between AA6061 and AA7075 Al alloys effects of process parameters. *Mater. Des.* **2014**, *56*, 185–192. [CrossRef]

8. Kurakake, Y.; Farazila, Y.; Miyashita, Y.; Otsuka, Y.; Mutoh, Y. Effect of molten pool shape on tensile shear strength of dissimilar materials laser spot joint between plastic and metal. *J. Laser Micro Nanoeng.* **2013**, *8*, 161–164. [CrossRef]

9. Chatterjee, S.; Abinandanan, T.A.; Chattopadhyay, K. Phase formation in Ti/Ni dissimilar welds. *Mater. Sci. Eng. A* **2008**, *490*, 7–15. [CrossRef]

10. Schubert, E.; Klassen, M.; Zerner, I.; Walz, C.; Sepold, G. Light-weight structures produced by laser beam joining for future applications in automobile and aerospace industry. *J. Mater. Process. Technol.* **2001**, *115*, 2–8.

11. Seretsky, J.; Ryba, E. Laser-welding of dissimilar metals—Titanium to nickel. *Weld. J.* **1976**, *55*, 208–211.

12. Chatterjee, S.; Abinandanan, T.A.; Chattopadhyay, K. Microstructure development during dissimilar welding: Case of laser welding of Ti with Ni involving intermetallic phase formation. *J. Mater. Sci.* **2006**, *41*, 643–652.

13. Carpinteri, A.; Brighenti, R.; Huth, H.J.; Vantadori, S. Fatigue growth of a surface crack in a welded T-joint. *Int. J. Fatigue* **2004**, *27*, 59–69. [CrossRef]

14. Padovani, C.; Fratini, L.; Squillace, A.; Bellucci, F.E. Electrochemical analysis on friction stir welded and laser welded 6xxx aluminium alloys t-joints. *Corros. Rev.* **2007**, *25*, 475–489. [CrossRef]

15. Oliveira, J.P.; Panton, B.; Zeng, Z.; Andrei, C.M.; Zhou, Y.; Miranda, R.M.; Fernandes, F.M.B. Laser joining of niti to Ti6Al4V using a niobium interlayer. *Acta Mater.* **2016**, *105*, 9–15. [CrossRef]

16. Miranda, R.M.; Assunção, E.; Silva, R.J.C.; Oliveira, J.P.; Quintino, L. Fiber laser welding of NiTi to Ti–6Al–4V. *Int. J. Adv. Manuf. Technol.* **2015**, *81*, 1533–1538. [CrossRef]

17. Fratini, L.; Buffa, G.; Shivpuri, R. Influence of material characteristics on plastomechanics of the fsw process for T-joints. *Mater. Des.* **2009**, *30*, 2435–2445. [CrossRef]

18. Rebak, R.B.; Crook, P. Nickel alloys for corrosive environments. *Adv. Mater. Process.* **2000**, *157*, 37.

19. Badini, C.; Pavese, M.; Fino, P.; Biamino, S. Laser beam welding of dissimilar aluminium alloys of 2000 and 7000 series: Effect of post-welding thermal treatments on T joint strength. *Sci. Technol. Weld. Join.* **2009**, *14*, 484–492. [CrossRef]

20. Enz, J.; Khomenko, V.; Riekehr, S.; Ventzke, V.; Huber, N.; Kashaev, N. Single-sided laser beam welding of a dissimilar AA2024-AA7050 T-joint. *Mater. Des.* **2015**, *76*, 110–116. [CrossRef]

21. Unt, A.; Salminen, A. Effect of welding parameters and the heat input on weld bead profile of laser welded T-joint in structural steel. *J. Laser Appl.* **2015**, *27*. [CrossRef]

22. Ribic, B.; Palmer, T.A.; DebRoy, T. Problems and issues in laser-arc hybrid welding. *Int. Mater. Rev.* **2009**, *54*, 223–244. [CrossRef]

23. Gao, X.L.; Liu, J.; Zhang, L.J.; Zhang, J.X. Effect of the overlapping factor on the microstructure and mechanical properties of pulsed Nd:Yag laser welded Ti6Al4V sheets. *Mater. Character.* **2014**, *93*, 136–149.

24. Norris, J.T.; Robino, C.V.; Hirschfeld, D.A.; Perricone, M.J. Effects of laser parameters on porosity formation: Investigating millimeter scale continuous wave Nd:Yag laser welds. *Weld. J.* **2011**, *90*, 198.

25. Donachie, M.J. *Titanium: A Technical Guide*; ASM International: Novelty, OH, USA, 2000; p. 381.

26. Sudarshan, T.S. *Surface Modification Technologies: Proceedings of the 20th International Conference on Surface Modification Technologies*; ASM International: Vienna, Austria, 2007.

27. Hiraga, H.; Inoue, T.; Shimura, H.; Matsunawa, A. Cavitation erosion mechanism of niti coatings made by laser plasma hybrid spraying. *Wear* **1999**, *231*, 272–278. [CrossRef]

28. Richman, R.H.; Rao, A.S.; Kung, D. Cavitation erosion of NiTi explosively welded to steel. *Wear* **1995**, *181–183*, 80–85. [CrossRef]

29. Kocich, R.; Szurman, I.; Kursa, M. The methods of preparation of Ti–Ni–X alloys and their forming. In *Shape Memory Alloys-Processing, Characterization and Applications*; Fernandes, F.M.B., Ed.; Press: Ostrava, Czech Republic, 2013.

30. Brantley, W.A.; Eliades, T. *Orthodontic Materials: Scientific and Clinical Aspects*; Thieme: Stuttgart, Germany, 2011.

31. Atabaki, M.M.; Nikodinovski, M.; Chenier, P.; Ma, J.; Liu, W.; Kovacevic, R. Experimental and numerical investigations of hybrid laser arc welding of aluminum alloys in the thick T-joint configuration. *Opt. Laser Technol.* **2014**, *59*, 68–92. [CrossRef]

metals

MDPI

Article

Gas-Solid Reaction Route toward the Production of Intermetallics from Their Corresponding Oxide Mixtures

Hesham Ahmed [1,2,*], R. Morales-Estrella [3], Nurin Viswanathan [4] and Seshadri Seetharaman [5]

[1] Division of Minerals and Metallurgical Engineering, Department of Civil, Environmental and Natural Engineering, Luleå University of Technology, 97187 Luleå, Sweden
[2] Department of Minerals Technology, Central Metallurgical Research and Development Institute, Box 87-Helwan, Cairo, Egypt
[3] Instituto de Investigación en Metalurgia y Materiales, Universidad Michoacana de San Nicolás de Hidalgo, Ciudad Universitaria, C.P. 58030, Morelia, México; rmorales@umich.mx
[4] Centre of Excellence in Steel Technology (CoEST), Indian Institute of Technology Bombay, 400076 Mumbai, India; vichu@iitb.ac.in
[5] Royal Institute of Technology (KTH), S-100 44 Stockholm, Sweden; raman@kth.se
* Correspondence: hesham.ahmed@ltu.se; Tel.: +46-920-491-309

Academic Editor: Ana Sofia Ramos
Received: 29 June 2016; Accepted: 10 August 2016; Published: 17 August 2016

Abstract: Near-net shape forming of metallic components from metallic powders produced in situ from reduction of corresponding pure metal oxides has not been explored to a large extent. Such a process can be probably termed in short as the "Reduction-Sintering" process. This methodology can be especially effective in producing components containing refractory metals. Additionally, in situ production of metallic powder from complex oxides containing more than one metallic element may result in in situ alloying during reduction, possibly at lower temperatures. With this motivation, in situ reduction of complex oxides mixtures containing more than one metallic element has been investigated intensively over a period of years in the department of materials science, KTH, Sweden. This review highlights the most important features of that investigation. The investigation includes not only synthesis of intermetallics and refractory metals using the gas solid reaction route but also study the reaction kinetics and mechanism. Environmentally friendly gases like H_2, CH_4 and N_2 were used for simultaneous reduction, carburization and nitridation, respectively. Different techniques have been utilized. A thermogravimetric analyzer was used to accurately control the process conditions and obtain reaction kinetics. The fluidized bed technique has been utilized to study the possibility of bulk production of intermetallics compared to milligrams in TGA. Carburization and nitridation of nascent formed intermetallics were successfully carried out. A novel method based on material thermal property was explored to track the reaction progress and estimate the reaction kinetics. This method implies the dynamic measure of thermal diffusivity using laser flash method. These efforts end up with a successful preparation of nanograined intermetallics like Fe-Mo and Ni-W. In addition, it ends up with simultaneous reduction and synthesis of Ni-WN and Ni-WC from their oxide mixtures in single step.

Keywords: gas-solid reactions; fluidization reaction; nanosized structures

1. Introduction

Intermetallics are well-suited for applications in high technology, where there is a strong need for materials that can withstand high temperatures. Intermetallics are suitable materials for the manufacture of microstructured tools because of their excellent mechanical properties in regard to wear and mechanical durability. Ni-W alloys for example exhibit enhanced properties such as corrosion resistance and wear resistance. This kind of alloys also can be used for magnetic heads, bearings, magnetic relays, etc. The problem in the utilization of intermetallics is their brittleness which calls for grain refinement. The grain size needed to produce ductility is very small and is difficult to achieve. In this aspect, the gas-solid reaction route is of great advantage in controlling the nano-sized structures. On the other hand, near-net shape forming of metallic powders produced in situ from reduction of corresponding pure metal oxides has not been explored to large extent. Such a process can be probably termed in short as "Reduction-Sintering" process. This methodology can be especially effective in producing components containing refractory metals. Additionally, in situ production of metallic powder from complex oxides containing more than one metallic element may result in in situ alloying during reduction, possibly at lower temperatures. With this motivation, in situ reduction of complex oxides mixtures containing more than one metallic element has been investigated intensively over a period of years in the Department of Materials Science and Engineering, Royal Institute of Technology, Stockholm, Sweden. The strategy adopted by the present authors was to initially study the hydrogen reduction of thin beds of oxide powders leading to intermetallics and refractory metals. In order to produce the intermetallic phases in bulk, fluidized bed technique was adopted in view of the excellent contact between the reactant solid and the gas with achievable high reaction efficiencies; the inter-particle contact would be minimum and the temperature of the reaction would be low. Therefore, both sintering and grain growth in the produced intermetallic phase will be minimum. Moreover, carburization and nitridation of nascent intermetallics could be successfully carried out. A novel method based on material thermal property was explored to track the reaction progress and estimate the reaction kinetics. This method implies the dynamic measure of thermal diffusivity using laser flash method. Gases like H_2, CH_4 and N_2, with low negative impact on the environment were used for simultaneous reduction, carburization, and nitridation, respectively. Thus, the present results are likely to lead to the synthesis of an entirely new series of materials with interesting properties; for example, production of Fe-Mo and Ni-Wi intermetallics with nano-grained structures along with Ni-WN and Ni-WC composites produced by simultaneous reduction from their oxide mixtures in a single step. This novel method was further developed to produce intermetallic coatings on copper surfaces. Moreover, other intermetallics with superior structure produced from their corresponding oxides have been reported elsewhere [1].

2. Materials and Methods

This section describes relevant details of the experimental techniques and procedures involved in this work. The entire experimental work was carried out within The Department of Materials Science and Technology, Royal institute of technology (KTH), Sweden. The experimental procedures described below do not represent the order in which this work was conducted.

2.1. Materials and Sample Preparation for Kinetic Studies

Table 1 shows the starting materials used for the present work (reduction, reduction-carburization and reduction-nitridation). These studies can be divided in to 3 categories; (1) thermogravimetric studies, (2) fluidized bed studies and (3) thermal diffusivity measurements. In the case of thermogravimetric and fluidized bed studies systems studied were viz., Fe-Mo-O and Ni-W-O. In the case of thermal diffusivity measurements NiO-WO_3 powder was studied.

In order to produce stoichiometric Fe_2MoO_4, powders of Fe, Fe_2O_3, and MoO_3, with mole ratio 4:1:3, were mixed thoroughly using an eccentric oscillator at 200 round per minute. Then the mixture was placed into an iron crucible with 45 mm inner diameter. An iron lid was then welded to the top of the crucible to make it gas tight. Thereafter, the crucible was heated under argon atmosphere at 1173 K for 24 h followed by a similar period of time at 1373 K. The crucible was removed from the hot zone at the end and quenched in water. The Fe_2MoO_4 thus synthesized was submitted to X-Ray diffraction (XRD) analysis to verify it against its reference pattern corresponding to Powder Diffraction File 00-025-1403.

Table 1. Starting materials, their purity and corresponding supplier.

Compound	Purity %	Supplier
MoO$_3$	99.95	Alfa Aesar; Karlsruhe, Germany
Fe$_2$O$_3$	99.8	Alfa Aesar; Karlsruhe, Germany
Fe	99.95	Merck; Darmstadt, Germany
Fe	98	Merck; Darmstadt, Germany
NiO	99	Sigma-Aldrich (St. Louis, MO, USA)
WO$_3$	99.9	Atlantic Equipment Engineering (AEE) (Bergenfield, NJ, USA)
NiWO$_4$	99	Johnson Matthey Inc. (London, UK)
		Iron with 98 pct was used for the fluidized bed experiments

On the other hand, the excess of oxygen in the nickel oxide was removed by heating the powder to 1273 K in argon and then left to cool down in the furnace. Stoichiometric NiO and WO$_3$ were then mixed in predetermined different ratios to produce Ni-W-O mixtures with different Ni and W content. The oxides were then mixed thoroughly and pressed into briquettes (10 mm in diameter and 5 mm in height), heated up to 873 K, and kept at this temperature overnight (24 h). Then the temperature was raised to 1273 K, and the samples were left to sinter at this temperature for 72 h.

2.2. Methods (Techniques and Procedures)

Both isothermal and non-isothermal experiments were carried out by means of thermogravimetric unit (SETARAM TGA 92, SETARAM instrumentation, Caluire, France) having a detection limit of 1 μg. Complete details of the experimental set up are given elsewhere [2]. Nevertheless, the experimental conditions were adjusted as to obtain the rate of the chemical reaction as the rate controlling mechanism. That is to say, the following parameters were carefully optimized; a hydrogen flow above the starvation rate, a very thin layer of powder (10–40 mg), and an average particle size of about 1–5 μm. Additionally, preliminary experiments were conducted to ensure that there is no external mass transfer effect through the sample bed.

Fluidized bed experiments were conducted in an electrical resistance furnace. A quartz tube with dimensions 1000 mm long and an inner diameter of 15 mm was vertically positioned in the furnace. A porous quartz disc (2 mm thick) was fused in the middle of the reactor, as sample supporter as well as gas distributer. The water content of the off-gases was monitored using a Shimadzu Gas Chromatograph (GC), model GC-2014 with Thermal conductivity Detector (TCD) (Shimadzu Corp., Kyoto, Japan). The fluidized bed reactor was connected to the gas chromatograph by a stainless steel tube of 5 mm inner diameter. Minimum fluidization velocity (U^*_{mf}) at room temperature was firstly determined experimentally and corresponding U_{mf} values at higher temperatures were calculated according to Equation (1). More details of the experimental setup can be found elsewhere [3]. The fluidized-bed reduction experiments were conducted isothermally. The sample was allowed to rest on the porous disc in the reactor. The powder bed was kept under a continuous flow of argon gas during heating segment. When the desired temperature was reached and stabilized, the inert gas was replaced by hydrogen.

$$U_{mf} = U*_{mf} \frac{\rho_r u_r}{\rho_T u_T} \tag{1}$$

where ρ_r, u_r, ρ_T and u_T stand for the properties of the gas phase, viz. densities and viscosities at room temperature and high temperature, respectively.

A laser flash unit model TC-7000H/MELT provided by Sinku-Rico, Inc., Yokohama, Japan was used for thermal diffusivity measurements. The laser beams irradiate the top side of the sample and provides an instantaneous energy pulse. The laser energy is then absorbed by the top surface of the sample and diffuses through the sample down to the other side. Immediately after the laser flash, the temperature of the other side (the rear face) is recorded using a photovoltaic infrared detector.

The increase in temperature of the rare surface of the sample was plotted against time. Further details of instrument and procedure are reported in an earlier publication [4].

3. Results and Discussion

In the present section the data obtained for Fe-Mo-O and Ni-W-O systems will be shown and discussed separately.

3.1. Fe-Mo-O System

3.1.1. Isothermal Reduction of Fe_2MoO_4

Figure 1 shows the reduction fraction (X) as a function of time for the reduction of iron molybdate by hydrogen in the temperature range of 823–1073 K. The fractional reduction, X, is defined as the ratio of the instant mass loss, Δmt, over the theoretical final mass loss, $\Delta m\infty$, (calculated based on the loss of four oxygen atoms per Fe_2MoO_4 unit). It is clearly seen that, under the prevailing experimental conditions, the reduction process is sensitive to temperatures, which confirms that the rate of the chemical reaction is the rate controlling step. Moreover, the reduction curves suggested a single step reaction. XRD analyses on partially reduced samples revealed only Fe_2MoO_4 and Fe_2Mo phases. The completely reduced product was established to be a homogeneous Fe_2Mo intermetallic phase; the existence of which had been a controversy over the years [5–7].

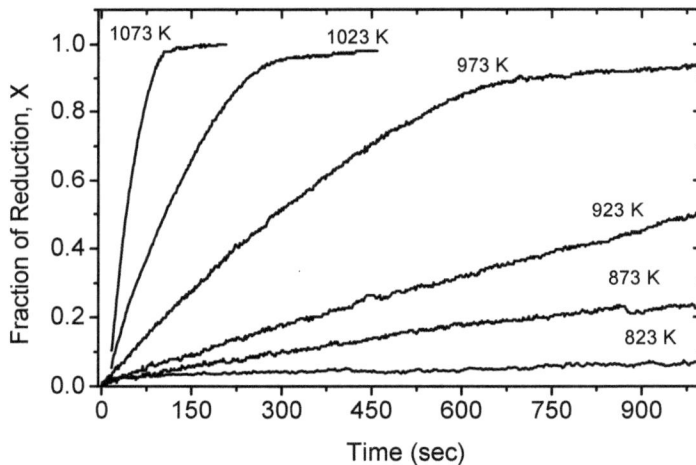

Figure 1. The isothermal reduction curves of shallow powder beds $FeMoO_4$ by hydrogen [2].

Hence, the chemical reaction for the reduction of Fe_2MoO_4 by hydrogen gas can be expressed as follows

$$1/4Fe_2MoO_4(s) + H_2(g) = 1/4Fe_2Mo(s) + H_2O(g) \qquad (2)$$

The kinetic analysis of the gas-solid reaction was worked out using the shrinking core model. Such model was combined with the Arrhenius rate law leading to the following expression [2]:

$$\frac{\left[1 - (1-X)^{1/3}\right]}{t} = \frac{M_{Fe_2MoO_4} \cdot P_{H_2} \cdot k_0}{\rho_{Fe_2MoO_4} \cdot r_0} \exp\left(-\frac{Q}{RT}\right) \qquad (3)$$

where, t is instant time, r_0 is the particle initial radius, $\rho_{Fe_2MoO_4}$ and $M_{Fe_2MoO_4}$ are the density and molecular weight of Fe_2MoO_4, respectively, k_0 is the frequency factor from the Arrhenius plot, P_{H_2} is the partial pressure of hydrogen, Q is the activation energy of the reaction, T is the temperature in K, and R is the gas constant. The plot of left hand side of Equation (2) as a function of $1/T$ is given in Figure 2. From the slopes of the plot, the corresponding activation energy is found to be 173 kJ/mol.

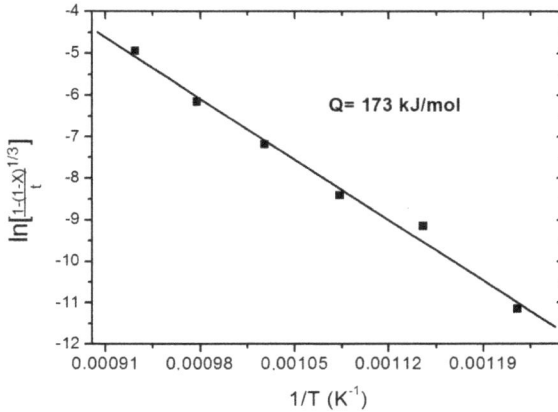

Figure 2. Arrhenius plot for the isothermal reduction of shallow powder beds of Fe_2MoO_4 [2].

3.1.2. Nonisothermal Reduction of Fe_2MoO_4

Figure 3 shows the non-isothermal reduction curves of Fe_2MoO_4 at three different heating rates, viz., 10, 12 and 15 K/min. It clearly shows that the reaction rates are sensitive to the heating rate. At a given temperature, the higher the heating rate the lower the reduction fraction is reached. To calculate the activation energy from the nonisothermal experimental data, a mathematical model derived earlier [8] was used. This model assumes that the rate of the chemical reaction is the rate-controlling mechanism and the reduced particles follow a shrinking core mode.

$$\ln\left(\frac{dX}{dt}\right) + \ln\left(T\right) - \ln\left(1 - X\right)^{2/3} = \ln\left(\frac{A_0 k_0}{R}\right) - \frac{Q}{RT} \qquad (4)$$

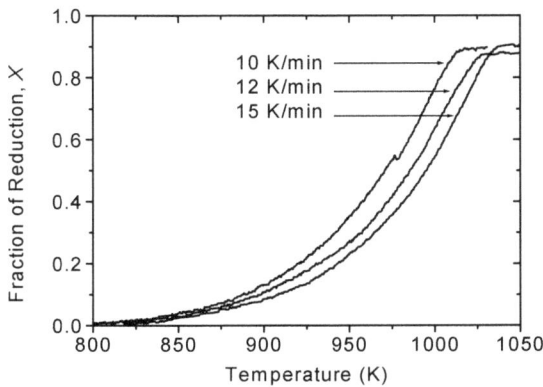

Figure 3. The non-isothermal reduction curves of shallow powder bed of Fe_2MoO_4 [2].

In Equation (4), the terms on the left hand side can be evaluated based on the reaction rate, conversion degree and temperature obtained from the non-isothermal curves in Figure 3. An Arrhenius plot, using Equation (4) for different heating rates, is presented in Figure 4.

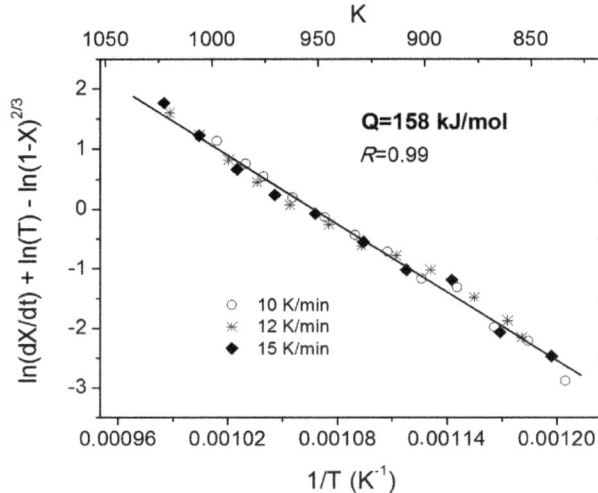

Figure 4. Arrhenius plot for the non-isothermal reduction of shallow powder beds of Fe_2MoO_4 [2].

The activation energy for Reaction (2) obtained from the regression line, in Figure 4, is 158.3 kJ/mol. Note that high correlation factor obtained suggests that the activation energy is independent of the heating rate, which in turn indicates that activation energy is a real function of the reacted fraction at a given temperature. The observed dependence implies that the Equation (4) provides accurate values of activation energies. In fact, the value of 158 kJ/mol is in good agreement with the value obtained from the isothermal experiments, 173 kJ/mol.

3.1.3. Characterization of Fe_2Mo Intermetallic

Figure 5 shows a Scanning electron microscope (SEM) image of reduced Fe_2MoO_4 isothermally at 1173 K [2]. The sponge-like structure is the result of the removal of oxygen which increases the specific surface area. The X-ray diffraction spectrum of the same sample is given in Figure 6 [2]. Two sharp peaks could be identified which correspond to the Fe_2Mo phase. Another broad Bragg peak was also detected which is an indication of an amorphous phase in the sample. However, Transmission Electron Microscopy (TEM) studies performed on a sample, pressed at 1 GPa, confirmed that the sample did not contain amorphous structure but indicated the existence of grains in both nano and micro scale. The TEM microstructural details are presented in Figure 7a–c. The small size of the domains along with the remarkable angle of disorientation among them (see Figure 7a) do diffract the incident beam of X-rays in larger deviated directions causing peak broadening. Selected Area diffraction Patterns (SAD) in Figure 7b,c indicate the existence of a hexagonal structure of Laves phase type Fe_2Mo on indexing.

The streaks shown in Figure 7c reveal that the lattice deformation present in the Fe_2Mo compact is due to the application of high compaction pressure at localized regions. To the best knowledge of the authors, these results represent the first documented evidence in successfully synthesizing the Fe_2Mo intermetallic powder which can be attributed to the advantages of the gas-solid reaction technique.

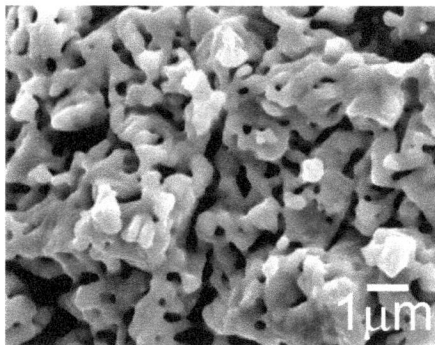

Figure 5. SEM micrograph of sponge-like porous Fe$_2$Mo powder particle used for unidirectional compaction.

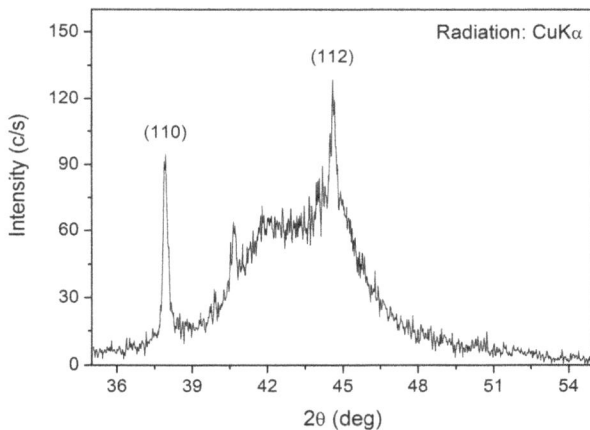

Figure 6. XRD pattern of the powder sample reduced by H$_2$ gas showing the sharpest Bragg peaks corresponding to the Miller indices of Fe$_2$Mo phase.

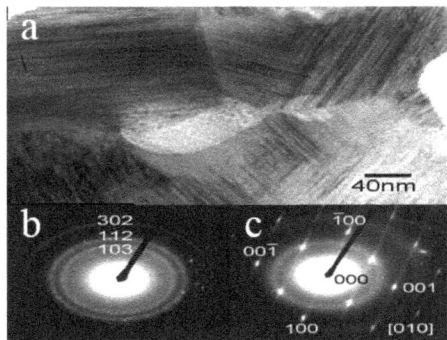

Figure 7. TEM micrographs of a Fe$_2$Mo pellet pressed at 1 GPa showing: (**a**) domains of different orientations with perfect coherency at the particle interface; (**b**) SAD pattern showing microcrystalline structure; and (**c**) SAD pattern showing satellite reflection superimposed on microcrystalline pattern of Fe$_2$Mo [9].

3.1.4. Fluidized Bed Reduction of Fe_2MoO_4

In view of the results obtained using shallow powder beds, it was decided to produce the intermetallic phase in bulk using a laboratory-scale fluidized bed reactor due to the excellent contact between the reactant solid and the gas. The reduction experiments were carried out isothermally and the rate of the reaction was followed by monitoring the rate of evolution of the product gas, viz. water vapor, using a gas chromatograph. The reduction rate curves at several temperatures are shown in Figure 8. Here, the times to complete the reaction are larger than in the thermogravimetric experiments due to the larger average particle size of Fe_2MoO_4 (100 μm versus < 1 μm). Despite the larger average particle size, it can be seen that the reduction curves are sensitive to temperature increase which is an indication that the process is controlled by the rate of the chemical reaction.

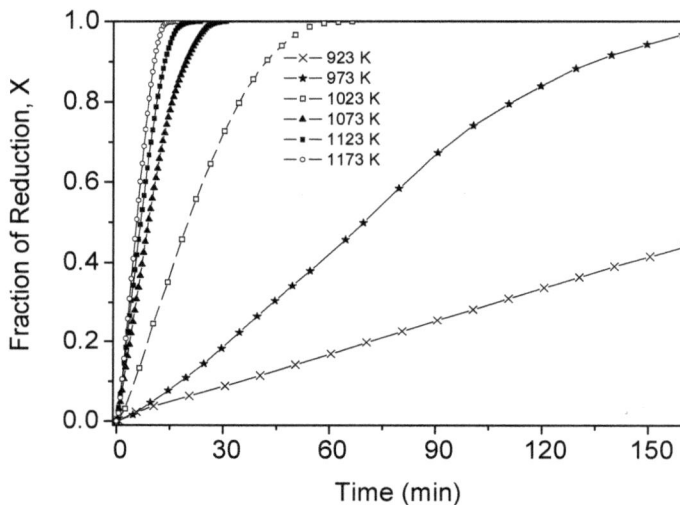

Figure 8. Experimental values of the fractional reduction of Fe_2MoO_4 by hydrogen in a fluidized bed reactor.

The same mathematical model (Equation (3)) was used to calculate the activation energy, of Reaction (2), from the slope of the Arrhenius plot. In this case, the range in particle size distribution was considered instead of taken a fix value of r_0. Thus, the value of the activation energy for the Reaction (2) was 158 ± 17 kJ/mol. This value is close to the activation energies calculated in isothermal and non-isothermal studies of fine shallow powder beds.

Figure 9a–d present the SEM images of the reduced samples at 923, 1023, 1073 and 1173 K, respectively. The images clearly show the effect of temperature on the morphology of the samples after being reduced. As shown in Figure 9a, the crystals are well below 100 nm. On the other hand, the crystal size is much bigger for samples exposed to higher temperatures (1173 K). The production of Fe_2Mo particles by gas-solid route in a fluidized bed reactor is clearly shown from the present results. Thus, the gas-solid reaction route with fluidization appears to be a very promising route towards the production of the Fe_2Mo phase with nano-crystalline structure.

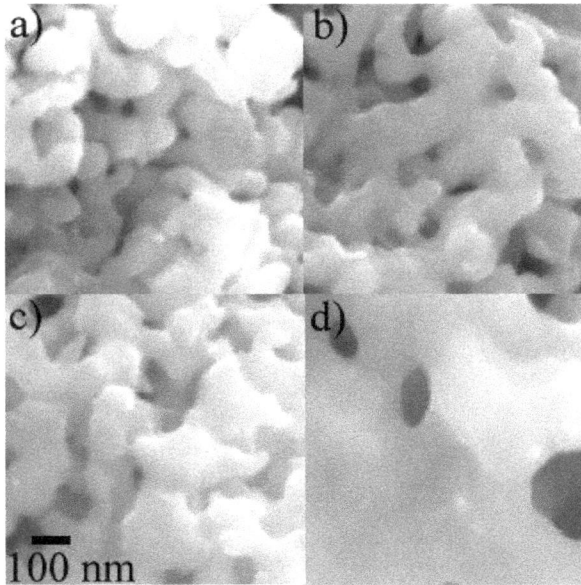

Figure 9. SEM micrographs showing the effect of reduction temperature on the microstructure of fluidizing powder; (**a**) 923 K, (**b**) 1023 K, (**c**) 1073 K, (**d**) 1173 K (at the same magnification) [9].

3.2. Ni-W-O System

3.2.1. Reduction of Ni-W-O System

Reduction of mixtures of NiO and WO_3 were conducted by means of thermogravimetric analyzer (TGA) to understand the intrinsic reduction kinetics. Subsequently, to explore the possibility of designing a process for the reduction, experiments were conducted using a fluidized bed reactor (FB).

Reduction experiments of four different compositions with different Ni/(Ni + W) molar ratios were first conducted by theromgravimetric means in the temperature range from 923 to 1173 K with a continuous hydrogen flow at rate of 0.5 L/min. Figure 10 shows the corresponding reduction fraction as function of temperature and time for the studied mixtures.

Generally, as the temperature increases the rate of reduction was found to increase. TG reduction curves manifest break points, which indicate change in the reaction mechanism (Figure 10a–d). The break points in the reduction curves and the XRD analysis of partially reduced sampled reveal that the reduction of NiO-WO_3 mixtures proceeds through successive steps, which could be represented as follows;

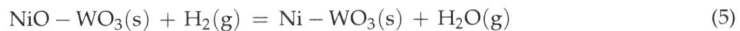

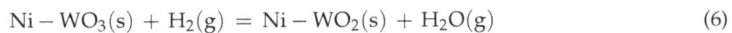

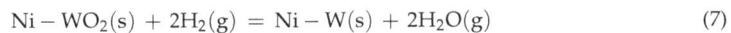

$$NiO - WO_3(s) + H_2(g) = Ni - WO_3(s) + H_2O(g) \tag{5}$$

$$Ni - WO_3(s) + H_2(g) = Ni - WO_2(s) + H_2O(g) \tag{6}$$

$$Ni - WO_2(s) + 2H_2(g) = Ni - W(s) + 2H_2O(g) \tag{7}$$

NiO-WO_3 mixtures were further reduced by hydrogen in a fluidized bed reactor in the temperature range from 973 to 1273 K. Figure 11a–c shows the curves resulted from these experiments.

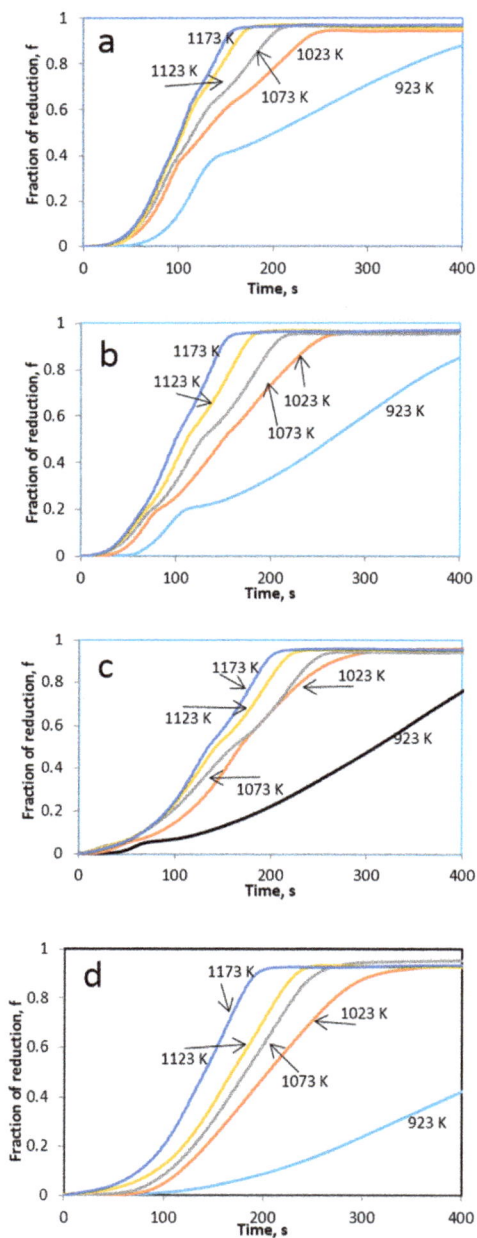

Figure 10. The mass changes for the reduction of the oxide precursors as a function of time. (**a**) Ni/(Ni + W) = 0.7; (**b**) Ni/(Ni + W) = 0.6; (**c**) Ni/(Ni + W) = 0.46; (**d**) Ni/(Ni + W) = 0.4 molar ratio [10].

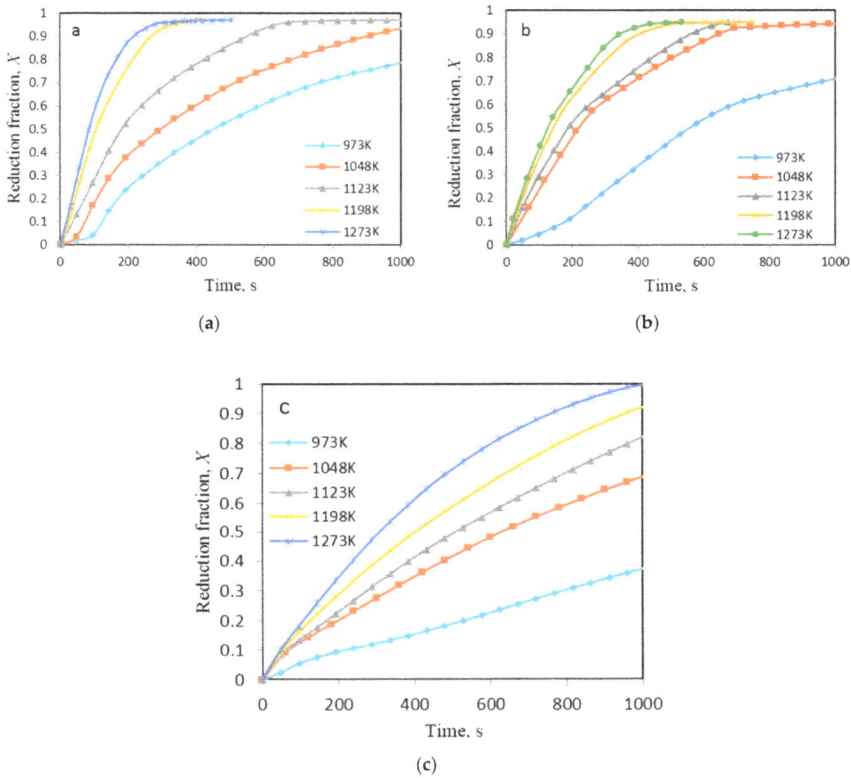

Figure 11. Experimental results for fractional reduction as a function of time. (**a**) Ni/(Ni + W) = 0.7;
(**b**) Ni/(Ni + W) = 0.5; (**c**) Ni/(Ni + W) = 0.4 molar ratio [3].

The symbol X in Equation (8) can be explained as the ratio of area under the curve, at any time *t*,
$A_P(t)$ to the area at the time when the reaction approaches completion. Therefore, fractional reduction
can be expressed as follows;

$$X = \frac{\int_0^t A_P(t)\,dt}{\int_0^\infty A_P(t)\,dt} \tag{8}$$

It can be seen clearly that the reduction curves show break points at different parts of the reduction
curves, which is in agreement with thermogravimetric results. Irrespective of the applied technique Ni
content seems to have a significant effect on the reduction rate. The higher the Ni/(Ni + W) molar
ratio, the higher was the reduction rate. On comparing the reduction rates obtained by TGA and FB,
the former was found to be faster. In order to correlate the obtained results and to understand the
mechanism behind the fluidized bed reduction process, a modeling approach was developed.

The developed model was based on the following assumptions;

I   The system is considered to be isothermal.
II The gas flow is plug flow.
III The mass transfer resistance for the reaction is small compared to the intrinsic reaction rate.
IV The particle sizes are small enough that diffusive transport of gas through the product particles
    can be neglected.

Based on these assumptions a mathematical description of this model is presented. The fluidized bed system is represented schematically in Figure 12.

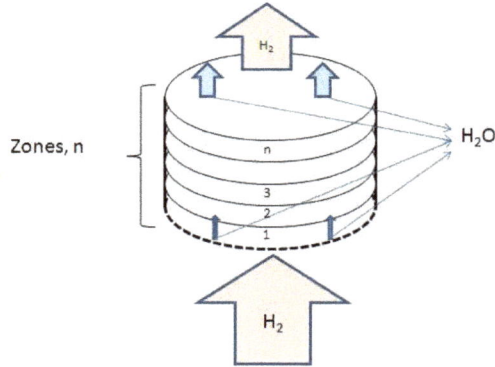

Figure 12. Schematic representation of fluidized powder bed [3].

An equation based on this representation can be written as follows;

$$\frac{d\left(\dot{n}\, x_{H_2O}\,(z)\right)}{dz} = A\,\dot{R}\,(z) \tag{9}$$

where A is the area of cross section of the reactor in m², z is axial co-ordinate, $\dot{n}$ is the molar flux of gas through the reactor in mol/s, x_{H_2O} is mole fraction of water vapor in the gas and $\dot{R}$ is the generation of water vapor due to chemical reaction in mol/m³.s.

The term $\dot{R}$ can be calculated as

$$\dot{R} = \frac{N_0^O}{AL} k_f \left(x_{H_2} - \frac{x_{H_2O}}{K_e} \right) \tag{10}$$

where N_0^O is the total number of moles of reducible oxygen in the Ni-W-O powder, L is height of the fluidized bed, k_f is the intrinsic reaction rate constant and K_e is the equilibrium constant for the reaction. The term $\frac{N_0^O}{AL}$ refers to the moles of reducible oxygen present per unit volume of the fluidized bed and $\dot{R}$ refers to reduction rate per unit volume of the fluidized bed. The intrinsic reaction rate k_f can be determined from dX^{TGA} which refers to the extent of reduction in TGA. Further details about model derivation and assumption can be found elsewhere [3]

$$\frac{dX^{TGA}}{dt} = k_f x_{H_2} \tag{11}$$

With a set of calculated values of rate constants, the model was used to predict the progress of reduction under the experimental conditions. The computed as well as the experimental results for reaction rate constant of NiWO₄ reduction by hydrogen are shown in Table 2.

Table 2. Calculated and experimentally obtained reaction rate constants of NiWO$_4$ [3].

Temp., K	Computed values		Experimental values *	
	2nd stage	3rd stage	2nd stage	3rd stage
973	1.81×10^{-3}	0.60×10^{-3}	1.09×10^{-3}	0.60×10^{-3}
1048	2.26×10^{-3}	0.87×10^{-3}	2.17×10^{-3}	0.88×10^{-3}
1123	2.67×10^{-3}	1.14×10^{-3}	2.56×10^{-3}	1.09×10^{-3}
1198	3.00×10^{-3}	1.43×10^{-3}	3.06×10^{-3}	1.49×10^{-3}
1273	3.28×10^{-3}	1.72×10^{-3}	3.34×10^{-3}	1.72×10^{-3}

* First stage was not possible to determine experimentally.

As seen from Table 2, the computed reduction rates of NiWO$_4$ by hydrogen based on TGA results are in good agreement with the experimental values of fluidized bed technique. The reduction kinetics was then estimated using Arrhenius plots. The calculated activation energies were found to follow the trend that indicates greater nickel content in the precursor would lead to greater activation energy (Table 3).

Table 3. Activation energy for different NiO-WO$_3$ mixtures [3,10].

(Ni/Ni+W) molar ratio	Activation energy kJ/mol		
	1st stage *	2nd stage	3rd stage
	TGA experiments		
0.7	17.9	62	51
0.6	17.5	51	43.9
0.5	18	37.9	35.5
0.46	20.6	38.2	34.5
0.4		40.3 **	
	Fluidized bed experiments ***		
0.7	—	58.6	50.8
0.5	—	36.3	35
0.4	—	46 **	

* It was not able to distinguish the 1st stage in fluidized bed. ** No clear discontinuity was found in the reaction rate, so it was difficult to calculate the activation energy for each step. *** Activation energy calculation based on surface chemical reaction model.

Investigation of reduced samples was further conducted by means of X-ray diffractometer (Siemens D5000 X-Ray diffractometer, Siemens Co., Munich, Germany). Corresponding peaks to metallic nickel phase were found slightly shifted from those that correspond to the pure metal (Figure 13). It was observed that as the WO$_3$ content increased in the mixture the shift increased. Unlike nickel peaks, peaks corresponding to metallic tungsten in the reduced samples overlapped with those peaks for pure W. This trend can be explained by the slight solubility of tungsten in nickel and the negligible solubility of nickel in tungsten. These results are in good agreement with the Ni-W binary phase diagram

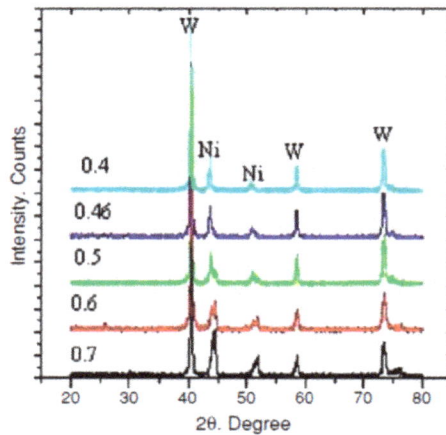

Figure 13. XRD pattern for synthesized Ni-W alloy phases at 1023 K, where 0.7, 0.6, 0.5, 0.46 and 0.4 are Ni/Ni + W molar ratio [10].

Figure 14 represents the SEM images of reduced samples (0.4 Ni/(Ni + W) molar ratio) at 1173 K. The sample is extremely porous. This porosity is similar to that observed earlier in case of Fe-Mo-O system after getting reduced by hydrogen [9]. Moreover, microstructural investigation of product samples was done by Scanning Electron Microscope (A JOEL JSM-840 SEM, Japan Electron Optics Ltd., Tokyo, Japan). Agglomerates of small particles (more common when W content is higher) could be clearly seen from SEM images. The small particles are spherical in shape and the large particles are more elongated.

Figure 14. SEM image of 0.4 Ni/(Ni + W) molar ratio at 1173 K, magnification 2000×.

3.2.2. Reduction-Carburization of Ni-W-O Mixed Oxides

In the present study, reduction-carburization of Ni-W mixed oxides using methane-hydrogen gas mixture was studied isothermally using thermogravimetric analyzer. The main advantage of carburizing metal oxides with methane is the high carbon activity of deposited solid carbon, which provides thermodynamic conditions to produce corresponding cemented carbides at relatively low temperature. The experiments were conducted in the presence of 5 vol.% methane-95 vol.% hydrogen gas mixture at temperatures from 973 K to 1237 K with 50 K interval. The targeted composition for this cemented carbide was WC-10 wt. pct Ni. The reaction progress as function of time and temperature is given in Figure 15. It can be seen clearly from the curves that the reaction proceeds through initially mass loss then followed by mass gain in most cases. The mass loss continues down to 20% which is corresponding to reduction of input sample. The afterwards mass gain resulted from carburization reaction and formation of corresponding cemented carbides.

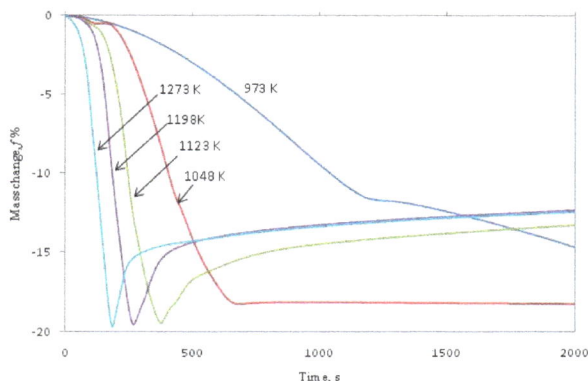

Figure 15. Mass change percentage of the oxide mixture 10.67 wt. pct NiO and 89.33 wt. pct WO_3 vs. time [11].

As can be seen from Figure 15, as long as the temperature is below 1048 K there was no observed mass gain. At temperature higher than 1048 K the TGA curves showed significant increase in weight, which is corresponding to carburization of nascent formed NiW intermetallic. As the temperature increases, the rate and the carburization extent increase. The carburization was observed to go through two consecutive steps. The first one goes up to f = −18.7% which corresponds to formation of the intermediate W_2C. The second step proceeds up to −15.5% mass change, which corresponds to complete formation of WC. The activation energy was calculated based on the initial rates and found to be 96 kJ/mol.

Mineralogical investigation revealed that carburization at 973 K was far from being complete. W metal phase was the predominant detected phase with only traces of the intermediate W_2C phase. This observation is in contradiction with an earlier investigation where it was stated that no carbide phase could form at such low temperatures [12]. As the temperature increased, phases like W, W_2C and WC were detected. The XRD pattern of W_2C is similar to that of standard W_2C peaks but broader .It was reported that nano-crystalline W_2C has been restricted from further development but instead it proceeds to the more stable WC phase [13]. The above observations agree very well with the thermogravimetric results. There are no signs of the presence of intermediate W_2C phase in the completely carburized samples.

Further evaluation of the above findings points to the fact that carburization can slowly start before complete reduction especially at lower temperatures. Similar observations have been reported earlier for the $CoWO_4$ system [14]. Microstructural investigation of product sample (reduced and carburized) shows the existence of agglomerates of hemispherical small particles (Figure 16).

Figure 16. SEM images of a reduced-carburized 0.27 Ni/(Ni + W) molar ratio sample at 1273 K.

3.2.3. Reduction-Nitridation of Ni-W-O Mixed Oxides

The reduction–nitridation reactions of Ni-W-O powders was carried out isothermally at 973–1273 K in a flow of 50% H_2 and 50% N_2 gas mixture using a fluidized bed reactor. In these experiments, H_2 gas was the reducing agent, while N_2 in the gas mixture was applied for the nitridation reactions. Similar to previously reported observations, it is expected that these precursors will first get reduced in H_2 gas to produce Ni–W intermetallics followed by the nitridation reaction of the reduced product. Because there is no reaction product during nitridation in the gas phase, analysis of the off-gases could not indicate the reaction progress. However, XRD results of reacted NiO-WO$_3$ precursors revealed the presence of WO$_2$ phase in NiO-WO$_3$ precursor as a main phase formed at 1048 K together with W, Ni, WN$_2$ and WN. This phase resulted from the stepwise reduction of WO$_3$. With further rise in temperature, the WO$_2$ phase is subsequently reduced to W metal, which is then reacted with N_2 gas to produce tungsten nitrides (WN and WN$_2$). The reduction-nitridation reactions of the stoichiometric NiWO$_4$ precursor proceed faster than that of NiO-WO$_3$, and tungsten nitrides are formed even at relatively lower. The extent of formation of WN, as the main reaction product at 1198 K, increases with rise in temperature. The higher the reaction temperature, the higher is the rate of formation of WN in the reaction products. Further, a higher degree of crystallinity was developed as indicated from the sharpening of WN peak at high temperatures. It is worth mentioning that, with an increase in the reaction time, the amount of WN formed increases and it becomes the predominant phase in NiWO$_4$ precursor [15].

3.2.4. Thermal Diffusivity Measurements

Isothermal thermal diffusivity measurements of pressed pellets of NiWO$_4$ were carried out in the temperature range from 973 to 1273 K under hydrogen using laser flash unit. Figure 17 shows the change of thermal diffusivity values as a function of time. In view of the shrinkage caused by sintering, the measured thermal diffusivity values were corrected according to the calculated thicknesses. The corrected values are plotted as solid lines in the same Figure. NiWO$_4$ thermal diffusivity values were affected by the shrinking caused by sintering. Corrected thermal diffusivity curves deviate from the experimental points at the later stages.

As the sample bed get heated up in hydrogen, the chemical composition changes from the oxidic phase to metallic phase, which corresponds to initial fast increase in thermal diffusivity values. After complete reduction, the bed starts to sinter, which is explained by the sluggish increase in thermal diffusivity values in the later stages of the experiments. This explanation is further confirmed by the fact that heat is conducted by phonons in oxides and by electrons in metallic phases [16].

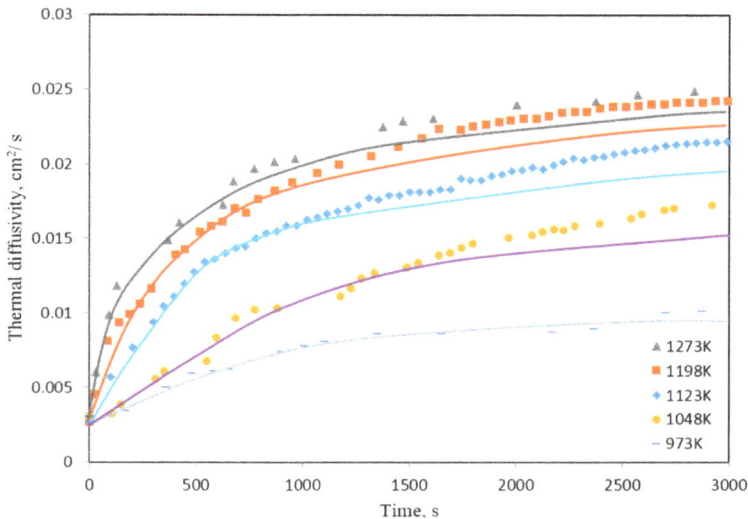

Figure 17. Effect of reaction progress on the thermal diffusivity values of pressed NiWO$_4$. Symbols represent the experimental raw values while solid lines represent the corrected values [4].

Thermal conductivity of a bed is expressed as a function of thermal conductivity of the gas, the solid and the void fraction of the bed. Conductivity can be considered as an over-all result of three mechanisms, namely; conduction, convection and radiation. Sun and Lu [17] reported that the heat transfer by convection is negligible for packed beds of fine particles. Moreover, it was reported that radiation became significant for 1 mm particle size above 673 K and 0.1 mm particles above 1773 K [18]. The average particle size of the studied NiWO$_4$ powder was 2.5 μm. In view of these arguments, the radiation effect can be ruled out under the present experiment conditions.

Effective thermal conductivity is calculated based on $\lambda = \alpha C_p \rho$ where α is the thermal diffusivity cm^2/s, C_p is the heat capacity J/Kg and ρ is the density g/cm^3. In this, the apparent densities of the samples were calculated based on their volume and mass. The temperature dependence of specific heat was calculated by the principle of addition. Temperature and chemical composition dependence of the sample volume were measured manually by stopping the reaction at different points and rapidly cooling the sample in a stream of argon. Figure 18 shows the experimental values of effective thermal conductivity of pellets of NiWO$_4$.

In the case of the reduction of an oxide by a gas, the outer layer of the oxide pellet is likely to be reduced first before the reduction proceeds towards the core. The reduction will occur along the sides as well. An outer metallic layer is likely to result in the conduction of heat along the sides of the pellet deviating from the axial heat flow across the pellet in the vertical direction. The possibility of such a conduction mechanism was examined in the present experimental series. It was found that, in the case of partially reduced pellets of NiWO$_4$, there were grains of reduced oxides. However, microscopic examination of these pellets revealed that the metallic grains formed were largely unconnected with each other, possibly due to the loss of oxygen from the oxide leaving voids. In the early stages, heat conduction along the sides will also have a resistance due to the fact that the metallic layer would be very thin and the heat flow would not be substantial. As the reaction proceeds further, a significant amount of oxide in the bulk of the pellet would be reduced. In such a case, surface conduction effects would not have a serious impact on the results.

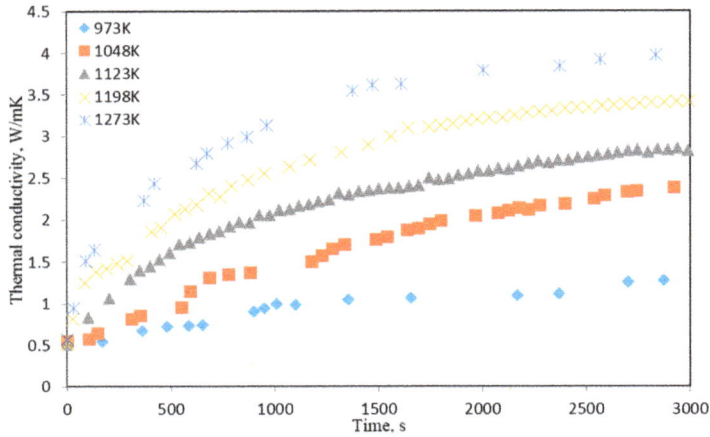

Figure 18. Effect of reaction progress on the effective thermal conductivity values of pressed $NiWO_4$.

The activation energy for the reduction $NiWO_4$ was calculated by an Arrhenius plot using the isothermal reduction rates at the initial stages of the reaction, which can be represented by the change in thermal diffusivity, at different temperatures. A plot of ln (s), where s = $d\alpha/dt$, against $1/T$ is presented in Figure 19.

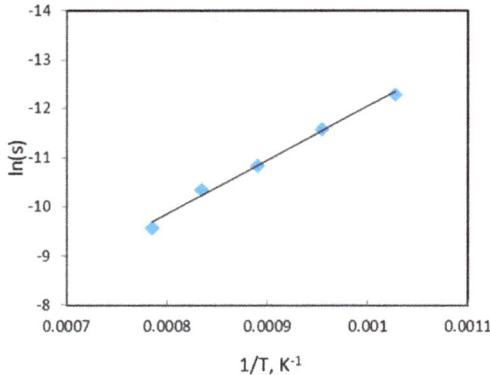

Figure 19. Arrhenius plot for the reduction of $NiWO_4$ by hydrogen using laser flash unit.

Based on the slope, activation energy value 91.4 kJ/mol was calculated. The calculated value was found to be higher than that obtained previously based on TGA experiments (37 kJ/mol). These inconsistencies required further investigation on the structural changes and reaction mechanism through the pellets. The difference can be explained, to some extent, on the basis of physical and structural changes during the process, since, thermal diffusivity is affected by change of porosity.

Sintering is a rate process, similar to reduction, influenced by temperature. Since the sintering has the dominant effect on diffusivity changes at later stages of the process, the degree of sintering may be expressed in terms of thermal diffusivity. In such a case, the activation energy of sintering can be calculated directly from the slopes of thermal diffusivity curves during the final stages (Figure 20). From the slopes of these curves an activation energy value of 36 kJ/mol was obtained.

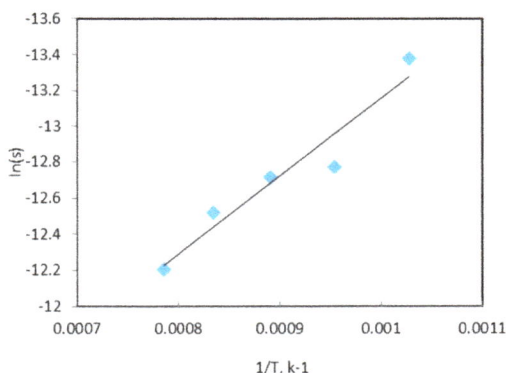

Figure 20. Arrhenius plot for sintering of NiWO$_4$ using laser flash unit.

The calculated activation energy of sintering was found to be less than that reported in the literature (160 kJ/mol [19] for Ni-W systems). One possible explanation is that the reduced metallic particles in the nascent state would have a significant population of active centers on the surface of each particle enhancing thereby the sintering process. Further, the presence of reducing atmosphere would prevent any formation of surface oxide coating and minimize the diffusion barrier due to these surface layers. Thus, a reducing atmosphere would promote sintering and is likely to lower the activation energy compared to inert atmosphere. Further, hydrogen molecules would diffuse into the pores and thereby facilitate the densification process via reaction with surface layers [20]. The physical capability of hydrogen to act as a lattice defect catalyst [21] would facilitate the movement of point defects and dislocations during sintering and thereby would lower the activation energy [22].

4. Conclusions

The present paper presents the work carried out at the Division of Materials Process Science, Royal Institute of Technology, Stockholm, Sweden on the hydrogen reduction of mixed oxides containing the refractory metals Mo and W. The main experimental technique used was TGA, fluidized bed reactor supplemented by thermal diffusivity measurements in one case. In all the cases, the products formed were characterized by standard methods.

The kinetics of reduction of the pure oxides were estimated and production of nano to micro grains of intermetallics was demonstrated. Up-scaling of thermo-gravimetric experiments using fluidized bed technique was successfully carried out. The up-scaling facilitates the mass production of intermetallics and also opens the possibility to produce other alloys in nano-range region. A varied range of Fe-Mo alloys could be produced utilizing the gas-solid reaction route in one step.

On the other hand, cemented carbide and nitrides were successfully produced utilizing friendly environmentally gases through short production route. The corresponding reaction kinetics and mechanism were identified. Thermal property of a material was used as a novel tool to track the reaction progress and evaluate the associated physical changes.

Therefore, it can be stated that the gas-solid reactions provide a very suitable process route to the production in bulk of intermetallics with suitable interstitials like carbon or nitrogen or both.

Acknowledgments: The financial support from the Swedish Research Council and CAMM–Centre of Advanced Mining and Metallurgy at Luleå University of Technology, Sweden is acknowledged.

Author Contributions: Hesham Ahmed, Ricardo Morales-Estrella and Seshadri Seetharaman conceived and designed the experiments; Hesham Ahmed and Ricardo Morales-Estrella performed the experiments; Hesham Ahmed, Ricardo Morales-Estrella, Nurni Viswanathan and Seshadri Seetharaman have analyzed the data; Hesham Ahmed and Ricardo Morales-Estrella wrote the paper; Nurni Viswanathan and Seshadri Seetharaman revised the paper and made in final form.

Conflicts of Interest: The authors declare no conflict of interest.

References

1. Yang, M.; MacLeod, M.J.; Tessier, F.; DiSalvo, F.J. Mesoporous metal nitride materials prepared from bulk oxides. *J. Am. Ceram. Soc.* **2012**, *95*, 3084–3089. [CrossRef]
2. Morales, R.; Sichen, D.; Seetharaman, S.; Arvanitidis, I. Reduction of Fe$_2$MoO$_4$ by hydrogen gas. *Metall. Mater. Trans. B* **2002**, *33*, 589–594. [CrossRef]
3. Ahmed, H.M.; El-Geassy, A.A.; Viswanathan, N.N.; Seetharaman, S. Kinetics and Mathematical Modeling of Hydrogen Reduction of NiO-WO$_3$ Precursors in Fluidized Bed Reactor. *ISIJ Int.* **2011**, *51*, 1383–1391.
4. Ahmed, H.M.; Seetharaman, S. Isothermal dynamic thermal diffusivity studies of the reduction of NiO and NiWO$_4$ precursors by hydrogen. *Int. J. Mater. Res.* **2011**, *102*, 1336–1344. [CrossRef]
5. Heijwegen, C.; Rieck, G. Determination of the phase diagram of the Mo-Fe system using diffusion couples. *J. Less Common Met.* **1974**, *37*, 115–121. [CrossRef]
6. Kleykamp, H.; Schauer, V. Phase equilibria and thermodynamics in the Fe-Mo and Fe-Mo-O systems. *J. Less Common Met.* **1981**, *81*, 229–238. [CrossRef]
7. Rawlings, R.; Newey, C. Study of the Iron-Molybdenum System by Means of Diffusion Couples. *J. Iron Steel Inst.* **1968**, *206*, 723.
8. Morales, R.; Arvanitidis, I.; Seetharaman, S. Interinsic Reduction of MoO$_3$ by Hydrogen. *Z. Metall.* **2000**, *91*, 589.
9. Morales, R.; Seetharaman, S.; Agarwala, V. Mechanical and structural characterization of uniaxially cold-pressed Fe$_2$Mo powders. *J. Mater. Res.* **2002**, *17*, 1954–1959. [CrossRef]
10. Ahmed, H.M.; El-Geassy, A.A.; Seetharaman, S. Kinetics of Reduction of NiO-WO$_3$ Mixtures by Hydrogen. *Metall. Mater. Trans. B* **2010**, *41*, 161–172. [CrossRef]
11. Ahmed, H.M.; Seetharaman, S. Reduction-carburization of NiO-WO$_3$ under isothermal conditions using H$_2$-CH$_4$ gas mixture. *Metall. Mater. Trans. B* **2010**, *41*, 173–181. [CrossRef]
12. Bondarenko, V.; Pavlotskaya, E. High-temperature synthesis of tungsten carbide in a methane-hydrogen gas medium. *Powder Metall. Met. Ceram.* **1996**, *34*, 508–512. [CrossRef]
13. Gao, L.; Kear, B. Synthesis of nanophase WC powder by a displacement reaction process. *Nanostructured Mater.* **1997**, *9*, 205–208. [CrossRef]
14. Lebukhova, N.; Karpovich, N. Carbothermic reduction of cobalt and nickel tungstates. *Inorg. Mater.* **2006**, *42*, 310–315. [CrossRef]
15. El-Geassy, A.; Nassir, N.A.; Ahmed, H.; Seetharaman, S. Simultaneous reduction nitridation for the synthesis of tungsten nitrides from Ni-W-O precursors. *Powder Metall.* **2013**, *56*, 411–419. [CrossRef]
16. Taylor, R. Construction of apparatus for heat pulse thermal diffusivity measurements from 300–3000K. *J. Phys. E Sci. Instrum.* **1980**, *13*, 1193. [CrossRef]
17. Sun, S.; Lu, W.K. Mathematical Modelling of Reactions in Iron Ore/Coal Composites. *ISIJ Int.* **1993**, *33*, 1062–1069. [CrossRef]
18. Tsotsas, E.; Martin, H. Thermal conductivity of packed beds: A review. *Chem. Eng. Process. Process Intensif.* **1987**, *22*, 19–37. [CrossRef]
19. Toth, I.; Lockington, N. The kinetics of metallic activation sintering of tungsten. *J. Less Common Met.* **1967**, *12*, 353–365. [CrossRef]
20. Liu, L.; Loh, N.; Tay, B.; Tor, S.; Murakoshi, Y.; Maeda, R. Micro powder injection molding: Sintering kinetics of microstructured components. *Scr. Mater.* **2006**, *55*, 1103–1106. [CrossRef]
21. Dominguez, O.; Bigot, J. Material transport mechanisms and activation energy in nanometric Fe powders based on sintering experiments. *Nanostructured Mater.* **1995**, *6*, 877–880. [CrossRef]
22. Paul, B.; Jain, D.; Bidaye, A.; Sharma, I.; Pillai, C. Sintering kinetics of submicron sized cobalt powder. *Thermochim. Acta* **2009**, *488*, 54–59. [CrossRef]

metals

MDPI

Article

Devising Strain Hardening Models Using Kocks–Mecking Plots—A Comparison of Model Development for Titanium Aluminides and Case Hardening Steel

Markus Bambach [1],*, Irina Sizova [1], Sebastian Bolz [2] and Sabine Weiß [2]

[1]   Chair of Mechanical Design and Manufacturing, Brandenburg University of Technology
     Cottbus-Senftenberg, Konrad-Wachsmann-Allee 17, Cottbus D-03046, Germany; sizova@b-tu.de
[2]   Chair of Physical Metallurgy and Materials Technology, Brandenburg University of Technology
     Cottbus-Senftenberg, Konrad-Wachsmann-Allee 17, Cottbus D-03046, Germany;
     sebastian.bolz@b-tu.de (S.B.); sabine.weiss@b-tu.de (S.W.)
*   Correspondence: bambach@b-tu.de; Tel.: +49-355-69-3108

Academic Editor: Hugo F. Lopez
Received: 29 June 2016; Accepted: 22 August 2016; Published: 29 August 2016

Abstract: The present study focuses on the development of strain hardening models taking into account the peculiarities of titanium aluminides. In comparison to steels, whose behavior has been studied extensively in the past, titanium aluminides possess a much larger initial work hardening rate, a sharp peak stress and pronounced softening. The work hardening behavior of a TNB-V4 (Ti–44.5Al–6.25Nb–0.8Mo–0.1B) alloy is studied using isothermal hot compression tests conducted on a Gleeble 3500 simulator, and compared to the typical case hardening steel 25MoCrS4. The behavior is analyzed with the help of the Kocks-Mecking plots. In contrast to steel the TNB-V4 alloy shows a non-linear course of θ (i.e., no stage-III hardening) initially and exhibits neither a plateau (stage IV hardening) nor an inflection point at all deformation conditions. The present paper describes the development and application of a methodology for the design of strain hardening models for the TNB-V4 alloy and the 25CrMoS4 steel by taking the course of the Kocks-Mecking plots into account. Both models use different approaches for the hardening and softening mechanisms and accurately predict the flow stress over a wide range of deformation conditions. The methodology may hence assist in further developments of more sophisticated physically-based strain hardening models for TiAl-alloys.

Keywords: titanium aluminides alloy; single phase steel; hot forming; dynamic recrystallization; modeling

1. Introduction

Intermetallic titanium aluminides (TiAl) alloys are relatively new class of lightweight materials which can be applied in automotive applications as well as in aviation gas turbines where they have to withstand temperatures up to 800 °C. The high potential of TiAl for these applications stems from the fact that the TiAl-based alloys have remarkable mechanical properties, such as a low specific weight of about 4 g/cm^3, a high strength at elevated temperatures and a very good oxidation/corrosion resistance [1–3]. In recent years, forgeable β-solidifying TiAl alloys have been developed with the goal of exploiting the benefits of the hot-worked microstructure, which allows for better performance than the as-cast state. However, as many other intermetallic alloys titanium aluminides are brittle, even at elevated temperatures.

An important task in processing TiAl to products consists in overcoming problems associated with the limited workability of TiAl-based alloys and the robust design of forging processes. With the rapid development of computing techniques, Finite Element (FE) simulation is widely applied nowadays to study metal forming processes. In metal forming, the accuracy of FE simulations depends crucially on the constitutive model. It is hence very important to establish a precise constitutive model for the flow stress which predicts the dependence of flow stress on deformation temperature, strain rate, and strain. In the past, various analytical, phenomenological [4–6] and empirical models [7–9] have been devised to model the strain hardening behavior of ordinary single phase steels. For steels, hardening is usually modeled using semi-empirical material models. For softening caused by dynamic recrystallization, the Johnson-Avrami-Mehl-Kolmogorov (JMAK) equation is commonly used.

The existing models have been successfully applied to steel and also transferred to TiAl. Extensive studies have been carried out to describe the flow stress as a function of temperature, strain rate and strain of the TiAl-alloys. Cheng et al. [10] adopt the model proposed by Laasraoui and Jonas [11] to TiAl-alloys with a nominal composition of Ti–42Al–8Nb–0.2W–0.1Y (at %). Godor et al. [12] applied two phenomenological constitutive models, the Sellars–McTegart model [13] and the Hensel-Spittel model [14], for the description of the deformation behavior of two γ-TiAl alloys with nominal compositions Ti–41Al–3Mo–0.5Si–0.1B and Ti–45Al–3Mo–0.5Si–0.1B. He et al. [15] and Pu et al. [16] established a constitutive model based on Arrhenius-type equations and the sine hyperbolic description for the Zener-Hollomon parameter for a Ti–45Al–8.5Nb–(W,B,Y) alloy and for a TiAl–Cr–V alloy, respectively. Deng et al. [17] proposed a flow stress model of Ti–47Al-alloy based on a regression analysis using Hensel-Spittel model. All existing descriptions of the flow behavior of TiAl-alloys are based on empirical models, which were derived primarily for steels. Steels are typically hot-worked in the austenitic phase, i.e., the models are designed for microstructure evolution in a single phase material.

In contrast to steel, TiAl-alloys have a complex multi-phase microstructure during hot working. The understanding of the flow behavior of TiAl alloys under hot deformation conditions thus has a great importance for the design of hot working processes. The majority of studies show that the flow curves of TiAl-based intermetallic alloys exhibited a sharp peak stress at a relatively low strain. Then, the flow stress decreased rapidly into a steady-state stress. Previous research [18–21] reported that dynamic recrystallization contributes to flow softening. Several other influencing factors have been reported, involving most notably flow localization that is partly initiated by dynamic recrystallization and adiabatic heating during deformation [21]. Hot compression tests with various TiAl-alloys with different primary structures have shown that the primary structure strongly influences the material behavior during hot forming [22]. Semiatin et al. [23,24] and Schaden et al. [25] concluded that the presence of lamellar colonies is responsible for a strong softening during hot forming (i.e., the pronounced maximum in the course of the flow curves).

It is known that the strain hardening rate $\theta = \partial\sigma/\partial\varepsilon$ as a function of flow stress (this type of plot is often referred to as Kocks-Mecking plot) reveals different stages of microstructure evolution during hot deformation as characteristic stages [26]. Ordinary single phase alloys such as steels in the austenitic phase exhibit at first a linearly decreasing strain hardening rate (stage III hardening) followed by a slowly decreasing hardening rate called stage IV, before attaining stage V where the values of θ approach zero [4,27]. Steels, which are hot-compression-tested in the austenitic phase, show inflection points in Kocks-Mecking plots that correspond to the onset of DRX [28]. In contrast to the ones of steel, the Kocks-Mecking plots of TiAl-alloys initially decrease monotonically, but not with a linear rate. Recent research of the authors has shown that TiAl-alloys initially show a non-linear, concave down course of θ (i.e., no stage-III hardening) and exhibit neither a plateau (stage IV hardening) nor an inflection point under all tested conditions [29].

The aim of this study is to investigate the peculiarities of the work hardening behavior of a Ti–44.5Al–6.25Nb–0.8Mo–0.1B (in at %) TiAl-alloy by direct comparison to a typical 25MoCrS4 steel used for forging by means of isothermal hot compression tests. It is shown how the features of the TiAl-alloy affect the set-up of strain hardening models. Special attention is paid on the accuracy of the strain hardening model in θ-σ-space, which is typically not taken into account in model development.

The paper is structured as follows: Section 2 gives an overview of the TNB-V4 alloy and 25MoCrS4 steel and the performed hot compression tests. Section 3 details the flow curves, Kocks-Mecking plots and metallographic results. The methodology for designing strain hardening models based on the behavior observed in the Kocks-Mecking plots is presented in Section 4, which also details the parameter identification procedure. In Sections 5 and 6, a comparison of the model results to experimental data and final conclusions are presented.

2. Materials and Hot Compression Testing

We studied a high Nb containing TiAl alloy with a nominal composition of Ti–44.5Al–6.25Nb–0.8Mo–0.1B (in at %) and a 25MoCrS4 steel with the chemical composition of 0.25 wt % C, 0.25% Si, 0.70% Mn, 1.05% Cr, 0.25% Mo, balance Fe. In order to start with chemically homogeneous samples the TiAl alloy was prepared by VAR melting. Residual porosity was removed by hot isostatic pressing at 1260 °C for 4 h with a pressure of 200 bars. A Gleeble 3500 thermo-mechanical simulator (DSI Europe GmbH, Weissenhorn, Germany) was used to simulate the hot forging process by means of hot compression tests. Strain measurement was performed with the standard LVDT (Linear variable differential transformer) of the Gleeble machine (DSI Europe GmbH, Weissenhorn, Germany). The testing temperatures ranged from 1150 °C to 1300 °C and the strain rates ranged from 0.001 s^{-1} to 0.5 s^{-1}. Compression tests were conducted in vacuum of 10^{-4} mbar. The specimens were heated with a heating rate of 10 K/s and held for 5 min to obtain a homogeneous microstructure before compression testing.

In order to investigate the microstructure evolution under hot working conditions, each specimen was quenched immediately after upsetting. Metallographic examination was carried out on cross sections parallel to the compression direction of the deformed specimens. The sections were also analyzed using scanning electron microscopy (Tescan Orsay Holding, Brno-Kohoutovice, Česká Republika).

The 25MoCrS4 steel was analyzed by compression testing in previous work of the first author of this paper [30]. Cylindrical specimens with a diameter of 10 mm and a height of 15 mm were deformed at temperatures ranging from 700 °C to 1200 °C, and strain rates from 0.01 s^{-1} to 100 s^{-1}. In all tests, the specimens were first heated from room temperature to 1250 °C and homogenized for 10 min, then placed into the furnace of the compression testing machine (Servotest Testing Systems Ltd., Surrey, UK) and cooled to the specific deformation temperature. The maximum true strain obtained in all tests was ~0.8. All specimens were quenched after testing and examined for grain size and recrystallized volume fractions.

3. Experimental Results

3.1. Hot Deformation Behavior

Figure 1 illustrates the true stress–strain curves at deformation temperature of 1200 °C and different strain rates in the isothermal compression of the 25MoCrS4 steel (Figure 1a) and the TNB-V4-alloy (Figure 1b). It can be seen that the hot deformation behavior of the TNB alloy is in strong contrast to the behavior of the 25MoCrS4 steel, which is fully austenitic under the testing conditions.

It is observed that the isothermal compression behavior of the 25MoCrS4 steel is approximately divided into three stages: (i) The flow stress increases to a peak value (σ_p) with increasing strain; (ii) The flow stress decreases to a steady state value. In this stage the softening becomes increasingly predominant due to the onset of DRX; (iii) The flow stress keeps a steady state due to the dynamic balance between work hardening and softening induced by DRX. It is apparent that the flow curves of steel display the typical characteristic of softening by DRX. Typical Kocks-Mecking plots (θ–σ curves) of the 25MoCrS4 steel at 1200 °C and various strain rates are shown in Figure 1c. The steel exhibits a characteristic work-hardening behavior. At first, θ decreases linearly (so called stage-III hardening) until reaching an inflection point, which determines the critical stress (σ_c) and accordingly the onset of DRX [28]. Just before the inflection point, the working hardening rate decreases to a lower slope (stage IV hardening). After the onset of DRX, θ decreases rapidly towards the peak stress (σ_p), at which θ = 0. The θ–σ curves can be used to determine the onset of DRX.

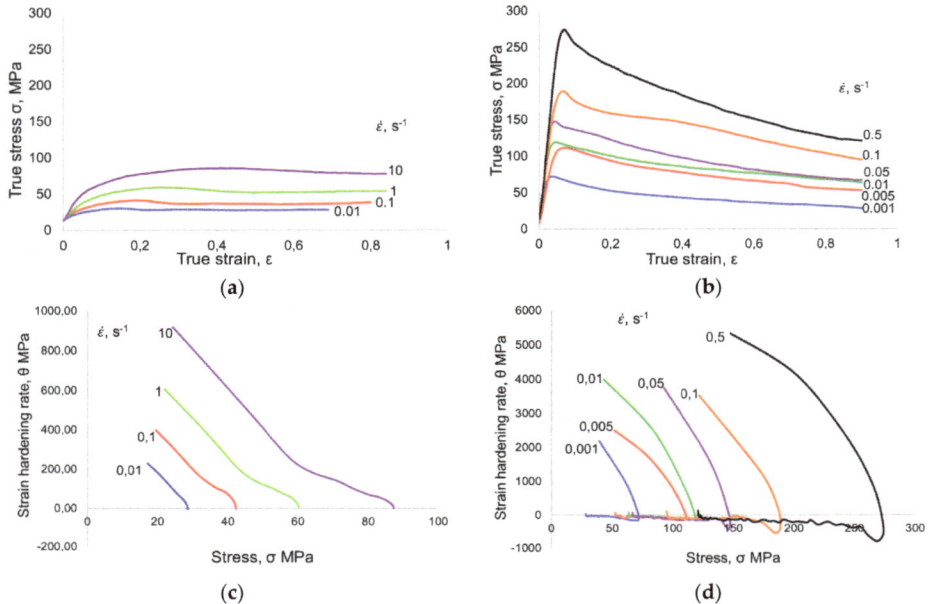

Figure 1. The flow stress curves at temperature 1200 °C and various strain rates: (**a**) for single phase steel 25MoCrS4; (**b**) for TNB-V4-alloy; (**c**) Kocks–Mecking plots of 25MoCrS4; and (**d**) Kocks-Mecking plots of TNB-V4-alloy.

In contrast to the flow curves of the 25MoCrS4 steel, the TNB-V4 alloy has a high initial hardening with a pronounced flow stress maximum, which is much sharper than for steel in the austenitic range. Then, the flow stress decreases rapidly to a steady-state stress with increasing strain, albeit with a different curvature than steel. The measured flow curves of TNB-V4 show a very strong dependence on strain rate and temperature. The flow stress increases very rapidly with increasing strain rate and decreasing temperature. In contrast to steel, it can be seen (Figure 1d) that the Kocks-Mecking plots of TNB-V4 look absolutely different. The alloy under investigation neither shows a linear initial decrease of θ (i.e., no stage-III hardening) nor does it exhibit a plateau (stage IV hardening) or an inflection point at all forming conditions. The work-hardening behavior of the TNB-V4-alloy can be explained by the complex multiphase microstructure, and many factors that may affect the hot deformation behavior of TiAl-alloys, such as flow localization, additional interior adiabatic heating, rotation or buckling and breakdown of the lamellar colonies during hot forming, cf. [31].

It has to be mentioned here that measurement accuracy, adiabatic heating, inhomogeneous deformation as well as further processing of the recorded data will affect the computation of stress-strain curves and the derivatives of the stress strain curves, which form the basis of the Kocks-Mecking plots used in this paper. In previous work by Lohmar and Bambach [32], the processing of recorded data was treated in detail with respect to the computation of Kocks-Mecking plots. The effect and correction of adiabatic heating was treated in Xiong et al. [33]. Data processing in the present paper was performed according to these findings. The stress-strain curves and Kocks-Mecking plots should hence well represent the behavior of the material.

3.2. Microstructure Evolution

The microstructure of the latest industrially used generation of Titanium aluminides consists of globular γ-grains, α + γ-lamellar colonies and an increased amount of β/β0-grains along the boundaries of lamellar colonies. Compared with two phase alloys, the stabilization of the β/β0-phase leads to a better strength-ductility ratio. Especially the workability of these TiAl-alloys is better than the one of previous generations of TiAl-alloys [34]. The deformation in two-phase α2 + γ- alloys is mainly based on dislocation glide and twinning in the γ-phase combined with deformation anisotropy of the lamellar colonies. The anisotropy of the lamellar colonies leads to soft or hard deformation modes depending on the direction of the applied stress. The yield and fracture stresses are low if the deformation occurs parallel to the lamella plane. In the opposite case the stresses are high if the deformation occurs perpendicular to the lamellae [35]. Under load hard mode, colonies rotate into flow direction or a kinking of the lamellas occurs. As can be seen in Figure 2b, after kinking both parts of the colonies subsequently rotate into flow direction during forming. It can be assumed that, depending on the bcc structure of the β-phase and its ordered counterpart β0, both phases have a sufficient amount of independent slip modes. That leads, combined with the limited deformability of the α- and γ-phase, to a high deformation gradient between the different phases. As a consequence, the β/β0-phase acts as a lubricant for the whole microstructure and supports the rotation of hard mode lamellar α2 + γ-colonies [36]. It is hence assumed that the rotation of the α2 + γ-colonies from hard into soft orientations leads to a non-linear course of strain hardening rate in the Kocks-Mecking plots.

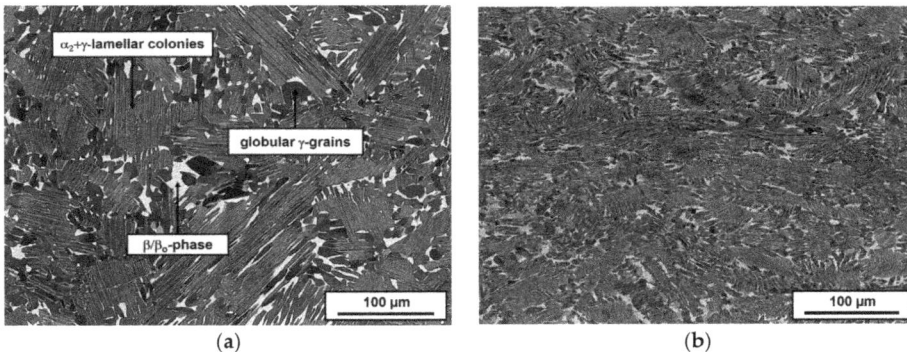

Figure 2. Typical microstructures of TNB-V4: (**a**) as-cast + HIP (hot isostatic pressing) state and (**b**) compressed at 1250 °C with a strain rate of 0.25 s^{-1} to a strain of 60%.

4. Microstructure-Based Strain Hardening Models for Hot Working

4.1. Derivation of Model Equations

Various microstructure-based strain hardening models have been derived in the past (see, e.g., Bariani et al. [37] for a review). For steel, dynamic recrystallization (DRX) is a dominant softening mechanism which restores the deformed microstructure during forming and allows for large plastic deformations without failure of the workpiece. Most strain hardening models rely on Johnson-Mehl-Avrami-Kolmogorov (JMAK) kinetics for DRX. Poliak and Jonas [28] showed that the onset of DRX corresponds to an inflection point in the course of the strain hardening rate θ as a function of stress σ, i.e., the critical stress is a root of the second derivative of the strain hardening rate with respect to stress:

$$\frac{\partial}{\partial \sigma}(-\frac{\partial \theta}{\partial \sigma}) = 0 \tag{1}$$

Recently, two problems in modeling the strain hardening behavior under hot working conditions were discovered: (i) inconsistency of JMAK kinetics with the Poliak-Jonas criterion; and (ii) absence of an inflection point in practical experiments despite occurrence of DRX [38].

Since the criterion by Poliak and Jonas was derived from principles of irreversible thermodynamics and has been confirmed with experimental data, a flow stress model that takes DRX into account should adhere to this criterion, not only to be consistent with experimental data but also to make sure that the model is sound in a thermodynamical sense. Hence, the flow stress σ should be three times continuously differentiable with respect to ε. It was recently shown by Bambach [38] that the Poliak-Jonas criterion requires that the Avrami exponent exceeds a value of 3 for the model to be consistent with the criterion. Hence, generalized transformation kinetics was considered to find the root cause of the inconsistency and to alleviate them. As mentioned in the introduction, TiAl alloys do not seem to show an inflection point in θ in spite of the occurrence of DRX [29].

A flow stress model that includes DRX essentially consists of four types of equations: (i) a model for strain hardening and dynamic recovery of the non-recrystallized material; (ii) an initiation criterion for DRX; (iii) DRX kinetics and an equation for the recrystallized grain size; and (iv) a rule of mixture to determine the macroscopic flow stress when recrystallized and non-recrystallized grains coexist. In addition, the evolution of grain size is expressed by an additional equation, which is mostly not coupled to the flow stress model.

Various model choices for the individual parts of the model are available. Due to the large differences in strain hardening behavior for different materials, setting up a model for a certain material requires to take the peculiarities of the strain hardening and softening of the material into account. As a consequence, there does not seem to be a unified model that would be applicable to all kinds of materials. In the following, we describe the features of typical strain hardening models that take DRX into account. We then show how to devise and select the model equations so that a suitable model is attained, at the example of the case hardening steel and the TNB alloy.

Model equations for characteristic points. Semi-empirical material models are the most commonly used models in industrial applications. They are often preferred to physically-based models due to their fast computing times. The basic idea behind semi-empirical models is that characteristic points of the flow curves can be expressed as a function of the Zener-Hollomon parameter Z. A typical semi-empirical model based on the equations proposed by Luton and Sellars [39] and Beynon and Sellars [40] is given in Figure 3 together with an illustration of a flow curve, the strain hardening rate and their characteristic points.

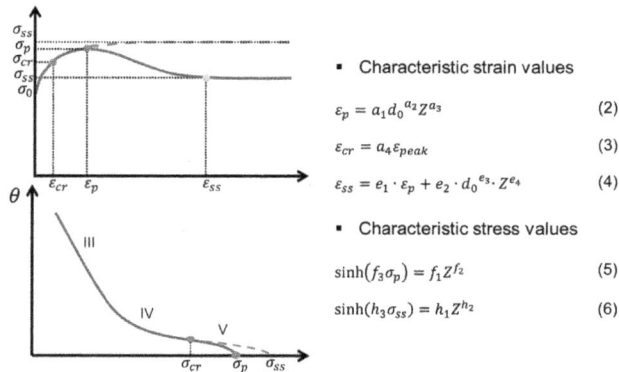

- Characteristic strain values

$$\varepsilon_p = a_1 d_0^{a_2} Z^{a_3} \tag{2}$$

$$\varepsilon_{cr} = a_4 \varepsilon_{peak} \tag{3}$$

$$\varepsilon_{ss} = e_1 \cdot \varepsilon_p + e_2 \cdot d_0^{e_3} \cdot Z^{e_4} \tag{4}$$

- Characteristic stress values

$$\sinh(f_3 \sigma_p) = f_1 Z^{f_2} \tag{5}$$

$$\sinh(h_3 \sigma_{ss}) = h_1 Z^{h_2} \tag{6}$$

Figure 3. Flow curve and strain hardening rate with characteristic point and model equations for the Beynon and Sellars model [40].

The strain at which the maximum value of the flow stress is attained is given by Equation (2). The critical strain for the initiation of DRX is assumed to be a constant fraction a_4 of ε_p by Equation (3). The strain at which a steady state deformation is reached is given by Equation (4), which uses four fit parameters, e_1–e_4. The peak and steady state flow stress σ_p and σ_{ss} are modeled as functions of the Zener-Hollomon parameter Z via Equations (5) and (6), with material-dependent parameters f_1, f_2, f_3 and h_1, h_2, h_3.

Strain hardening. A plenitude of models is available to model strain hardening. Recent research has shown that different kinds of inconsistencies with the Poliak-Jonas criterion can occur depending on the model choice [38]. If the strain hardening models do not show an inflection point, they do not reach the critical point. If the strain hardening model shows an inflection point, care must be taken that the inflection point coincides with the critical conditions of DRX nucleation models. In addition, the model might just not capture the course of the strain hardening rate in a Kocks-Mecking plot in the right way, which shows that the underlying physics are not taken into account.

Typical dislocation density models are formulated as ordinary differential equations, which contain terms for the generation and annihilation of dislocations. These models were derived from physical mechanisms of dislocation theory. The Taylor equation correlates dislocation density to the macroscopic flow stress. It can be used to transform the model equations into θ-σ-space (Kocks-Mecking plots), where the course of the strain hardening rate can be analyzed. None of the dislocation-based models detailed in Table 1 are appropriate for the TNB alloy since they cannot reproduce the curvature in a Kocks-Mecking plot. The phenomenological model by Cingara and McQueen, however, possesses the right curvature for $C \geq 1$. The constant C determines the curvature of the flow curve up to the peak. The location of the peak is explicitly taken into account by the strain hardening model. Softening will be superimposed after the onset of DRX and will be discussed below.

In contrast, the model by Cingara and McQueen is not the best choice for the 25MoCrS4 steel. The 25MoCrS4 steel shows all hardening stages, i.e., a linear decrease of $\theta(\sigma)$ initially, which is referred to as stage-III hardening, an stage IV hardening with a reduced softening rate and an inflection point, and a subsequently increased softening rate due to DRX.

In physical theories of crystal plasticity, the dislocation density is commonly used as the structural parameter for macroscopic descriptions of plastic flow, with the flow stress governed by Taylor-like hardening following:

$$\sigma = \sigma_0 + \alpha G b \sqrt{\rho} \tag{7}$$

where σ_0 is the yield stress, at which point $\rho = \rho_0$ and $\varepsilon = \varepsilon_0$, α is a constant of 0.5–1, G is the shear modulus, and b is the Burger's vector. The Taylor equation can be utilized to transform the model equations into Kocks-Mecking plots. Equally, models can be formulated directly in the θ-σ-space and then transferred into stress–strain curves based on the Taylor equation. The strain hardening model proposed in [38] reproduces the course of the experimentally observed θ–σ-curves using a linear function for the initial stage-III hardening, a transition to a second linear function which represents stage IV and then, starting from the critical point, a cubic function of stress is used to model stage V hardening. Hence, the model is designed to mimic the hardening behavior observed for steel (cf. Figure 1c), but it would not be applicable to TNB-V4.

Recrystallization kinetics. In the original work of Luton and Sellars [39], JMAK kinetics was proposed for DRX, which were formulated as function of time. Later, a strain-dependent version of the kinetics was introduced by Sellars [41]. Recent research has shown that JMAK kinetics is only consistent with the Poliak-Jonas criterion if the Avrami exponent exceeds a value of 3 [38]. As a consequence, DRX kinetics based transformation kinetics that go back on Cahn were proposed, which are detailed in Table 2. In contrast, the softening of TNB is not only determined by DRX but also by the rotation and break-down of lamellar colonies, adiabatic heating etc. Since consistency to the Poliak-Jonas criterion is not an issue for TNB-V4 (it shows no inflection point), ordinary JMAK kinetics may be used, albeit with no physical meaning.

Table 1. Comparison of different strain hardening models and their course in Kocks-Mecking plots $\theta(\sigma)$.

Evolution of ρ	Flow Stress	$\theta(\sigma)$	Curvature	Form/Reference
$\frac{d\rho}{d\varepsilon} = k_1\sqrt{\rho} - k_2\rho$	$\sigma(\varepsilon) = \sigma_0 + (\sigma_{ss} - \sigma_0)(1 - e^{-C\varepsilon})$	$\theta(\sigma) = C(\sigma_{ss} - \sigma)$	None (linear course)	[4]
$\frac{d\rho}{dt} = \frac{\dot\varepsilon}{bl} - 2M\tau\rho^2$	$\sigma(\varepsilon) = \sigma_0 + \alpha Gb\sqrt{\rho}$ $\rho = \rho_s\tanh(2M\tau\rho_s\varepsilon/\dot\varepsilon)$	$\theta(\sigma) = \frac{\alpha^2 G^2 b}{2l(\sigma-\sigma_0)} - \frac{M\tau(\sigma-\sigma_0)^3}{\alpha^2 G^2 b^2\dot\varepsilon}$	Concave up/concave down	[42] [43]
$\frac{d\rho}{d\varepsilon} = k_1 - k_2\rho$	$\sigma = \left[\sigma_{ss}^{*2} + (\sigma_0^2 - \sigma_{ss}^{*2})e^{-\Omega\varepsilon}\right]^{1/2}$	$\theta(\sigma) = \frac{\Omega}{2}\left(\frac{\sigma_{ss}^2}{\sigma} - \sigma\right)$	Concave up	[11]
-	$\sigma(\varepsilon) = \sigma_p\left[\frac{\varepsilon}{\varepsilon_p}\exp(1-\frac{\varepsilon}{\varepsilon_p})\right]^C$	N.A.	concave down for $C \geq 1$	[44]
-	N.A.	$\theta(\sigma) = \theta_{III}(1 - H_1(\sigma - \sigma_{IV})) +$ $\theta_{IV}H_1(\sigma - \sigma_{IV}) + \theta_V H_2(\sigma - \sigma_V)$ $\theta_{IV} = b_{IV} - m_{IV}\sigma$ $\theta_V = -C(\sigma - \sigma_c)^3$ $H_1(x) = \frac{1}{2} + \frac{1}{2}\tanh(cx) = \frac{1}{1+e^{-2cx}}$	linear (stage III)/linear (stage IV)/concave down	[38]

Table 2. Model equations for 25MoCrS4 and TNB.

Model Part	25MoCrS4 Steel	TNB
Strain hardening	$\theta(\sigma) = \quad \theta_{\mathrm{III}}(1 - H_1(\sigma - \sigma_{\mathrm{IV}})) +$ $\theta_{\mathrm{IV}} H_1(\sigma - \sigma_{\mathrm{IV}}) + \theta_V H_2(\sigma - \sigma_V)$ $\theta_{\mathrm{IV}} = b_{\mathrm{IV}} - m_{\mathrm{IV}}\sigma$ $\theta_V = -C(\sigma - \sigma_c)^3$ $H_{1,2}(x) = \frac{1}{2} + \frac{1}{2}\tanh(c_{(1,2)}x) = \frac{1}{1 + e^{-2c_{(1,2)}x}}$	$\sigma(\varepsilon) = \sigma_p \left[\frac{\varepsilon}{\varepsilon_p} \exp(1 - \frac{\varepsilon}{\varepsilon_p}) \right]^C$
Critical strain	$\varepsilon_{cr} = \alpha\varepsilon_p$ (Equation (3), s. Figure 3)	
Peak strain	$\varepsilon_p = a_1 \cdot d_0{}^{a_2} \cdot Z^{a_3}$ (Equation (2), s. Figure 3)	
Steady state stress	$\varepsilon_{ss} = e_1 \cdot \varepsilon_m + e_2 \cdot d_0{}^{e_3} \cdot Z^{e_4}$ (Equation (4), s. Figure 3)	
Peak stress	$\sinh(f_3 \cdot \sigma_p) = f_1 \cdot Z^{f_2}$ (Equation (5), s. Figure 3)	
Steady state stress	$\sinh(h_3 \cdot \sigma_p) = h_1 \cdot Z^{h_2}$ (Equation (6), s. Figure 3)	
DRX kinetics	$\dot{X}(t) = (1 - X(t))\left[\begin{array}{l} (1 - H(t_{ss}, t_s))\frac{4\pi}{3}I_s S v^3 t^3 \\ + H(t_{ss}, t_s)AX(t) \end{array} \right]$	$X = 1 - \exp\left[k \cdot (\frac{\varepsilon - \varepsilon_{cr}}{\varepsilon_{ss} - \varepsilon_{cr}})^q \right], \quad \varepsilon \geq \varepsilon_{cr}$
Flow stress	$\sigma_Y = \begin{cases} \sigma_0 & \text{if } \varepsilon < \varepsilon_{cr} \\ \sum\limits_{i=0}^{n-1}(X_i - X_{i+1})\sigma_i + X_n\sigma_n & \text{if } \varepsilon \geq \varepsilon_{cr} \end{cases}$	

Dynamically recrystallized grain size. The dynamically recrystallized grain size is independent of the initial grain size and only depends on the Zener-Hollomon parameter via

$$d_{\mathrm{DRX}} = b_1 \cdot Z^{b_2} \tag{8}$$

This equation is used in almost all semi-empirical models for DRX. In this work, it is applied to the 25MoCrS4 steel and to the phases of the TNB alloy that show recrystallization.

Strain hardening models. Based on the observations made about the strain hardening behavior of TNB-V4 and 25MoCrS4, two strain hardening models have been set up, which are detailed in Table 2. Both models are designed to meet the requirements of the materials

Using non-linear regression by least-squares, the models have been fitted to a single flow curve of TNB-V4 and 25MoCrS4, respectively. Figure 4 shows that for a single flow curve, the models reproduce not only the flow curves but also the behavior in the Kocks-Mecking plot with very high accuracy. When the model is applied to a larger range of temperatures and strain rates, parameter identification by non-linear regression may fail. To allow for successful parameter fitting, most of the model equations are first fitted to the experimental data using the characteristic points extracted from the flow curves. Regression is thus performed separately for each model equation. In the model, all equations are coupled. Hence, a non-linear regression of the entire model is performed subsequently, which will improve the initial parameter values obtained by fitting the model equations individually. The procedure is described in detail in the subsequent sections.

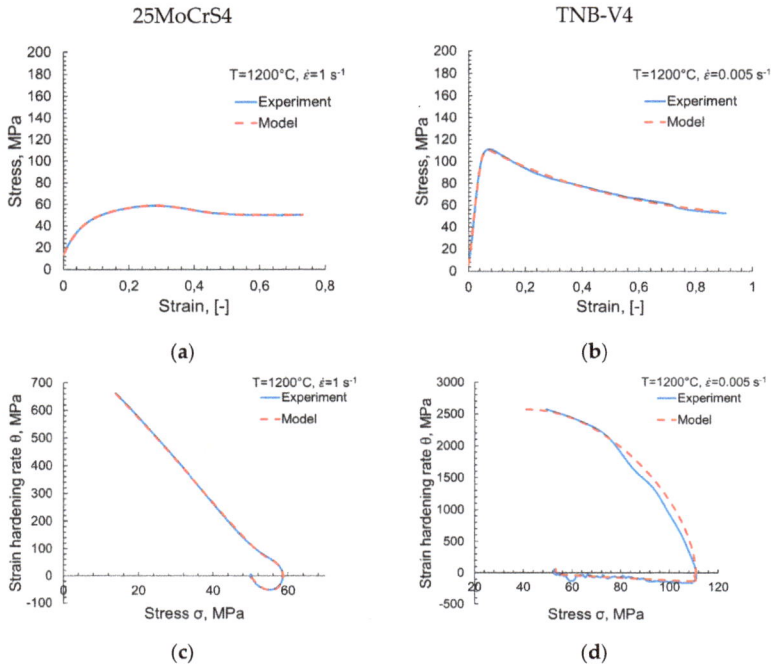

Figure 4. Model prediction vs. experimental data: (**a**) flow curve of 25MoCrS4; (**b**) flow curve of TNB-V4; (**c**) Kocks-Mecking plot of 25MoCrS4; and (**d**) Kocks-Mecking plot of TNB-V4.

4.2. Evaluation of Experimental Data and Parameters Identification

The procedures for parameter identification depend largely on the model equations. However, some steps are common for the two models defined in Table 2 above. As a first step, all characteristic points such as ε_p, σ_p, ε_{ss}, and σ_{ss} have to be determined from the experimental flow curves. Using σ_p, the activation energy and hence the Zener-Hollomon parameter Z, sometimes referred to as temperature-compensated strain rate, can be determined,

$$Z = \dot{\varepsilon} \exp\left(\frac{Q_w}{RT}\right) = A\left[\sinh(\alpha\sigma)\right]^n \tag{9}$$

where A and α are constants, T is the deformation temperature, σ is the flow stress (MPa), $\dot{\varepsilon}$ is the strain rate (s^{-1}), n is the stress exponent, R is the gas constant (kJ·mol^{-1}·K^{-1}) and Q_w is the activation energy of deformation (kJ·mol^{-1}).

For some model parameters (predominantly those with a physical background), dedicated procedures exist for parameter determination. The activation energy of hot working Q_w, for instance, can be determined from the following relation:

$$Q = R\left\{\frac{\partial\ln\dot{\varepsilon}}{\partial\ln\left[\sinh(\alpha\sigma)\right]}\right\}_T \left\{\frac{\partial\ln\left[\sinh(\alpha\sigma)\right]}{\partial(1/T)}\right\}_{\dot{\varepsilon}} \tag{10}$$

According to Poliak and Jonas [28], the critical strain ε_{cr} for the onset of DRX can be detected from the inflection point in the course of the strain hardening rate $\theta = \partial\sigma/\partial\varepsilon$ as a function of flow stress σ. To this end, the strain hardening rate has to be computed from flow stress data. All other model parameters of the equations in Figure 3 can be determined for each equation independently by regression analysis.

A specific problem with the TNB alloy is that it shows no inflection point in the Kocks-Mecking plot. Hence, no critical stress or strain can be determined from flow stress data only.

To describe the relation between the dynamically recrystallized volume fraction X and strain an Avrami-type equation is used for TNB-V4 while different kinetics are used for 25MoCrS4 to ensure consistency with the Poliak-Jonas criterion. The model parameters are not determined from the flow curves but from the recrystallized fractions determined by quantitative metallography of quenched specimens. Based on the phase fraction and grain size of the different phases in the not deformed state a histogram for every phase was calculated. These histograms were used as basis to determine the differences of the phase fractions and grain sizes after deformation.

In the same way, the coefficients b_1 and b_2 in Equation (8), which describes the dynamically recrystallized grain size, were identified from metallographic evaluation of the specimens that show DRX.

4.3. Determination of Model Parameters Common to Both Models

In this section, the experimental data from the cylinder compression tests and the metallographic analysis are used to determine the parameters for the strain hardening models. For this purpose, firstly the characteristic points such as ε_p, σ_p, ε_{ss}, and σ_{ss} were determined from the flow curves shown in Figure 1.

Based on Equation (10), the slope of the plots of $\ln[\sinh(\alpha\sigma)]$ versus $\ln\dot{\varepsilon}$ and $1/T$ can be utilized for calculating the value of Q_w. The required plots are shown in Figure 5. Values of 481.3 kJ·mol^{-1} for 25MoCrS4 and 321.2 kJ·mol^{-1} for TNB-V4 were obtained for the activation energy Q_w.

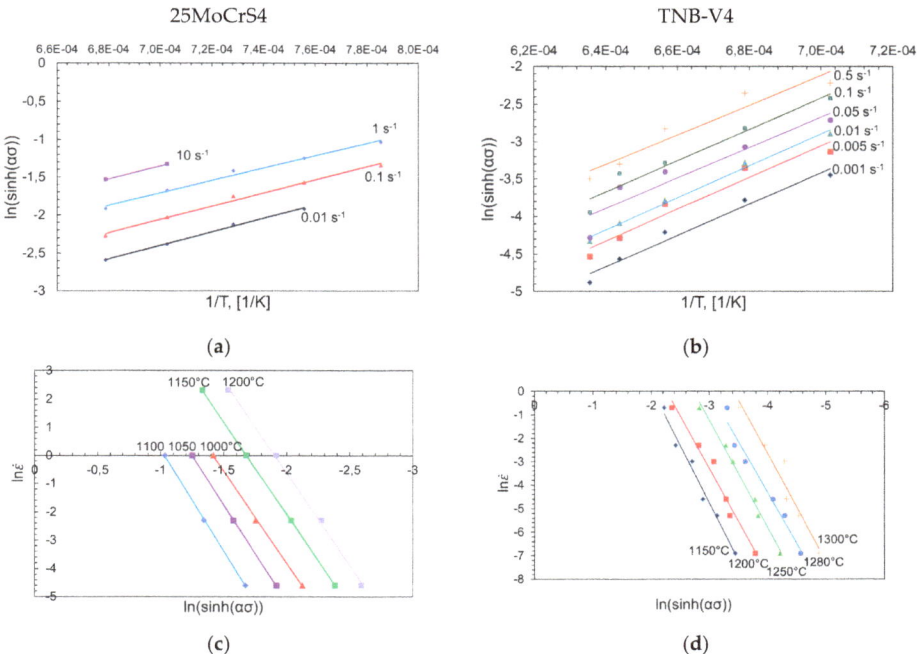

Figure 5. Evaluating Q_w: $\ln[\sinh(\alpha\sigma)]$ versus $1/T$ (**a**) for 25MoCrS4 and (**b**) for TNB-V4; and $\ln\dot{\varepsilon}$ versus $\ln[\sinh(\alpha\sigma)]$ (**c**) for 25MoCrS4 and (**d**) for TNB-V4.

In Figure 6a, a plot of the peak strain as a function of the Zener-Hollomon parameter is shown together with the results of regression analysis. The steady state strain strongly influences the shape of the flow curves. The extraction of values for the steady state strain is complicated for TNB-V4 since only a few flow curves attain a steady state regime. To determine the parameters e_1, e_2, and e_3 in Equation (4), which defines the dependence of ε_{ss} on the Zener-Hollomon parameter, a regression analysis was performed as shown in Figure 6b.

Since the TNB-V4-alloy shows no inflection point in the Kocks-Mecking plots no critical stress/strain can be determined from flow stress data only and α was set to 0.5 from experience. For 25MoCrS4 previous work showed that a value of $0.7\varepsilon_p$ is suitable.

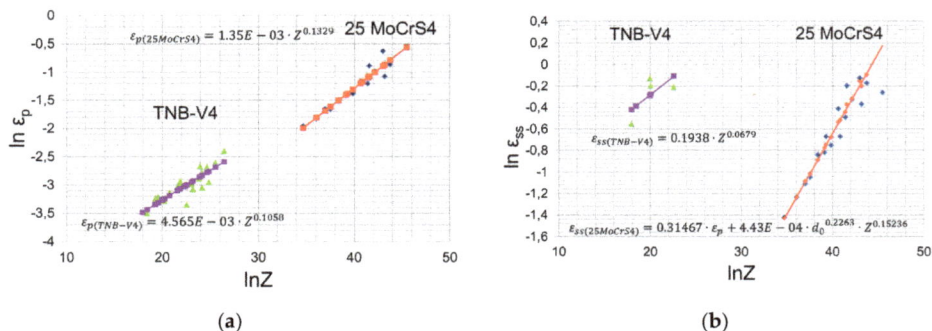

Figure 6. Strain depending on Zener-Hollomon-Parameter: (**a**) relationships between ln ε_p and lnZ; and (**b**) relationships between lnε_{ss} and lnZ.

The dependence of peak and steady state stress on the Zener-Hollomon parameter was determined by non-linear regression using Equations (5) and (6). The fitted and measured values are shown in Figure 7. As mentioned above, the equation of Cingara et al. [44] was employed to describe the work-hardening condition of TNB-V4-alloy up to the peak stress. Hence, the fitted peak stress values can be used directly.

The dynamically recrystallized grain size of 25MoCrS4 increases with decreasing Zener-Hollomon parameter, i.e., by increase of temperature and decrease of strain rate the resulting grain size increases. In Figure 8a, the grain sizes determined for various Z values are shown along with the best fit of Equation (8). The kinetics of dynamic recrystallization is displayed in Figure 8b.

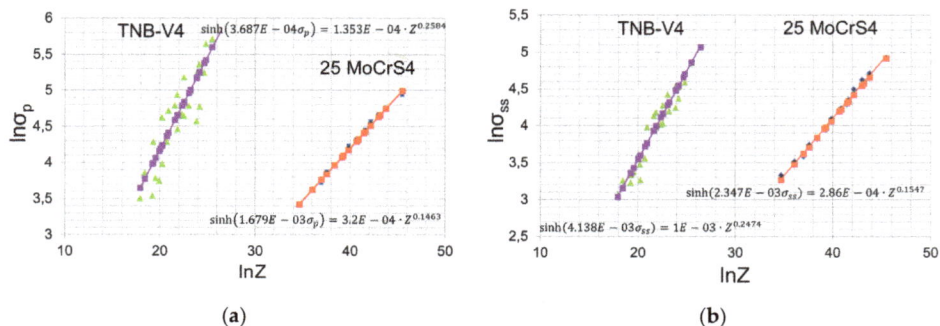

Figure 7. Stress depending on Zener-Hollomon-Parameter: (**a**) relationships between lnσ_p and lnZ; and (**b**) relationships between lnσ_{ss} and lnZ.

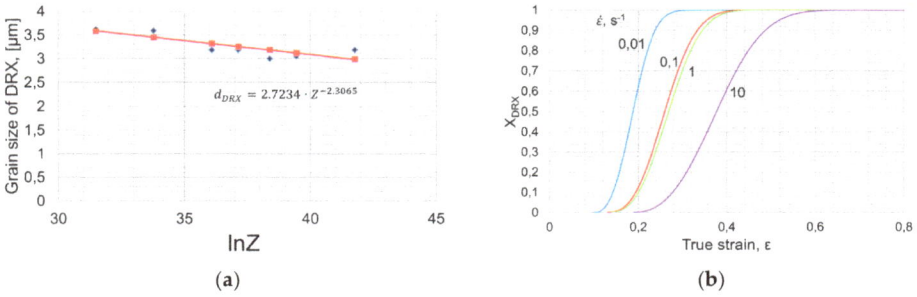

Figure 8. (**a**) Relationship between DRX grain size and Zener-Hollomon-Parameter: and (**b**) volume fraction of DRX obtained at a temperature of 1200 °C and different strain rates.

The relationship between the dynamically recrystallized grain size of the TNB-V4-alloy and the Zener-Hollomon-Parameter is shown in Figure 9. In the multiphase TiAl-alloys, solely the β / β_0- (Figure 9a) and γ-phase (Figure 9b) show recrystallization. The α-phase only forms during hot forming. The typical microstructure of the TNB-V4-alloy after forming can be seen in Figure 9c.

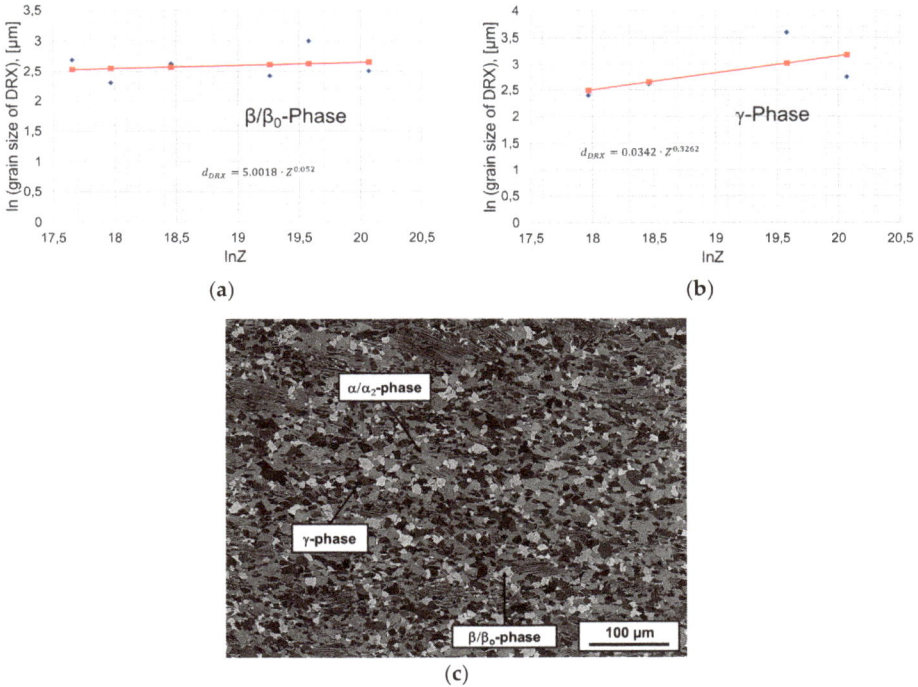

Figure 9. Relationship between DRX grain size and Zener-Hollomon-Parameter of TNB-V4: (**a**) for β / β_0-Phase; and (**b**) γ-Phase. (**c**) Typical microstructure of TNB-V4 compressed at 1250 °C with a strain rate of 0.001 s^{-1} to a strain of 60%.

5. Final Calibration and Validation of the Model

The direct determination of model parameters detailed in section 4 is not applicable to all model parameters. In addition, since the interplay of the individual parts of the model determines the accuracy of the model, an inverse parameter identification is favorable as final step, i.e., after starting values of most model parameters are available from the direct determination performed in Section 4. Inverse parameter identification is essentially the solution of an optimization problem. The cost function is the sum of squares of the differences between measured and calculated flow stress values for all available values from the experiments:

$$F(\beta) = \sum_i \sum_j \sum_k \left(\sigma_f^{(exp)}(T_i, \dot{\varepsilon}_j, \varepsilon_k) - \sigma_f^{(sim)}(T_i, \dot{\varepsilon}_j, \varepsilon_k, \beta) \right)^2 \tag{11}$$

The cost function $F(\beta)$ is minimized using an optimization algorithm such as the Levenberg-Marquardt method to determine the material parameters, which are assembled in the parameter vector β.

Figure 10 shows a comparison between the predicted and measured flow stress values for the 25MoCrS4 steel for temperatures ranging from 900 °C to 1200 °C and five different strain rates. The model parameters were determined in a similar way as in the previous study of Konovalov et al. [30]. However, in the present work, a new model formulation was used. A very good agreement is obtained between the calculation and measurement in view of the shape of the flow curves and the absolute level of the calculated flow stress over the entire range of strain rates and temperatures.

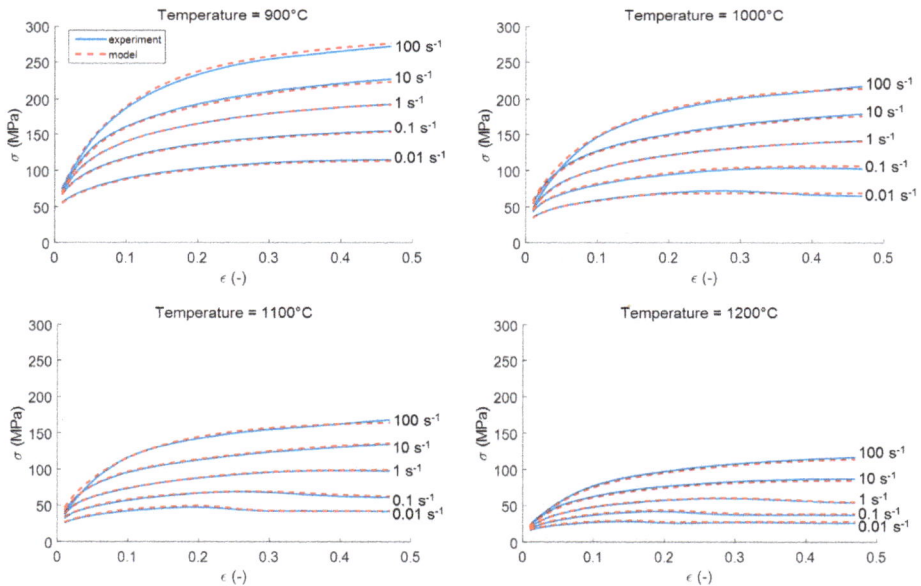

Figure 10. Model prediction vs. experimental data of 25MoCrS4 at various temperatures und strain rates.

The model shows a very high accuracy for a wide range of temperatures and strain rates. In Figure 11, the comparison between the calculated and measured flow stresses is shown for the TNB-V4-alloy for strain rates varying from 0.001 s^{-1} to 0.5 s^{-1} and for a temperature of 1200 °C suitable for isothermal forging.

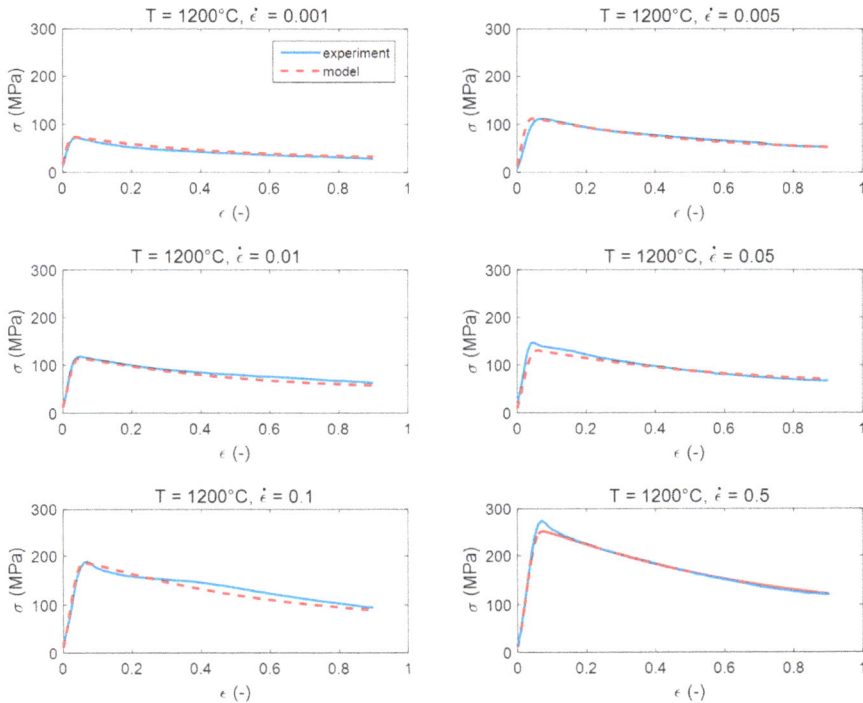

Figure 11. Model prediction vs. experimental data of TNB-V4-alloy at 1200 °C and various strain rates.

The results indicate that the proposed strain hardening model can properly describe and predict the flow stress of the TiAl-alloy under isothermal forging conditions. The model accurately predicts the flow stress up to the peak stress and to some extent also the transition to the steady state.

The results show that for both 25MoCrS4 and TNB-V4, building a model from the behavior observed in Kocks-Mecking plots has the advantage that the model is consistent with the observed strain hardening stages and shows a high accuracy. If only curve fitting in terms of stress strain curves is performed and valuable information provided by the Kocks-Mecking plots is neglected, inconsistencies and larger uncertainties of the model performance may be the consequence. For the 25MoCrS4, an intricate model was designed which takes the physics of dynamic recrystallization into account. For TNB-V4, both the strain hardening before the initiation of DRX and the softening kinetics are entirely phenomenological. For TiAl alloys in general, a deeper insight into the deformation mechanism leading to the observed behavior seems necessary. Reflecting the model behavior with the experimental results not only in stress-strain space but also in θ-σ-space, as proposed in this paper, yields important information to guide the model development.

6. Conclusions

In the present study the peculiarities of the work hardening and softening behavior of a Ti–44.5Al–6.25Nb–0.8Mo–0.1B (in at %) TNB-V4-alloy were investigated and compared to a typical case-hardening steel 25MoCrS4. The flow curves were analyzed using Kocks-Mecking plots in order to illustrate the difference in deformation behavior between a typical steel and the high Nb containing TiAl-alloy. Specific model equations for 25MoCrS4 and Ti–44.5Al–6.25Nb–0.8Mo–0.1B were derived based on the course of the Kocks-Mecking plots. The empirical equation proposed by Cingara and

McQueen was used to model the work-hardening region for TNB-V4, since this model is able to show a concave down course in the Kocks-Mecking plot, as observed experimentally. A new model formulation capable of showing an inflection point was used for 25MoCrS4. For the DRX kinetics, a new formulation consistent with the Poliak-Jonas criterion was applied to 25MoCrS4. For TNB, a JMAK type kinetics was used, which describes all softening mechanisms without taking into account the physical background.

The following conclusions can be drawn from the presented studies:

- Kocks-Mecking plots reveal that in contrast to 25MoCrS4 steel the Ti–44.5Al–6.25Nb–0.8Mo–0.1B alloy neither shows a linear decrease of θ (i.e., no stage-III hardening) nor does it exhibit a plateau (stage IV hardening) or an inflection point (marking the onset of DRX) at all forming conditions.

- The information obtained from Kocks-Mecking plots should be taken into account in the development of strain hardening models. Otherwise, inconsistencies, e.g., with the Poliak-Jonas criterion, may result.

- Both models show a high accuracy. They may hence be used in finite element simulations of metal forming processes. For TNB-V4, however, the complex microstructure evolution, i.e., the recrystallization of the individual phases, the reorientation of the lamellar colonies and flow localization effects during deformation need to be taken into account in continuative work.

Acknowledgments: The model development was performed as preliminary study in the project BA4253/2-1 funded by the German Research Foundation. M. Bambach and I. Sizova gratefully acknowledge the funding obtained.

Author Contributions: Bambach, M. devised the constitutive models, wrote and edited the manuscript; Sizova, I. analyzed the experimental data, performed the parameter identification and wrote parts of the manuscript; and Bolz, S. and Weiß, S. performed the microstructure analysis and interpretation of the obtained results and wrote the respective sections of the manuscript.

Conflicts of Interest: The authors declare no conflict of interest.

References

1. Appel, F.; Wagner, R. Microstructure and deformation of two-phase γ-titanium aluminides. *Mater. Sci. Eng. Rep.* **1998**, *22*, 187–268. [CrossRef]
2. Dimiduk, D. Gamma titanium aluminide alloys—An assessment within the competition of aerospace structural materials. *Mater. Sci. Eng. A* **1999**, *263*, 281–288. [CrossRef]
3. Kabir, M.R.; Chernova, L.; Bartsch, M. Numerical investigation of room-temperature deformation behavior of a duplex type γTiAl using a multi-scale modeling approach. *Acta Mater.* **2010**, *58*, 5834–5847. [CrossRef]
4. Mecking, H.; Kocks, U.F. Kinetics of flow and strain-hardening. *Acta Metall.* **1981**, *29*, 1865–1875. [CrossRef]
5. Estrin, Y.; Mecking, H. A unified phenomenological description of work hardening and creep based on one-parameter models. *Acta Metall.* **1984**, *32*, 57–70. [CrossRef]
6. Bergström, Y. A dislocation model for the stress-strain behaviour of polycrystalline α-Fe with special emphasis on the variation of the densities of mobile and immobile dislocations. *Mater. Sci. Eng.* **1970**, *5*, 193–200. [CrossRef]
7. Baragar, D.L. The high temperature and high strain-rate behaviour of a plain carbon and an HSLA steel. *J. Mech. Work. Technol.* **1987**, *14*, 295–307. [CrossRef]
8. Davenport, S.B.; Silk, N.J.; Sparks, C.N.; Sellars, C.M. Development of constitutive equations for modelling of hot rolling. *Mater. Sci. Technol.* **2013**, *16*, 539–546. [CrossRef]
9. Rao, K.P.; Hawbolt, E.B. Development of constitutive relationships using compression testing of a medium carbon steel. *J. Eng. Mater. Technol.* **1992**, *114*, 116. [CrossRef]
10. Cheng, L.; Xue, X.; Tang, B.; Kou, H.; Li, J. Flow characteristics and constitutive modeling for elevated temperature deformation of a high Nb containing TiAl alloy. *Intermetallics* **2014**, *49*, 23–28. [CrossRef]
11. Laasraoui, A.; Jonas, J.J. Prediction of steel flow stresses at high temperatures and strain rates. *Metall. Trans. A* **1991**, *22*, 1545–1558. [CrossRef]
12. Godor, F.; Werner, R.; Lindemann, J.; Clemens, H.; Mayer, S. Characterization of the high temperature deformation behavior of two intermetallic TiAl–Mo alloys. *Mater. Sci. Eng. A* **2015**, *648*, 208–216. [CrossRef]

13. Sellars, C.M.; McTegart, W.J. On the mechanism of hot deformation. *Acta Metall.* **1966**, *14*, 1136–1138. [CrossRef]

14. Hensel, M.; Spittel, T. *Ver- u. Entfestigung bei Warmumformung*; Dt. Verl. für Grundstoffindustrie: Leipzig, Germany, 1982.

15. He, X.; Yu, Z.; Liu, G.; Wang, W.; Lai, X. Mathematical modeling for high temperature flow behavior of as-cast Ti–45Al–8.5Nb–(W,B,Y) alloy. *Mater. Des.* **2009**, *30*, 166–169. [CrossRef]

16. Pu, Z.J.; Wu, K.H.; Shi, J.; Zou, D. Development of constitutive relationships for the hot deformation of boron microalloying TiAl–Cr–V alloys. *Mater. Sci. Eng. A* **1995**, *192–193*, 780–787. [CrossRef]

17. Deng, T.-Q.; Ye, L.; Sun, H.-F.; Hu, L.-X.; Yuan, S.-J. Development of flow stress model for hot deformation of Ti-47%Al alloy. *Trans. Nonferr. Met. Soc. China* **2011**, *21*, s308–s314. [CrossRef]

18. Nobuki, M.; Hashimoto, K.; Takahashi, J.; Tsujimoto, T. Deformation of cast TiAl intermetallic compound at elevated temperatures. *Mater. Trans. JIM* **1990**, *31*, 814–819. [CrossRef]

19. Fukutomi, H.; Nomoto, A.; Osuga, Y.; Ikeda, S.; Mecking, H. Analysis of dynamic recrystallization mechanism in γ-TiAl intermetallic compound based on texture measurement. *Intermetallics* **1996**, *4*, S49–S55. [CrossRef]

20. Kim, H.Y.; Sohn, W.H.; Hong, S.H. High temperature deformation of Ti–(46–48)Al–2W intermetallic compounds. *Mater. Sci. Eng. A* **1998**, *251*, 216–225. [CrossRef]

21. Fröbel, U.; Appel, F. Hot-workability of gamma-based TiAl Alloys during severe torsional deformation. *Metall. Mater. Trans. A* **2007**, *38*, 1817–1832. [CrossRef]

22. Kim, H.Y.; Hong, S.H. Effect of microstructure on the high-temperature deformation behavior of Ti–48Al–2W intermetallic compounds. *Mater. Sci. Eng. A* **1999**, *271*, 382–389. [CrossRef]

23. Semiatin, S.L.; Frey, N.; Thompson, C.R.; Bryant, J.D.; El-Soudani, S.; Tisler, R. Plastic flow behavior of Ti–48Al–2.5Nb–0.3Ta at hot-working temperatures. *Scr. Metall. Mater.* **1990**, *24*, 1403–1408. [CrossRef]

24. Wiezorek, J.M.K.; Deluca, P.M.; Mills, M.J.; Fraser, H.L. Deformation mechanisms in a binary Ti–48 at % Al alloy with lamellar microstructure. *Philos. Mag. Lett.* **1997**, *75*, 271–280. [CrossRef]

25. Schaden, T.; Fischer, F.D.; Clemens, H.; Appel, F.; Bartels, A. Numerical modelling of kinking in lamellar γ-TiAl based Alloys. *Adv. Eng. Mater.* **2006**, *8*, 1109–1113. [CrossRef]

26. McQueen, H.; Ryan, N. Constitutive analysis in hot working. *Mater. Sci. Eng. A* **2002**, *322*, 43–63. [CrossRef]

27. Kocks, U.F.; Mecking, H. A mechanism for static and dynamic recovery. In *Strength of Metals and Alloys*; Haasen, P., Gerold, V., Kostorz, G., Eds.; Pergamon Press: Oxford, UK, 1980; pp. 345–350.

28. Jonas, J.J.; Quelennec, X.; Jiang, L.; Martin, É. The Avrami kinetics of dynamic recrystallization. *Acta Mater.* **2009**, *57*, 2748–2756. [CrossRef]

29. Bambach, M.; Sizova, I.; Bolz, S.; Weiß, S. Development of a dynamic recrystallization model for a β-solidifying titanium aluminide alloy using kocks-mecking plots. In Proceedings of the 19st ESAFORM Conference on Material Forming, Nantes, France, 27–29 April 2016.

30. Konovalov, S.; Henke, T.; Jansen, U.; Hardjosuwito, A.; Lohse, W.; Bambach, M.; Prahl, U. Test case gearing component. In *Integrative Computational Materials Engineering: Concepts and Applications of a Modular Simulation Platform*; Schmitz, G.J., Prahl, U., Eds.; John Wiley & Sons: Weinheim, Germany, 2012.

31. Semiatin, S.L.; Frey, N.; El-Soudani, S.M.; Bryant, J.D. Flow softening and microstructure evolution during hot working of wrought near-gamma titanium aluminides. *Metall. Mater. Trans. A* **1992**, *23*, 1719–1735. [CrossRef]

32. Lohmar, J.; Bambach, M. Influence of different interpolation techniques on the determination of the critical conditions for the onset of dynamic recrystallization. *Mater. Sci. Forum* **2013**, *762*, 331–336. [CrossRef]

33. Xiong, W.; Lohmar, J.; Bambach, M.; Hirt, G. A new method to determine isothermal flow curves for integrated process and microstructural simulation in metal forming. *Int. J. Mater. Form.* **2015**, *8*, 59–66. [CrossRef]

34. Bolz, S.; Oehring, M.; Lindemann, J.; Pyczak, F.; Paul, J.; Stark, A.; Lippmann, T.; Schrüfer, S.; Roth-Fagaraseanu, D.; Schreyer, A.; et al. Microsturcture and mechanical properties of a forged β-solidifying γ-TiAl alloy in different heat treatment conditions. *Intermetallics* **2015**, *58*, 71–83. [CrossRef]

35. Fujiwara, T.; Nakamura, A.; Hosomi, M.; Nishitani, S.; Shirai, Y.; Yamaguchi, M. Deformation of polysynthetically twinned crystals of TiAl with a nearly stoichiometric composition. *Philos. Mag. A* **1990**, *61*, 591–606. [CrossRef]

36. Schwaighofer, E.; Clemens, H.; Lindemann, J.; Stark, A.; Mayer, S. Hot-working behavior of an advanced intermetallic multi-phase γ-TiAl based alloy. *Mater. Sci. Eng. A* **2014**, *614*, 297–310. [CrossRef]

37. Bariani, P.F.; Negro, T.D.; Bruschi, S. Testing and modelling of material response to deformation in bulk metal forming. *CIRP Ann. Manuf. Technol.* **2004**, *53*, 573–595. [CrossRef]

38. Bambach, M. Implications from Poliak—Jonas criterion for the construction of flow stress models incorporating dynamic recrystallization. *Acta Mater.* **2013**, *61*, 6222–6233. [CrossRef]

39. Luton, M.; Sellars, C. Dynamic recrystallization in nickel and nickel-iron alloys during high temperature deformation. *Acta Metall.* **1969**, *17*, 1033–1043. [CrossRef]

40. Beynon, J.H.; Sellars, C.M. Modelling microstructure and its effects during multipass hot rolling. *ISIJ Int.* **1992**, *32*, 359–367. [CrossRef]

41. Sellars, C.M. Modelling of structural evolution during hot working processes. In Proceedings of the International Symposium on Annealing Processes: Recovery, Recrystallisation and Grain Growth, Sheffield, UK, 8–12 September 1986; pp. 167–187.

42. Sandström, R.; Lagneborg, R. A model for hot working occurring by recrystallization. *Acta Metall.* **1975**, *23*, 387–398. [CrossRef]

43. Sommitsch, C.; Mitter, W. On modelling of dynamic recrystallisation of fcc materials with low stacking fault energy. *Acta Mater.* **2006**, *54*, 357–375. [CrossRef]

44. Cingara, A.; McQueen, H.J. New formula for calculating flow curves from high temperature constitutive data for 300 austenitic steels. *J. Mater. Process. Technol.* **1992**, *36*, 31–42. [CrossRef]

![metals logo] *metals*

MDPI

Article

Study of the Isothermal Oxidation Process and Phase Transformations in B2-(Ni,Pt)Al/RENE-N5 System

Luis Alberto Cáceres-Díaz [1], Juan Manuel Alvarado-Orozco [2], Haidee Ruiz-Luna [3], John Edison García-Herrera [1], Alma Gabriela Mora-García [1], Gerardo Trápaga-Martínez [1,4], Raymundo Arroyave [5] and Juan Muñoz-Saldaña [1,6,*]

[1] Centro de Investigación y de Estudios Avanzados del IPN, Unidad Querétaro, Libramiento Norponiente 2000, Real de Juriquilla, 76230 Querétaro, Mexico; lcaceres@cinvestav.mx (L.A.C.-D.); johngarcia@cinvestav.mx (J.E.G.-H.); almamora@cinvestav.mx (A.G.M.-G.); gerardo.trapaga@ciateq.mx (G.T.-M.)

[2] Centro de Ingeniería Avanzada y Desarrollo Industrial, CIDESI-Querétaro, Av. Pie de la Cuesta No. 702, 76125 Santiago de Querétaro, Mexico; juan.alvarado@cidesi.edu.mx

[3] CONACYT-Universidad Autónoma de Zacatecas, Av. Ramón López Velarde 801, 98000 Centro Zacatecas, Mexico; hruizlu@conacyt.mx

[4] Centro de Tecnología Avanzada, CIATEQ A.C.-San Agustín del Retablo 150, Constituyentes Fovissste, 76150 Querétaro, Mexico

[5] Materials Science and Engineering Department, Texas A & M University, College Station, TX 77843, USA; rarroyave@tamu.edu

[6] German Aerospace Center (DLR), Institute of Materials Research, D-51170 Cologne, Germany

* Correspondence: jmunoz@cinvestav.mx; Tel.: +52-442-211-9924

Academic Editor: Hugo Lopez
Received: 8 June 2016; Accepted: 20 August 2016; Published: 1 September 2016

Abstract: Changes in composition, crystal structure and phase transformations of B2-(Ni,Pt)Al coatings upon isothermal oxidation experiments (natural and scale free oxidation) at 1100 °C, as a function of time beyond their martensitic transformation, are reported. Specifically, the analysis of lattice parameter and composition are performed to identify changes in the B2-(Ni,Pt)Al phase upon the chemically-driven $L1_0$-(Ni,Pt)Al and $L1_2$-(Ni,Pt)$_3$Al transformations. The B2-(Ni,Pt)Al phase was found to disorder and transform the martensite during the heat treatments for both oxidation experiments at approximately 36.3 and 40.9 at. % of Al, 47.7 and 42.9 at. % of Ni, 6.2 and 8.5 at. % of Pt, 4.2 and 2.9 at. % of Cr and 4.4 and 3.8 at. % of Co. The lattice constant and the long-range order parameter of the B2-(Ni,Pt)Al phase decreased linearly as a function of the elemental content irrespective of the nature of the oxidation experiments.

Keywords: B2-(Ni,Pt)Al bond coatings; oxidation; martensitic transformation; crystal structure

---

1. Introduction

B2-(Ni,Pt)Al intermetallics are widely used as bond coats (BC) in Thermal Barrier Coating systems (TBCs) for aeronautic applications due to their high temperature oxidation and corrosion resistance [1,2]. Such properties are achieved because these intermetallics serve as Al reservoirs for the formation of a Thermally Grown Oxide (TGO) α-Al$_2$O$_3$ layer upon oxidation at high temperature. The formed alumina scale acts as a barrier for the diffusion [3,4] and is the bonding material between the superalloy (SA) and the ceramic top coat (TC). NiAl has a CsCl crystal structure and can be defined by the sublattice model proposed by Ansara et al. [5]. In their work, two sublattices are used to define the NiAl system; Al is located in the first sublattice, β; Ni is located in the second sublattice, α. In this system, variations from stoichiometry are addressed as Ni antisites in the Ni-rich region, and as Ni vacancies (Va) in the Al-rich region of the B2 phase, according to the triple point defects model defined by Pike et al. [6]. The mechanical and oxidation behavior of NiAl varies as the Al content deviates from the 50:50 composition until the Ni-rich phase is achieved, but also with additions of ternary elements such as Cr, Co, Hf, Fe, Mo, among others. In the present case, the Pt content in the B2-(Ni,Pt)Al coatings reduces the detrimental effect of sulfur impurities on the scale adherence, prevents the formation of voids at the scale interface and improves the oxidation resistance because it directly diminishes the chemical activity of Al [7]. In the B2 phase, Pt occupies the Ni sites and has a solubility limit of 42 at. % [8,9]. Other elements, such as Cr and Co, are present in the bond coat since they diffuse from the superalloy during the heat treatments or high temperature exposures. Cr and Co are known to improve the oxidation and corrosion resistance and are assumed to occupy the Al and Ni sites, respectively [10,11].

On the other hand, the Al depletion during operation (cyclic thermal and isothermal exposure at high temperature) leads to changes in mechanical (Young modulus and other elastic constants [12]) structural (ordering and size of crystals [1,13]) thermodynamic (phase stability [14], enthalpy of formation, entropy [15]) and diffusion kinetics (mobility of atoms [16]) of the NiAl alloy, and subsequently affects the lifetime of the TBC system. A loss of stability of the B2 phase also leads to martensitic, or even gamma prime phase transformations, generating stresses in the coating beneath the oxide scale and consequently, leads to TBC failure. For instance, some reports suggest that the stress fields are the main intrinsic failure mechanisms of TBCs [4] because they contribute to the rumpling of the BC [4,17], although there has been some controversy about the role of the martensitic phase transformation on the failure of the TBCs [18]. The martensitic phase transformation in B2-(Ni,Pt)Al coatings has been studied elsewhere by XRD, SEM and in-situ TEM in samples with 28% and 100% of their Furnace Cycle Test (FCT) life. In these reports, changes in the crystal structure, such as volume fractions, linear and thermal strains, microstructure and their role in the development of internal stresses are presented and discussed [19,20]. Additionally, Glynn et al. [21] performed numerical simulations to study the effect of the thermal expansion coefficient on the stress and strains distributions of the bond coat during cooling-heating cycles; they concluded that elastic (heating) and inelastic (cooling) loadings of the bond coat develop a maximum tensile stress in the Top Coat.

The effect of ternary element additions on the nature and evolution of the martensitic transformation has also been previously reported. Pt additions, and other ternary elements such as Co and Cr, present an influence on the martensitic transformation temperature. For instance, Kainuma et al. [22] showed that the start temperature of the martensitic transformation, M_S increases with Pt or Co additions and decreases when Cr, Mo and Fe are added to the B2-NiAl phase. These studies have been carried out either by performing atomistic and numerical simulations or in high temperature cycled samples that already contain the martensite phase. However, the effect of the heat treatments on the stability of B2-(Ni,Pt)Al phase and its disordering until transformation to $L1_0$-(Ni,Pt)Al phase has not previously been reported.

The present contribution focuses on the effect of the isothermal exposure at 1100 °C of B2-(Ni,Pt)Al bond coats deposited on RENE N5 superalloys on the chemical composition, structure and microstructure of the bond coat, as well as on the chemically-driven B2 $\leftrightarrow$ L1$_0$ martensitic transformation. The oxidation experiments are designed to vary the diffusion rates so that the structural changes in the B2 phase, and early stages of the transformation to the martensite phase, are separately identified and linked to the local chemical content along the BC.

2. Materials and Methods

The samples used for the isothermal oxidation experiments were coupons of 25 mm diameter and 3 mm thickness, provided by GE-Aircraft Engines-Cincinnati, OH, USA. The coupons are composed of a low-activity B2-(Ni,Pt)Al bond coat on a RENE N5 superalloy. In this process, a Pt layer is electrolytically deposited over the substrate, whereas the Al is provided by a chemical vapor deposition process (CVD) and finally, the nickel required to form the B2 phase diffuses from the SA through vacuum heat treatments [2,23].

Two types of isothermal oxidation experiments were conducted in order to identify structural changes of the coatings related to the composition variations, regarding the possible effects of thermal and mechanical stresses as the alumina scale grows. Both experiments were conducted at 1100 °C in a single zone tube furnace (GSL-1100X, MTI Corporation, Richmond, CA, USA) in air (pO_2 ~ 0.21 atm), varying the growth conditions of the scale, as follows: In the first case, eight specimens were oxidized as a function of time for 0–18 h in 3 h intervals. This experiment stands for a regular diffusion-controlled oxidation process, where Al selectively oxidizes to form alumina (Al_2O_3) [24,25] at the surface. Since Al depletes in the bond coat, following the natural diffusion process controlled by the pO_2 present at the metal-scale interface, this first case has been called *Natural Oxidation* (hereafter NO) *experiments*. On the contrary, the second type of experiment is called *Scale-Free Oxidation* (hereafter SFO) *experiments* because it corresponds to isothermal oxidation keeping the same pO_2 (0.21 atm) with a minimum presence of Al_2O_3 scale at the surface. One sample was used throughout the SFO experiment and exposed to high temperature for 5 min intervals for a total of 110 min. The thin layer of alumina grown after every 5 min of exposure was removed mechanically with SiC paper, and the sample was exposed again to high temperature for 5 min. This process was repeated until 110 min of exposure of the samples was completed. The TGO scale was removed in 5 min intervals to increase, for instance, the Al depletion rate in order to address its effect on the crystal structure of the intermetallic.

For comparative purposes, the units of time for NO and SFO experiments are unified to hours in Figures 2, 4 and 5. All samples were cooled down from 1100 °C to room temperature (RT) with a constant cooling rate (100 °C/min). Characterization of samples was also all performed at RT. It is well known that metallographic polishing introduces mechanical stresses and distorts the B2 lattice shadowing the structural characterization of the coating. To avoid this effect, the crystal structure of the bond coat is first characterized after every exposure to high temperature, then the alumina scale is removed by metallographic polishing and characterized in terms of chemical content and microstructure.

The structural properties of the coatings are studied by X-ray diffraction (XRD) using a diffractometer (Dmax2100, Rigaku, Tokyo, Japan) with Cu-Kα radiation. All samples, irrespective of the oxidation process, are characterized before the oxidation treatments and right after every exposure to high temperature. The lattice parameter (a) of the B2 phase, as a function of oxidation time, is obtained using Bragg's law and applying Bradley and Jay's extrapolation [26]. Additionally, the long-range order parameter, S, is calculated using the following equation [27]:

$$S = \sqrt{\frac{\frac{I_{SL\_obs}}{I_{F\_obs}}}{\frac{I_{SL\_ref}}{I_{F\_ref}}}} \tag{1}$$

where I is the relative intensity of the peaks of the B2 phase. The subscripts correspond to the contributions of the superlattice (*SL*) and fundamental (*F*) peaks of the thermally treated (*obs*) and "reference" (*ref*) samples. The "as-received" sample for the NO and SFO experiments are used as the reference condition, since no crystallographic information JCPDS (Joint Committee on Powder Diffraction Standards) files are available for the B2-(Ni,Pt)Al phase. The determination of S is performed for those XRD patterns that allowed a reliable measurement of the superlattice/fundamental peaks ratio. This corresponds to the experiments from 0 to 1.25 h for SFO and from 0 to 12 h for NO oxidation experiments. After those times, superlattice peaks are not observed.

As mentioned before, after each XRD measurement, the samples were polished in order to remove the alumina scale (oxidation product), and the microstructure of the surface is studied by scanning electron microscopy (SEM) using a XL30 (Philips, Eindhoven, The Netherlands) with an incidence energy of 15 keV and a working distance of 10 mm. The microstructural details are revealed by etching the surface of the samples with an HCl/HNO$_3$ solution in a volume ratio of 5:1. The etching is carried out by dipping the sample 5 times for 5 s. This procedure was defined after several procedures in dummy samples to avoid excessive sample etching and/or pitting. The chemical content on the surface of the samples is measured by using Energy Dispersive X-ray spectroscopy (EDS) with the Philips, XL30, Eindhoven, The Netherlands and verified in some samples by electron probe microanalysis (JXA-8530F, JEOL, Tokio, Japan). The incidence energy was 20 keV and the working distance was 11 mm. Diffusing elements from the SA (Cr, Co, W, Ta, Mo and Re) were identified in the bond coat besides the Ni, Al and Pt, but only Ni, Al, Pt, Cr, and Co are considered in the analysis because they belong to the bond coat as solid solution substitutional atoms and also have high solubility in the B2 phase. Other elements such as W, Ta, Mo, and Re that diffuse from the superalloy are heterogeneously distributed either in the BC or in a so-called interdiffusion zone (IDZ); they are not considered in the scope of this work and further work has to be done to address their effect on the properties of B2-(Ni,Pt)Al coatings.

3. Results

Figure 1 shows the composition of Ni, Al, Pt, Cr and Co along the cross section of the as-received B2-(Ni,Pt)Al coatings used for the NO and SFO experiments, as a function of the distance from the surface covering the bond coat, the Interdiffusion Zone (IDZ) and the last 40 μm belongs to the superalloy. The major constituent elements in the BC (Al, Ni and Pt) varied almost linearly from 51.5, 37.5 and 9.0 at. % to 34.2, 44.2 and 3.7 at. %, respectively. A slight increase in Pt is observed on the first 13 μm of the BC and later decreased to 0 at the IDZ/SA interface. Co and Cr are also measured in the BC and vary from 1 and 0.3 at. % to 7 and 7 at. %, respectively. The IDZ showed heterogeneous concentrations of Ni, Al, Pt, Co and Cr distributed in γ and γ′ phases, as well as μ precipitates [7,28]. Minor content elements, mainly allocated in the IDZ such as W, Re, Ta, Mo (not shown in Figure 1) are also present even in the BC but are not considered in the analysis. Finally, the chemical content of the SA is in agreement with that previously reported in the literature [29] and showed a constant behavior along the measured section.

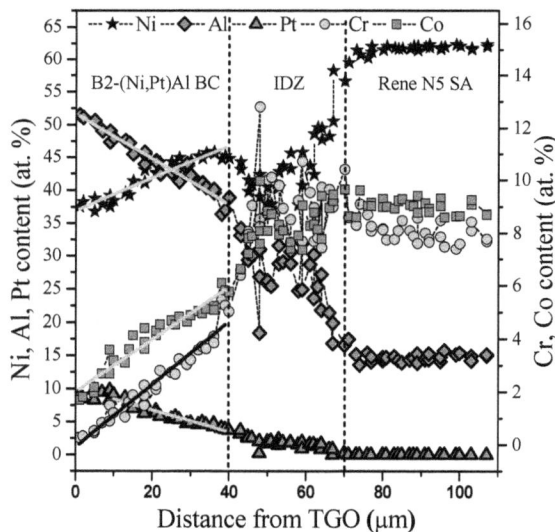

Figure 1. Concentration profiles of a cross section reference sample obtained from EPMA, starting from the BC surface, through the IDZ zone up to the SA.

Changes in the crystal structure as a function of oxidation time are obtained by XRD. The B2 crystal structure of the as-received coatings is unambiguously identified in Figure 2a,b for NO and SFO experiments, respectively, through the appearance of its fundamental and superlattice peaks.

The fundamental (110), (200) and (211), and the superlattice (100), (111) and (210) diffraction peaks of the B2-(Ni,Pt)Al phase are clearly identified in the as-received samples for both experiments. The fundamental peaks correspond to those diffraction planes that satisfy the selection rules for BCC lattices ($h + k + l = even$), and superlattice peaks correspond to the peaks that do not satisfy such rules. Since the martensite phase has a Face Centered Tetragonal crystal structure, the fundamental peaks correspond to the diffraction planes that satisfy the selection rules for FCC lattices (*unmixed $h + k + l$*), and superlattice for those that do not apply to these selection rules [27]. The Al_2Pt phase is also identified in some of the samples in the as-received conditions for the NO experiments and is represented with the symbol V. As expected, α-Al_2O_3 is also observed in all the samples exposed to NO experiments whereas it is weakly detected in SFO samples. A monotonic shifting to higher angles and a decrease in the superlattice peaks' intensity are evidence of the structural changes taking place during the NO and SFO experiments. The martensite $L1_0$-(Ni,Pt)Al phase is clearly identified after 6 h and 0.92 h for the NO and SFO samples respectively. The criterion to define the onset time of $L1_0$-phase formation is based on the appearance of the fundamental (111) peak at 43.3 degrees of 2θ. Superlattice peaks of the B2 phase are not observed after 15 h for NO and 1.83 h for SFO. However, small fundamental peaks indicate the presence of a partially disordered B2 phase. The gamma prime $L1_2$-(Ni,Pt)$_3$Al phase is also observed in the experiments after 18 h and 1.83 h for NO and SFO, respectively.

In the region of the mixture of phases, the martensite exhibited both superlattice and fundamental peaks during all the periods of high temperature exposure, while the austenite (B2 phase) partially decreased the superlattice peaks until they disappeared at 15 h and 1.83 h of NO and SFO experiments. It is important to remark that the superlattice peaks result after the contribution of the scattered X-rays by the atoms located in the α and β sublattices of the B2 phase, and therefore they depend on the difference in the Atomic Scattering factors (ASF) of the elements. Thus, the superlattice peaks allow to identify if the Ni and Al atoms occupy their corresponding sublattices in the B2-NiAl alloy or, if they are randomly arranged, promoting a disordered state of this intermetallic [27].

The decrease of the austenite superlattice peaks shows that this phase partially disorders as a function of the oxidation time and becomes completely disordered after 15 h and 1.25 h of NO and SFO, while the martensite phase remains ordered as the oxidation time increases.

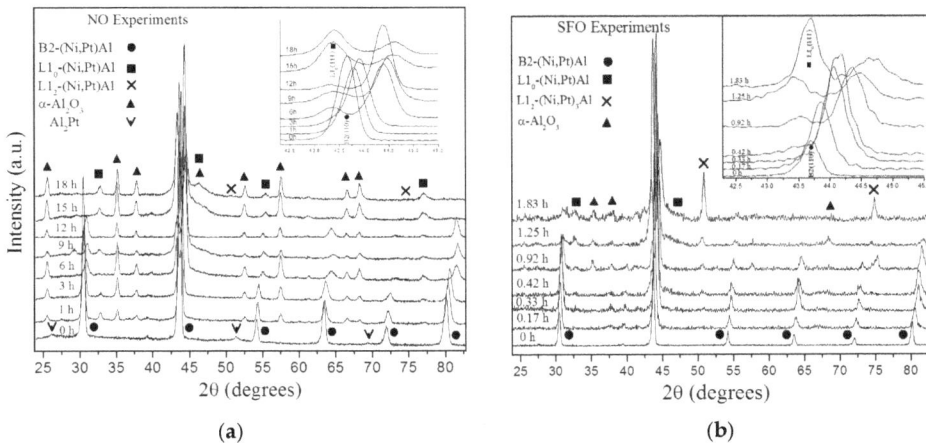

Figure 2. XRD patterns for samples processed by (**a**) natural oxidation for thermal exposure times of 0–18 h and (**b**) scale free oxidation for thermal exposure times of 0–110 min. The inset is a magnified view which better shows the evolution of the $L1_0$ (111) and B2 (110) peaks.

SEM micrographs of the microstructure of the NO samples in the as-received conditions, as well as those oxidized for 18 h, are presented in Figure 3a,b, respectively. The expected B2 phase (previously identified by XRD) shows equiaxed grains with a heterogeneous size in the range between 3 and 6 µm. For the 18 h - NO sample, B2-(Ni,Pt)Al, $L1_0$-(Ni,Pt)Al and $L1_2$-(Ni,Pt)$_3$Al phases are identified based on their morphology. Particularly, the $L1_0$-(Ni,Pt)Al phase is identified because the grains show the common habit planes of the martensite phase, which is the product of a diffusionless transformation of the B2 phase upon cooling [21]. This sample also shows a grain coarsening effect of both beta and the martensite grains (grain size between 8 and 12 µm).

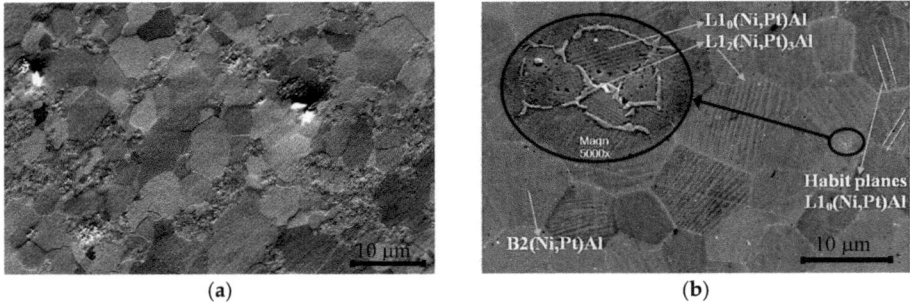

(a)   (b)

Figure 3. SEM micrographs of the bond coat surface (**a**) before and (**b**) after scale free oxidation for 110 min. B2-(Ni,Pt)Al, $L1_0$-(Ni,Pt)Al and $L1_2$-(Ni,Pt)$_3$Al phases are identified with arrows. The martensite phase is well identified in (**b**) by its habit planes.

The pores observed in the grains in Figure 3b are likely to be the result of either vacancy condensation (e.g., Kirkendall effect) and/or chemical etching attack during the metallographic sample preparation. Additionally, precipitates of γ' phase are observed in grain boundaries due to Al diffusion. This diffusion mechanism generates regions with a higher content of Ni and leads to the transformation to γ' (Ni,Pt)$_3$Al [23]. The identification of the γ' phase shows that the content of the coating exposed to 18 h is in the region of mixture of phases B2 + $L1_2$, which correlates well to XRD measurements.

Figure 4 shows variations in composition from the sample surfaces exposed to both oxidation experiments. Particularly, Figure 4a (Ni,Al) and 4b (Pt, Cr, Co) show the values corresponding to the NO experiments and Figure 4c (Ni,Al) and 4d (Pt, Cr, Co) correspond to the SFO experiments.

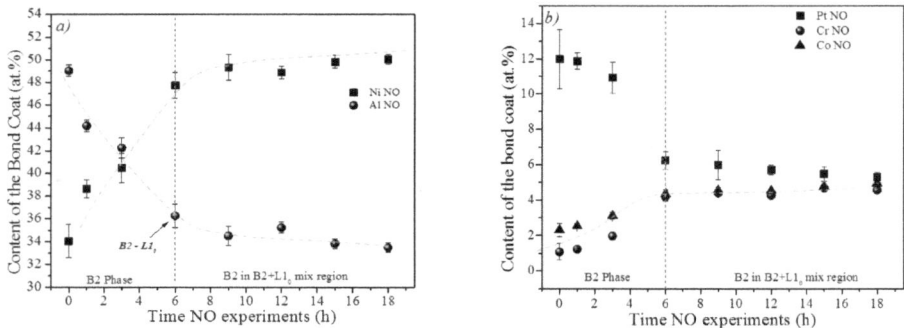

Figure 4. *Cont.*

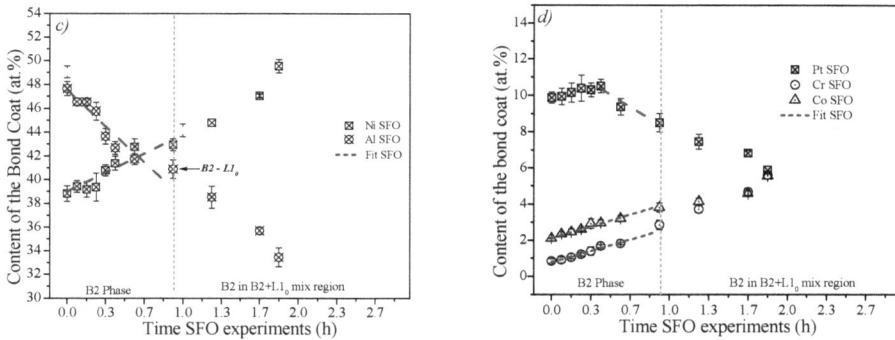

Figure 4. Variation in Al and Ni content as a function of (**a**) NO and (**c**) SFO heat treatment time. The martensitic transformation conditions are indicated at 36.9 at. % Al after 6 h of thermal exposure and at 40.9 at. % Al after 0.92 h of thermal exposure. Variation in Pt, Cr and Co content as a function of (**b**) NO and (**d**) SFO heat treatment time.

The Al content decreased as a function of heat treatment time from 49.0 to 33.5 at. % and from 47.6 to 33.4 at. %, whereas the Ni increased from 34.0 to 50.0 at. % and from 38.8 to 49.5 at. % for NO and SFO, respectively (Figure 4a,c). The Pt content decreased from 12.0 to 5.3 at. % and from 9.9 to 5.9 at. % for NO and SFO. In contrast, Co increased from 2.3 to 4.9 at. % and from 2.1 at. % to 5.6 at. % for NO and SFO, respectively. Similarly, Cr content increased from 1.0 to 4.6 at. % and from 0.8 to 5.6 at. %.

For comparative purposes, the X axes are matched from the as-received (0 h) up to the onset of martensitic transformation. These conditions correspond to 6 h and 0.92 h for NO and SFO, respectively. This criterion applies for composition and lattice constant and long-range ordering from Figures 4 and 5.

From Figure 4a,b, a quasi-parabolic composition profile as a function of the oxidation time is observed for the NO experiments, whereas an almost linear profile is observed for the SFO experiments. For NO experiments, two regions of variation that are defined by the onset of the martensitic transformation are observed. The approximate ratio of the depletion rates in these regions are 10%, 12%, 14%, 15% and 7% for Al, Ni, Pt, Cr and Co, respectively.

Thus, from 0 to 6 h, high depletion rates are observed for all the elements, keeping however the B2 composition; hereafter called B2 region, since it corresponds to the oxidation phenomena that occurs before the martensitic transformation. In the second region, the elements presented a lower depletion rate and are called "B2 + $L1_0$ region" because they correspond to the zone where beta and martensite phase coexist at RT.

On the other hand, the composition of samples exposed to SFO experiments followed only one tendency in the diffusion processes during all the experiments, even after the martensitic transformation (Figure 4d).

4. Discussion

Considering the three major elements (Al, Pt and Ni as a balance), the composition of the as-received coating surface corresponds to the stoichiometric B2-(Ni,Pt)Al phase of the Al-Ni-Pt phase diagram (see Figure 1). A composition gradient is observed from the surface through the coating thickness moving to the B2 + $L1_2$ region as it approaches the IDZ boundary [10]. The crystal structure from the surface of the as-received and heat-treated samples was verified by XRD (see Figure 2). As expected, during coating preparation both the B2-(Ni,Pt)Al and the underlying superalloy show interdiffusion processes leading to composition gradients with positive (Ni, Cr, Co) or negative (Al and Pt) trends [2,23].

The particular increase of Pt on the first 13 μm of the BC causes inversely proportional changes in the Ni content. This behavior is because Pt has a preference for the Ni sites in the B2 crystal structure since the bond strength of AlPt is stronger than that of AlNi. As a consequence, Pt decreases the chemical activity of Al [1,7].

It is worth noting that the oxidation time does not represent a reliable variable to correlate the structural and chemical properties in both experiments because it changes according to specific oxidation conditions. Instead, physical and/or thermodynamic considerations, as well as the phase transformations that occur in the coating, must be taken into account. To do so, and as mentioned before, the experiments are compared in Figures 4 and 5 from 0 h to the onset of the martensitic transformation.

The NO and SFO experiments represent different oxidation kinetics. Nevertheless, the boundaries for the B2 → L1$_0$ martensitic transformation in the NO experiments are 36.3 at. % Al, 6.2 at. % Pt, 4.2 at. % Cr, 4.4 at. % Co and 47.7 at. % Ni, and those for the SFO experiments are 40.9 at. % Al, 8.5 at. % Pt, 2.9 at. % Cr, 3.8 at. % Co and 42.9 at. % of Ni.

The increase of Pt during the first 0.42 h of SFO experiments (Figure 4d) seems to be associated with the fabrication process, since a similar trend is also observed in the as-coated sample, in which, as mentioned before, Pt increases during the first 13 μm of the BC (see Figure 1).

Another non-expected increment of Al is observed in the sample exposed to 12 h of NO due to the presence of the Al$_2$Pt phase. The characterization of all the specimens in the as-received conditions allowed the identification of the aforementioned phase in the samples that were later exposed to 0, 3 and 12 h. In this case, the Al$_2$Pt phase dissolves, which increases the Al content of the coating.

Figure 5a,b show the lattice constant *a* and long-range order parameter *S* of the B2-(Ni,Pt)Al phase obtained from the analysis of the XRD patterns for the NO and SFO experiments as a function of the oxidation time. The lattice constant of the as-received sample is 2.932 Å, which is in turn 1.6% higher than that expected for the stoichiometric binary NiAl alloy. This increase is due to Pt-additions (12 at. %) and is in agreement with results reported by Chen et al. [30]. In general, it is found that the lattice constant decreases as a function of the thermal exposure time and follows similar trends for NO and SFO experiments during the first stages of the oxidation.

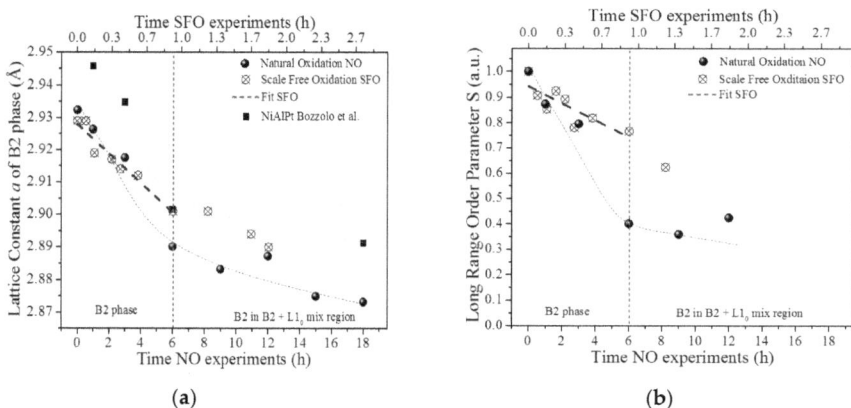

Figure 5. (a) Experimental and calculated lattice constant *a* of the B2 phase for NO and SFO experiments; (b) Long range ordering S of the B2 phase as a function of the exposure time for natural and scale free oxidation experiments.

The coupled behavior between composition and lattice constant has already been studied by Bozzolo [13] and Pike [31] for the binary NiAl systems, which follows the Vergard's law for substitutional solid solution alloys. The contribution of other elements to the lattice constant of the B2 phase depends on its size and specific location in every sublattice, according to the structural triple point defects generated in this alloy [6]. Based on the sublattice model for NiAl by I. Ansara [5] and on the site preference models for Pt, Cr and Co proposed by Jiang et al. [8,12], the B2-(Ni,Pt)Al BC can be represented by a two sublattice descriptions as B2-(Ni, Pt, Co)(Al, Cr, Ni). Given that Al depletes during the oxidation, the studied system will remain in the Ni-rich alloy region, where no vacancies are expected and Ni antisites are considered. Additionally, Pt, Cr and Co are taken as substitutional solid solution defects.

A detailed observation of the composition variations (Figure 4), a and S at the martensitic transformation boundary (6 h NO and 0.92 h SFO), shows a direct relation between the Al and Pt contents and these structural parameters of the B2-(Ni,Pt)Al phase. Unlike the other elements, the higher Al content the higher a and S is observed. A similar behavior is observed at 12 h of oxidation in the NO experiment, where the effect of Al content on the a and S is probably due to the dissolution of the Al_2Pt in the B2 phase at high temperature.

Del Grosso et al. [32] performed a theoretical calculation of the effects of Pt on the lattice constant of NiAl as a function of composition based on the Bozzolo-Ferrante-Smith method. For comparison, their results are plotted as squares in Figure 5a for 1, 3, 6 and 18 h. The compositions for these times are normalized to the NiAlPt system and correspond to the simulated values by Del Grosso et al. Although the theoretically calculated a values are higher than the experimental ones, there is a good agreement in their tendencies as a function of the oxidation time, which is proportional to the composition. Thus, changes in the crystal structure of B2-(Ni,Pt)Al coatings caused by the oxidation are prompt to be addressed through experimental and atomistic simulations approaches seeking a better understanding of their structural stability and phase transformations.

For instance, Sordelet et al. [33] studied the effects of Pt on the isothermal nature of the martensitic transformation of NiAl and found that Pt increases the start martensitic transformation temperature Ms. Complementary results are presented by Jiang et al., [34] based on Density Functional Theory calculations, and show that Pt additions decrease the shear C' and C_{44} moduli and thus destabilize the B2 phase. Additionally, the Al-Ni-Pt phase diagram reported by Hayashi et al. [10] proposes an extended region where the martensitic transformation may occur. Additions of 20 at. % Pt move the Al boundary for transformation up to approximately 40 at. % Al [33,35]. These observations are directly associated with the present results because the coatings exposed to SFO experiments contain higher Pt content and transformed at higher Al contents than the values required for martensitic transformation in the binary NiAl (about 37 at. %) alloy.

The behavior of S is found to be in accordance with the chemical content changes in the bond coat, following the trends of Al depletion in every region. Although S presents similar values for the B2 region, a significant difference is observed at the transformation boundary (6 h for NO and 0.92 h for SFO) which is explained in terms of the chemical content on each sample, and their role contributing to the scattering of the X-rays depending on the site occupancy of the sublattices in the B2 structure [12]. As can be seen in Figure 4c,d, the sample exposed to 0.92 h contained more Al and Pt than the sample exposed to 6 h (Figure 4a,b) and consequently less Ni. No significant differences are observed for Cr and Co between both types of samples. Remembering that the B2 phase can be described by the 2 sublattice model, where Pt and Co occupy the Ni sublattice and Cr occupies the Al sublattice, six binary systems (B2-XAl and B2-XCr, X: Ni, Pt and Co) can be compared to understand the role that the elemental composition of the alloy has on the superlattice peaks' intensity observed in the X-ray patterns. These peaks are a consequence of the difference in ASF in every sublattice and therefore define the long-range order parameter. Despite the simplicity of the binary models, these are useful to show the contributions of the elements to the superlattice peaks. For instance, the difference in ASF of the binary models at $2\theta = 30.9°$ that correspond to the superlattice (100) peak are: B2-PtAl $\rightarrow$ 44.86, B2-PtCr $\rightarrow$ 38.57, B2-NiAl $\rightarrow$ 8.97, B2-CoAl $\rightarrow$ 8.28, B2-NiCr $\rightarrow$ 2.68 and B2-CoCr $\rightarrow$ 1.99. It is clearly observed that the larger differences correspond to the Pt containing alloys due to the higher electronic density of Pt. This analysis is in good agreement with the content of Al and Pt at the transformation boundary.

Thus, composition, lattice parameter, as well as for long-range ordering, all follow similar behaviors, showing two regions of variation before and after the martensitic transformation for NO experiments, and an almost linear decrease for SFO experiments.

Given that the oxidation time is not a physically meaningful independent variable, nor does it allow direct comparison of natural and scale free oxidation results, the lattice constant and long-range ordering parameter are plotted in Figure 6 versus the corresponding variable, as a function of the Al content for NO and SFO experiments, since it is the most representative contributor to the crystal structure properties and ordering. The oxidation process is shown by the expressions "Oxidation starts" for 0 h and "Oxidation ends" for 18 h and 1.83 h of NO and SFO, respectively. Data from the literature for the binary NiAl system is also represented in Figure 6 for comparative purposes. The values reported by Bozzolo et al. [36] correspond to first principles calculations performed in B2-NiAl and applying Ni antisite defects for compositions out of the stoichiometry, in order to maintain the B2 crystal structure.

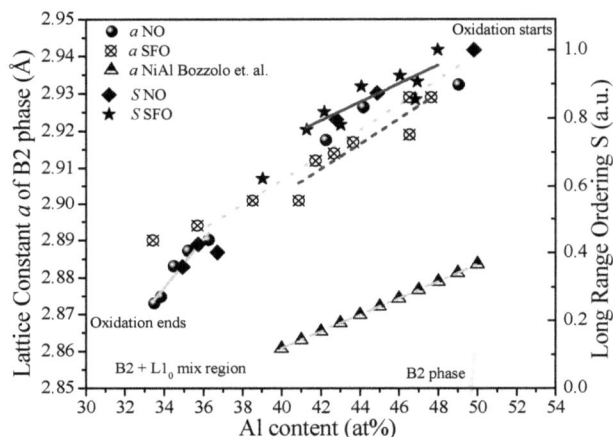

Figure 6. Variation of lattice constant as a function of Al content for NO and SFO experiments. Results of the analysis from a Binary (Bozzolo) and commercial systems are included.

The lattice constant of the B2-(Ni,Pt)Al phase shows a linear dependence with the Al content during the complete isothermal oxidation process, irrespective of the nature of the oxidation (NO or SFO). This behavior corresponds to substitutional solid solution alloys, where the elements are replaced by others following a model of hard spheres and where Vegard's law can be applied. Based on the slopes of the linear trends, we can see that the dependence of the lattice constant with Al content in the commercial coatings showed good agreement with the behavior of the simulated NiAl binary system. The dispersion of *a* values and slight differences in the slopes from 40 to 50 at. % Al are attributed to the change of Pt, Cr and Co in the Ni and Al sublattices from the as-received conditions (0 h) to the transformation boundaries (6 h and 0.92 h). The difference in the lattice constant between the simulations and experiments is mainly due to the effect of Pt substitutions in the Ni site. These results are in agreement with the work reported by Chen et al. [30]. The effect of Cr and Co on the lattice constant are not expected to be determinant on the size of the unit cell because their atomic radiuses are similar to that of Ni, and according to the triple point defect model and site preferences, they do not affect this parameter.

The linear dependence of the lattice constant as a function of Al content in the region of mixed B2 + L1$_0$ phases showed a higher slope than the one observed in the single B2 phase region. This behavior deserves further studies.

These findings show the importance of the Al content in the crystal structure of commercial B2-(Ni,Pt)Al bond coatings and the effect of the ternary additions on the size of the B2 unit cell (bigger intercept than for NiAl).

The long-range ordering S for NO and SFO experiments plotted in Figure 6 also exhibits a linear dependence with the Al content, independently of the nature of the oxidation. These results show how the B2 phase disorders as the Al is depleted during the oxidation until its transformation to the martensitic phase. The behavior of S is also affected by the presence of Pt in the alloy due to their difference in ASF with Al, and this affects the ordering and consequently the stability of the crystal structure, as shown before.

5. Conclusions

Two different isothermal oxidation experiments conducted at 1100 °C and pO_2 = 0.2 atm, on B2-(Ni,Pt)Al/Rene N5 systems were performed to study the effect of chemical content on the structural stability and the martensitic phase transformation of B2-(Ni,Pt)Al bond coats. The main conclusions are presented below:

- Natural (NO) and Scale Free (SFO) Oxidation experiments allowed us to study the stability and martensitic transformation of the B2-(Ni,Pt)Al phase as a function of the composition.
- The B2 phase destabilized and transformed to martensite for NO and SFO experiments at 36.3 and 40.9 at. % of Al, 47.7 and 42.9 at. % of Ni, 6.2 and 8.5 at. % of Pt, 4.2 and 2.9 at. % of Cr and 4.4 and 3.8 at. % of Co, respectively.
- The lattice constant and long range ordering parameter of the B2-(Ni,Pt)Al phase have a linear dependence on the Al content of the coating and the slopes are similar to the binary NiAl system.
- Diffusing Cr and Co elements from the SA were identified in the bond coat and did not cause significant effects on the size and ordering of the coating.
- Al and Pt were found to play a determinant role in the lattice constant and S of the B2 phase.

Acknowledgments: This project was funded by Conacyt FOMIX-QRO Project No. 10599 and Conacyt CIENCIA BASICA-2009 No. 130591 Projects. Authors thank D.G. Konitzer (General Electric Aircraft Engines, Cincinnati, OH, USA), J.L. Ortiz Merino and Mike Boldrick (Peace Corps) for his support in the helpful discussion and revision of the manuscript. RA recognizes partial support of NETL through grant No. FE0008719 and JMS of the Av Humboldt foundation for the HERMES fellowship. This research has been carried out partially at CENAPROT and LIDTRA national laboratories.

Author Contributions: L.A.C.-D., H.R.-L., J.M.A.-O. and J.M.-S. conceived and designed the experiments; L.A.C.-D., J.E.G.-H., A.G.M.-G. and J.M.A.-O. performed the experiments; L.A.C.-D., J.E.G.-H., H.R.-L., J.M.A.-O. and J.M.-S. analyzed the data; G.T.-M., R.A., J.M.A.-O. and J.M.-S. contributed reagents/materials/analysis tools; J.M.-S., L.A.C.-D., H.R.-L. and J.M.A.-O. wrote the paper.

Conflicts of Interest: The authors declare no conflict of interest.

References

1. Gleeson, B.; Wang, W.; Hayashi, S.; Sordelet, D.J. Effects of Platinum on the Interdiffusion and Oxidation Behavior of Ni-Al-Based Alloys. *Mater. Sci. Forum* **2004**, *461–464*, 213–222. [CrossRef]
2. Zhang, Y.; Lee, W.Y.; Haynes, J.A.; Wright, I.G.; Cooley, K.M.; Liaw, P.K. Synthesis and cyclic oxidation behavior of a (Ni,Pt)Al coating on a desulfurized Ni-base superalloy. *Metall. Mater. Trans. A* **1999**, *30*, 2679–2687. [CrossRef]
3. Clarke, D.R.; Levi, C.G. Materials design for the next generation thermal barrier coatings. *Annu. Rev. Mater. Res.* **2003**, *33*, 383–417. [CrossRef]
4. Evans, A.G.; Clarke, D.R.; Levi, C.G. The influence of oxides on the performance of advanced gas turbines. *J. Eur. Ceram. Soc.* **2008**, *28*, 1405–1419. [CrossRef]

5. Ansara, I.; Dupin, N.; Lukas, H.L. Thermodynamic Assessment of the Al-Ni System. *J. Alloys Compd.* **1997**, *247*, 20–30. [CrossRef]

6. Pike, L.M. Point Defect Concentrations and Hardening in Binary B2 Intermetallics. *Acta Mater.* **1997**, *45*, 3709–3719. [CrossRef]

7. Zhang, Y.; Haynes, J.A.; Lee, W.Y.; Wright, I.G.; Pint, B.A.; Cooley, K.M.; Liaw, P.K. Effects of Pt Incorporation on the Isothermal Oxidation Behavior of Chemical Vapor Deposition Aluminide Coatings. *Metall. Mater. Trans. A* **2001**, *32A*, 1727–1741. [CrossRef]

8. Jiang, C.; Besser, M.F.; Sordelet, D.J.; Gleeson, B. A combined first-principles and experimental study of the lattice site preference of Pt in B2 NiAl. *Acta Mater.* **2005**, *53*, 2101–2109. [CrossRef]

9. Bozzolo, G.; Noebe, R.D.; Honecy, F. Modeling of ternary element site substitution in NiAl. *Intermetallics* **2000**, *8*, 7–18. [CrossRef]

10. Hayashi, S.; Ford, S.I.; Young, D.J.; Sordelet, D.J.; Besser, M.F.; Gleeson, B. α-NiPt(Al) and phase equilibria in the Ni-Al-Pt system at 1150 °C. *Acta Mater.* **2005**, *53*, 3319–3328. [CrossRef]

11. Miracle, D.B. The Physical and Mechanical Properties of NiAl. *Acta Metall. Mater.* **1993**, *41*, 649–684. [CrossRef]

12. Jiang, C. Site preference of transition-metal elements in B2 NiAl: A comprehensive study. *Acta Mater.* **2007**, *55*, 4799–4806. [CrossRef]

13. Wang, T.; Zhu, J.; Chen, L.; Liu, Z.; Mackay, R.A. Modeling of lattice parameter in the Ni-Al system. *Metall. Mater. Trans. A* **2004**, *35*, 2313–2321. [CrossRef]

14. Meyer, R.; Entel, P. Computer simulations of martensitic transformations in NiAl alloys. *Comput. Mater. Sci.* **1998**, *10*, 10–15. [CrossRef]

15. Arroyave, R.; Shin, D.; Liu, Z.K. Ab initio thermodynamic properties of stoichiometric phases in the Ni-Al system. *Acta Mater.* **2005**, *53*, 1809–1819. [CrossRef]

16. Frank, S.; Divinski, S.V.; Södervall, U.; Herzig, C. Ni tracer diffusion in the B2-compound NiAl: Influence of temperature and composition. *Acta Mater.* **2001**, *49*, 1399–1411. [CrossRef]

17. Spitsberg, I.T.; Mumm, D.R.; Evans, A.G. On the failure mechanisms of thermal barrier coatings with diffusion aluminide bond coatings. *Mater. Sci. Eng. A* **2005**, *394*, 176–191. [CrossRef]

18. Tolpygo, V.K.; Clarke, D.R. On the rumpling mechanism in nickel-aluminide coatings part I: An experimental assessment. *Acta Mater.* **2004**, *52*, 5115–5127. [CrossRef]

19. Chen, M.W.; Livi, K.J.T.; Hemker, K.J.; Wright, P.K. Microstructural characterization of a platinum-modified diffusion aluminide bond coat for thermal barrier coatings. *Metall. Mater. Trans. A* **2003**, *34A*, 2289–2299. [CrossRef]

20. Chen, M.W.; Glynn, M.L.; Ott, R.T.; Hufnagel, T.C.; Hemker, K.J. Characterization and modeling of a martensitic transformation in a platinum modified diffusion aluminide bond coat for thermal barrier coatings. *Acta Mater.* **2003**, *51*, 4279–4294. [CrossRef]

21. Glynn, M.L.; Chen, M.W.; Ramesh, K.T.; Hemker, K.J. The Influence of a Martensitic Phase Transformation on Stress Development in Thermal Barrier Coating Systems. *Metall. Mater. Trans. A* **2004**, *35A*, 2279–2286. [CrossRef]

22. Kainuma, R.; Ohtani, H.; Ishida, K. Effect of Alloying Elements on Martensitic Transformation in the Binary NiAl (β) Phase Alloys. *Metall. Mater. Trans. A* **1996**, *27*, 2445–2453. [CrossRef]

23. Yu, Z.; Hass, D.D.; Wadley, H.N.G. NiAl bond coats made by a directed vapor deposition approach. *Mater. Sci. Eng. A* **2005**, *394*, 43–52. [CrossRef]

24. Grabke, H.J. Oxidation of NiAl and FeAl. *Intermetallics* **1999**, *7*, 1153–1158. [CrossRef]

25. Brumm, M.W.; Grabke, H.J. The oxidation behaviour of NiAl-I. Phase transformations in the alumina scale during oxidation of NiAl and NiAl-Cr alloys. *Corros. Sci.* **1992**, *33*, 1677–1690. [CrossRef]

26. Bradley, A.J.; Jay, A.H. A method for deducing accurate values of the lattice spacing from X-ray powder photographs taken by the Debye-Scherrer method. *Phys. Soc.* **1932**, *44*, 563–579. [CrossRef]

27. Cullity, B.D. *Elements of X-ray Diffraction*, 1st ed.; Addison-Wesley: Boston, MA, USA, 1956.

28. Zhang, L.C.; Heuer, A.H. Microstructural Evolution of the Nickel Platinum-Aluminide Bond Coat on Electron-Beam Physical-Vapor Deposition Thermal-Barrier Coatings during High-Temperature Service. *Metall. Mater. Trans. A* **2005**, *36*, 43–53. [CrossRef]

29. Reed, R.C. *The Superalloys Fundamentals and Applications*, 1st ed.; Cambridge University Press: Cambridge, UK, 2006.

30.   Chen, M.W.; Ott, R.T.; Hufnagel, T.C.; Wright, P.K.; Hemker, K.J. Microstructural evolution of platinum modified nickel aluminide bond coat during thermal cycling. *Surf. Coat. Technol.* **2003**, *163–164*, 25–30. [CrossRef]

31.   Pike, L.M.; Liu, C.T.; Anderson, I.M.; Chang, Y.A. Solute Hardening and Softening Effects in B2 Nickel Aluminides, 1998. Available online: http://www.osti.gov/scitech/biblio/676873 (accessed on 23 August 2016).

32.   Del Grosso, M.F.; Mosca, H.O.; Bozzolo, G. Atomistic modeling of Pt additions to NiAl. *Intermetallics* **2008**, *16*, 1305–1309. [CrossRef]

33.   Sordelet, D.J.; Besser, M.F.; Ott, R.T.; Zimmerman, B.J.; Porter, W.D.; Gleeson, B. Isothermal nature of martensite formation in Pt-modified β-NiAl alloys. *Acta Mater.* **2007**, *55*, 2433–2441. [CrossRef]

34.   Jiang, C.; Sordelet, D.J.; Gleeson, B. Effects of Pt on the elastic properties of B2 NiAl: A combined first-principles and experimental study. *Acta Mater.* **2006**, *54*, 2361–2369. [CrossRef]

35.   Zhang, Y.; Haynes, J.A.; Pint, B.A.; Wright, I.G.; Lee, W.Y. Martensitic transformation in CVD NiAl and (Ni,Pt)Al bond coatings. *Surf. Coat. Technol.* **2003**, *163–164*, 19–24. [CrossRef]

36.   Bozzolo, G.; Amador, C.; Ferrante, J.; Noebe, R.D. Modelling of the defect structure of β-NiAl. *Scr. Metall. Mater.* **1995**, *33*, 1907–1913. [CrossRef]

Article

Utilization of a Porous Cu Interlayer for the Enhancement of Pb-Free Sn-3.0Ag-0.5Cu Solder Joint

Nashrah Hani Jamadon [1], Ai Wen Tan [1], Farazila Yusof [1,*], Tadashi Ariga [2], Yukio Miyashita [3] and Mohd Hamdi [1]

[1] Centre of Advanced Manufacturing and Material Processing (AMMP), Department of Mechanical Engineering, University of Malaya, Kuala Lumpur 50603, Malaysia; nashrahhani@gmail.com (N.H.J.); aiwen_2101@hotmail.com (A.W.T.); hamdi@um.edu.my (M.H.)
[2] Department of Metallurgical Engineering, Tokai University, Hiratsuka 259-1292, Japan; ttariga@keyaki.cc.u-tokai.ac.jp
[3] Department of Mechanical Engineering, Nagaoka University of Technology, Nagaoka 940-2188, Japan; miyayuki@mech.nagaokaut.ac.jp
* Correspondence: farazila@um.edu.my; Tel.: +603-796777633; Fax: +603-79675317

Academic Editor: Ana Sofia Ramos
Received: 29 June 2016; Accepted: 5 September 2016; Published: 15 September 2016

Abstract: The joining of lead-free Sn-3.0Ag-0.5Cu (SAC305) solder alloy to metal substrate with the addition of a porous Cu interlayer was investigated. Two types of porous Cu interlayers, namely 15 ppi—pore per inch (P15) and 25 ppi (P25) were sandwiched in between SAC305/Cu substrate. The soldering process was carried out at soldering time of 60, 180, and 300 s at three temperature levels of 267, 287, and 307 °C. The joint strength was evaluated by tensile testing. The highest strength for solder joints with addition of P25 and P15 porous Cu was 51 MPa (at 180 s and 307 °C) and 54 MPa (at 300 s and 307 °C), respectively. The fractography of the solder joint was analyzed by optical microscope (OM) and scanning electron microscopy (SEM). The results showed that the propagation of fracture during tensile tests for solder with a porous Cu interlayer occurred in three regions: (i) SAC305/Cu interface; (ii) inside SAC305 solder alloy; and (iii) inside porous Cu. Energy dispersive X-ray spectroscopy (EDX) was used to identify intermetallic phases. Cu_6Sn_5 phase with scallop-liked morphology was observed at the interface of the SAC305/Cu substrate. In contrast, the scallop-liked intermetallic phase together with more uniform but a less defined scallop-liked phase was observed at the interface of porous Cu and solder alloy.

Keywords: porous Cu interlayer; Sn-3.0Ag-0.5Cu solder alloy; joint strength; fracture morphology

1. Introduction

At present, solder is widely used in the semiconductor packaging industry with the purpose of providing electronic connection. It also plays an important role in the joining of metal components in the transportation, aerospace and energy industries [1–3]. Driven by miniaturization and increased requirements for advanced electronic devices, it is crucial to develop a solder alloy that can remain stable under extreme environments and perform well under high temperature operating systems [4]. Thus, the characteristic performance of the solder alloys is critical since a good quality solder joint could ultimately determine the effective function and reliability of a semiconductor device.

Due to health and environmental concerns, the use of lead-free (Pb-free) solders has been proposed to substitute the lead containing solders in the electronic packaging process [5,6]. To date, Sn-3.0Ag-0.5Cu (SAC305) solder alloy has been proposed as a promising alternative to replace the lead containing solder alloys in electronic applications due to its favourable mechanical properties including superior resistance to creep and thermal fatigue [3,7,8]. This solder is generally applied in commercial microelectronic devices where the operating temperature usually does not exceed 260 °C. However, due to the higher melting temperature of SAC305 solder alloy (217 °C) compared to the conventional Sn-Pb solder alloy (183 °C), SAC305 solder alloy is thus more viable for middle-temperature-ranged applications [9].

Several studies using SAC305 solder alloy in the electronics assemblies operated at higher temperature have been successfully reported. In a Chellvarajoo et al. study, they reengineered the composite SAC305 solder alloy by adding nanoparticles to enhance the reliability of the solder joint. They found the intermetallic compound (IMC) growth stunted by increasing the percentage of nanoparticles of iron nickel oxide (Fe_2NiO_4) or nickel oxide (NiO) in the SAC305 solder matrix [10,11]. In addition, attempts have also been made in improving the solderability of SAC305 solder alloy in high temperature electronic applications by altering the joining methods through the utilization of an interlayer material in the soldering process, such as during the transient liquid phase (TLP) bonding process. TLP bonding is an established joining technique for die attach bonding in the application of high temperature electronic systems [2]. Generally, a thin layer of low melting point metal is utilized as the interlayer material for an effective bonding during the TLP process. Liu et al. study reported enhanced joint strength and restrained formation of the IMC layer was observed through the use of a composite preform consisting of an Ag based metal core layer with Sn-Ag based solder layers at both sides during the TLP bonding process [12].

More recently, it has been reported that lead-free solder alloys containing additional metal elements have decreased the melting temperature and improved the tensile strength of the solder joints. This was successfully accomplished by adding Cu nanoparticles into Sn-Ag solder alloys [13]. The examination of interfacial interaction during the soldering process showed that Cu particles were dissolved in the solder alloys and this influenced the chemical structure of the Sn-Ag solders, leading to lower melting temperatures and an increase in the solder joint strength. As a result of this promising finding, few studies have been reported to adopt similar techniques for various lead-free solders.

Furthermore, some achievement with enhanced joint quality was also found in the studies through the utilization of porous metal interlayer during the joining process in the field of metal-joining application. Recently, porous metal interlayers have been widely used in various engineering designs such as cooling electronic systems and in aerospace facilities due to its porosity, thermal conductivity and usage efficiency [14–16]. The pores in the porous metal interlayer is reported to assist during the residual stress absorption, which endows this unique feature a remarkable advantage for metal joining applications [17].

For instance, Zaharinie et al. investigated the addition of a porous Cu/Ni interlayer sandwiched in between the sapphire and inconel that were joined using vacuum brazing. The results showed that the Cu/Ni porous composite in the sapphire-inconel joint altered the thermodynamic activity near the ceramic portion of the sapphire, leading to the ductile IMC formation which further improved the joining characteristics [18]. Previously, we have worked on the utilization of porous Cu interlayer in the soldering process where the IMC growth and formation was affected by the porosity of the porous Cu [19]. However, the results are preliminary and warrant further in-depth studies. Since this technique is still in its infancy in the field of soldering technique for electronic assembly application, in this paper, we aim to investigate the influence of porous Cu on its joint strengths, fracture morphologies and the IMC formation. Therefore, modifying the physical soldering process of SAC305 solder alloy through the incorporation of a porous Cu interlayer in the soldering configuration was attempted.

In this paper, the effects of a porous Cu interlayer on the mechanical properties and microstructure of SAC305 solder joints were investigated. We hypothesized the porous Cu interlayer to function as a metal foam allowing the molten solder alloy to penetrate into its internal porous structure to achieve a better joint quality [20].

2. Experimental Procedure

2.1. Material Selection

The SAC305 solder paste alloy material used was provided by Nihon Handa Co., Ltd. (Tokyo, Japan). The chemical composition and general properties of the solder material are given in Tables 1 and 2 [21], respectively. Two types of open cells foam, namely porous Cu interlayer of 15 pore per inch (ppi) and 25 ppi, signified as P15 and P25, were used as the solder joint reinforcement. The interlayer porosity was determined using the mass displacement equation with water as the immersing medium as shown in Figure 1. The percentage of porosity was calculated by the following equation from Archimedes' principle:

$$P(\%) = \frac{W_{as} - W_d}{W_{as} - W_s} \tag{1}$$

where P is the percentage of porosity, W_{as} the weight of porous Cu after submerged in water, W_d represents the weight in dry condition, and W_s the weight of porous Cu while submerged in water. Note that the volume of water displaced is equal to the volume of the porous Cu being submerged. The solid part in the porous Cu is called solid cell walls (SCW) [22].

Table 1. Chemical composition of solder alloy (wt. %).

Solder Alloy	Cu	Ag	Bi	Fe	As	Ni	Pb	Sb	Sn
Sn-3.0Ag-0.5Cu	0.516	3.083	0.011	0.002	0.005	0.001	0.022	0.015	Bal.

Table 2. Mechanical properties of SAC305 [21].

Solder Alloy	Melting Temperature, °C	Tensile Strength, MPa	Young Modulus, MPa	Hardness, HV
Sn-3.0Ag-0.5Cu	217	50.6	54	13.3

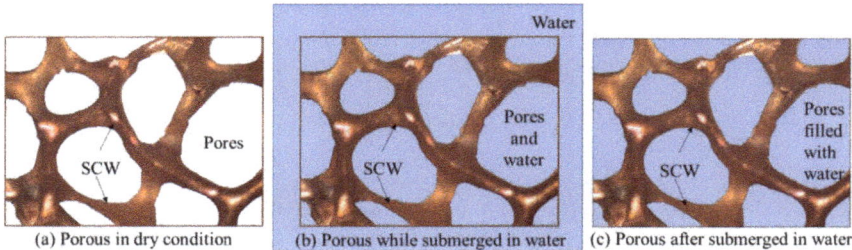

Figure 1. Archimedes' technique to calculate porosity of porous Cu interlayer. (**a**) Porous in dry condition; (**b**) porous while submerged in water; (**c**) porous after submerged in water.

2.2. Joining Procedure and Soldering Process Set Up

Two Cu cylindrical rods with dimensions of 30 mm in length and 8 mm in diameter were used as the substrates. They were soldered together using 0.1 g of SAC305 solder alloy with porous Cu in between the solder alloy, as configured in Figure 2. Prior to the soldering process, the porous Cu interlayers (P15 and P25) were rolled manually with a solid cylinder to obtain a uniform layer with 100 μm in thickness. The solder joint without porous Cu was prepared as a control specimen. The prepared specimen was clamped using a fabricated jig to hold the sample during the soldering process in the tube furnace, as shown in Figure 3. The tube furnace was set up in accordance with the Japanese Industrial Standard, JIS Z 3191. Briefly as shown in Figure 4, the specimen was heated inside a movable tube furnace under Argon gas atmosphere at a predefined temperature and time. The soldering process was conducted at temperatures of 267, 287, and 307 °C, with three different holding times of 60, 180, and 300 s as listed in Table 3.

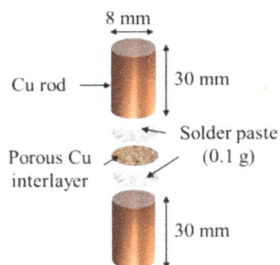

Figure 2. Configuration of a solder joint.

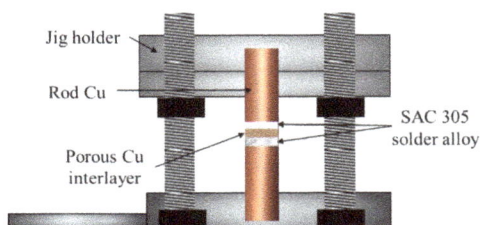

Figure 3. Assembled solder joint at fabricated jig.

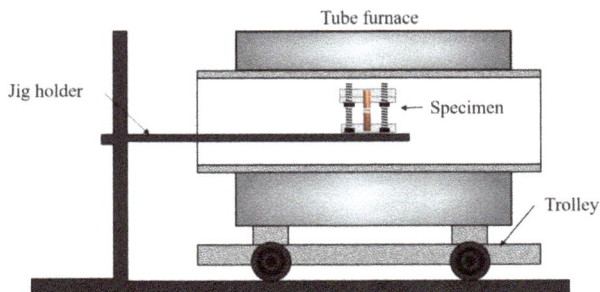

Figure 4. Schematic view of furnace set up as in Japanese Industrial Standard, JISZ 3191.

Table 3. Parameters setting for mechanical testing.

Parameters	Setting Value
Soldering temperature, °C	267, 287, 307
Soldering time, s	60, 180, 300
Porous Cu	0, P15, P25

2.3. Joint Strength Analysis

The strength of the solder joint was measured by tensile testing. The tensile test was carried out at room temperature using Instron® Corporation Universal Testing machine (Model No. 3369, Norwood, MA, USA), with crosshead speed of 0.5 mm/min. The specimen was initially machined at the testing section to avoid the effect of difference in edge shape caused by over flowing of molten solder during the soldering process. The results of this tensile strength were determined by the average value of three different samples for each condition.

2.4. Microstructural Analysis

After the soldering process, the microstructural observations of the fractured surface were observed under the optical microscope (OM, Olympus, Tokyo, Japan). The cross sections of the soldered joints were also prepared using standard metallorgraphical procedures (sectioning, grinding and polishing) for microstructural analysis. The elemental compositions of the IMC layer were analyzed using scanning electron micrograph (SEM, Crest System (M) Sdn. Bhd., Eindhoven, The Netherlands) equipped with energy dispersive X-ray spectroscopy (EDX, Crest System (M) Sdn. Bhd., Eindhoven, The Netherlands). The fractography (surface and cross section) of the solder joint after the tensile test was also examined by SEM and EDX to study their failure behaviour.

3. Results and Discussion

3.1. Verification of Porosity Percentage

Porous Cu interlayer is characterized by the size of the pore diameter which associates with the porosity [23]. The percentages of the porosity are presented in Figure 5, both experimentally and theoretically. The experimental value was measured by Archimedes' technique, while the theoretical value was calculated by the pore density formula. Figure 6 shows the surface structure of P15 and P25 porous Cu in its pre-rolled and post-rolled conditions. The average thickness of the porous Cu solid cell walls for P15 and P25 porous Cu was 0.23 mm and 0.06 mm, while the pore diameters for P15 and P25 after rolling were 0.3 mm and 0.1 mm, respectively. This showed that the pores were still distinct in shape even after being rolled into a very thin layer. The reason for rolling the porous Cu into a very thin layer was to reduce the joint gap and thereby promote a homogeneous joint layer. In fact, thick joints tend to reduce the joint strength [24].

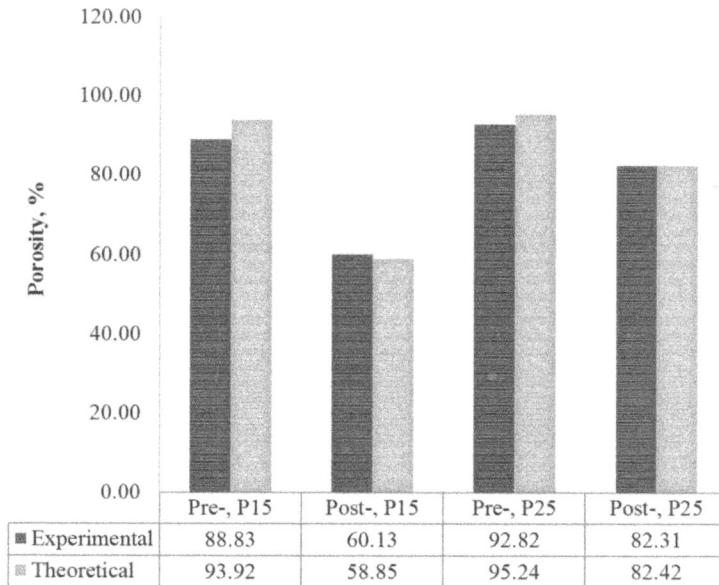

	Pre-, P15	Post-, P15	Pre-, P25	Post-, P25
Experimental	88.83	60.13	92.82	82.31
Theoretical	93.92	58.85	95.24	82.42

Figure 5. Experimental and theoretical measurement of porosity percentage of porous Cu interlayer.

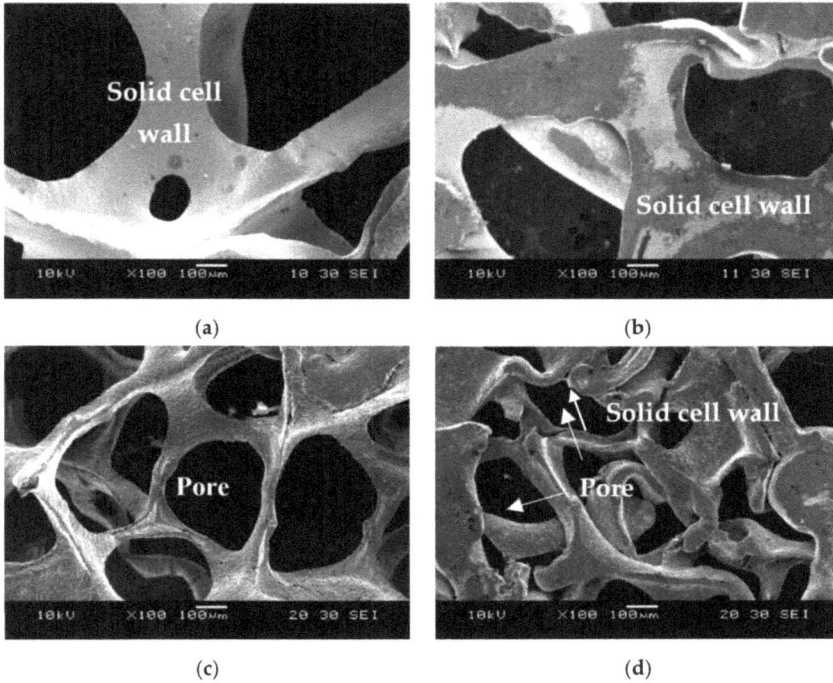

Figure 6. Structure of porous Cu interlayer of P15 and P25 before and after rolling. (**a**) P15, pre-rolled; (**b**) P15, post-rolled; (**c**) P25, pre-rolled; (**d**) P25, post-rolled.

3.2. Effect of Porous Cu Interlayer Addition on Joint Strength

Figure 7 shows the effect of soldering time on joint strength at three different temperatures of solder joint, with and without the two types of porous Cu interlayer. For the control sample without porous Cu, it was observed that the effect of soldering time and temperature on joint strength was minimal as similar joint strength values were recorded for all solder joints, ~40 MPa. In contrast, with the porous Cu addition samples, increasing the soldering time and temperature generally increased the joint strength of the solder joint. In addition, the joint strength of P25 porous Cu solder samples seemed higher compared to the samples with P15 porous Cu at each applied soldering time and temperature; with the exception for P15 porous Cu at soldering time of 60 s and 300 s and temperature of 307 °C. This is possibly due to the effect of differential reaction of porous Cu interface with molten solder in order to create bonding. Only a slight drop in the joint was observed for the P25 porous Cu sample when temperature increased from 287 °C to 307 °C.

The highest strength for the P25 porous Cu addition sample was recorded at 51 MPa with soldering time of 180 s and temperature of 307 °C, whereas, for the sample with P15 porous Cu addition, 54 MPa marked the highest observed strength at soldering time of 300 s and temperature of 307 °C. Both values are acceptable for the optimum joint strength of solder joint for the usage of electronics devices [25]. These values are, in fact, comparable to the joint strength of die attach using composite preform layer during joining as reported in Liu et al. study [12].

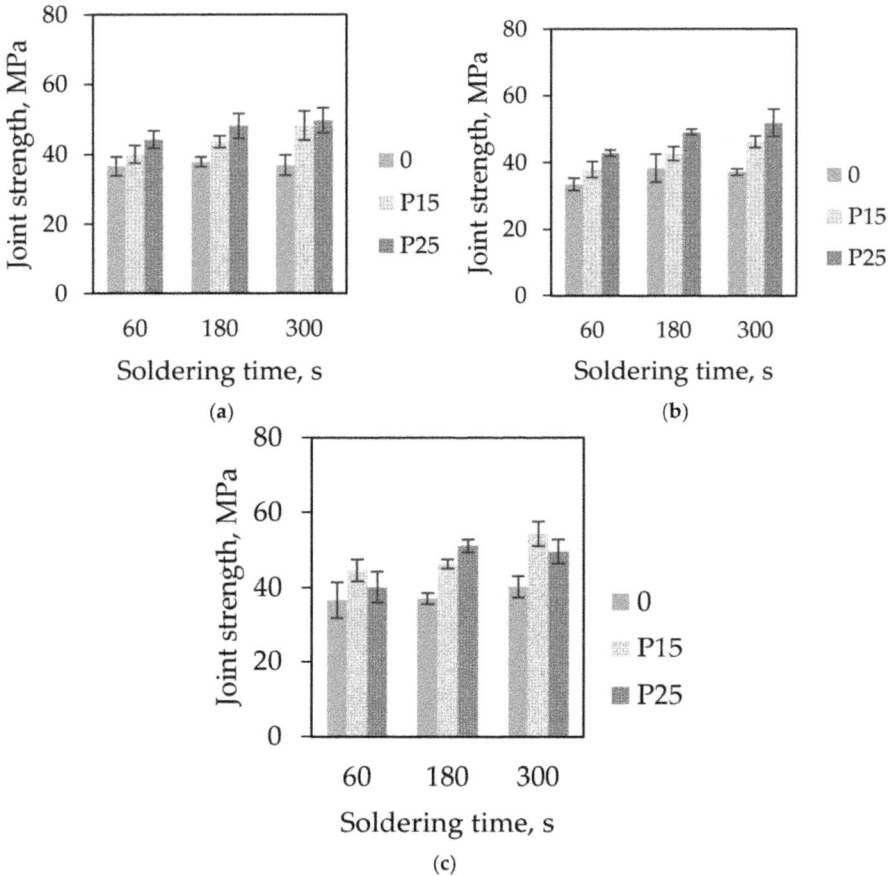

Figure 7. Effect of soldering time and porosity on joint strength at soldering temperature of (**a**) 267 °C; (**b**) 287 °C; and (**c**) 307 °C.

3.3. Effect of Porous Cu Interlayer Addition on Fracture Behaviour

3.3.1. Fractured Surface

The specimens at soldering temperature of 307 °C and 300 s were selected for the microstructure analysis due to the promising results obtained in the joint strength evaluation. The fractured surfaces of both samples, (control and with porous Cu interlayer) are illustrated in Figure 8a–c, while high magnification images of the selected area are shown in Figure 8d–f. The fracture of the solder joint without porous Cu interlayer occurred in between the Cu substrate and the solder, as well as inside the solidified SAC305 solder alloy, as shown in Figure 8a. At higher magnification, the surface was seen to be mostly flat, consisting of micro ductile dimples and brittle failure indicating a mixed failure that occurred in the solder of SAC305 (Figure 8d). Similar fracture behavior was also found on the other control sample with different soldering parameters. This observation is in accordance with the joint strength measurement obtained where they had approximately similar value of joint strength (Figure 7).

A rougher fractured surface was obtained on the sample soldered with P15 porous Cu, as shown in Figure 8b. Fractured form of the porous Cu interlayer was obviously detected at the broken surface after the tensile testing. This observation is similar to the brazing process involving metal foam where a cup and cone liked structure of metal foam was detected at fractured surface [26]. The failure locations were found at the solder interface close to the porous Cu interface, which corresponds to the IMC layer. This can be explained by the formation of Cu_6Sn_5 IMC phase at the boundary between the solder and porous Cu, as shown in Figure 8e. A larger solid cell wall for P15 will reduce the penetration of molten solder inside porous Cu, creating a segment unreachable by molten solder during the soldering process. Consequently, P15 bonding solid cell walls develop weakness hence explaining the observation for lower joint strength of P15 porous Cu solder joint compared to the P25 porous Cu at soldering temperatures of 267 °C and 287 °C (Figure 7a,b). However, at a higher soldering temperature (307 °C) with soldering time 60 s and 300 s, the joint strength of solder joint with P15 porous Cu obtained is slightly higher than the corresponding solder joint with P25 porous Cu.

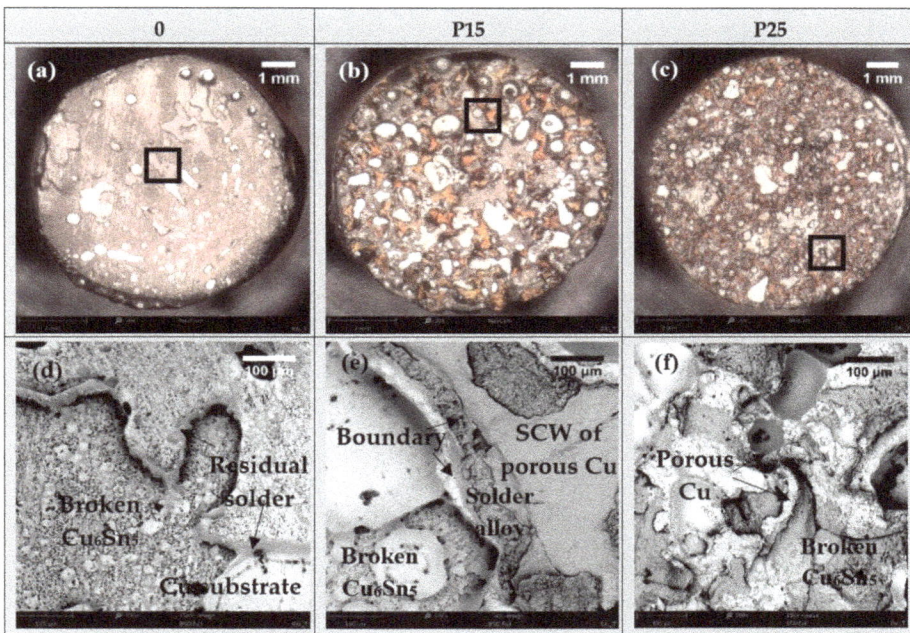

Figure 8. Fracture morphology of solder joint at soldering temperature of 307 °C with soldering time of 300 s; (**a–c**) optical microscope; (**d–f**) high magnification from SEM.

Figure 8c shows the fracture surface of the P25 porous Cu interlayer specimen. As shown in Figure 7, higher joint strength was obtained for the solder joint with P25 porous Cu. The bright dimple which arises from the cleavage surface was detected at the wall boundary of porous Cu and solder alloy (Figure 8f). This cleavage surface developed as a result of decohesion of IMC involving Cu from porous Cu and Sn from molten solder during slow elongation rate of tensile testing. As the amount of Cu atom increased in solder joint system of P25 porous Cu, the reaction of Cu-Sn increased as well to form the IMC phase. The IMC phase formation hindered the motion of atoms dislocation [27]. Subsequently, this strengthened the adhesive bonding of interfacial reaction to increase the joint strength of solder joint with P25 porous Cu, except for the solder joint formed at 307 °C with soldering time of 60 s and 300 s.

The relationship between the test load and displacement of solder joint of every sample at constant speed rate of 0.5 mm/min at room temperature is shown in Figure 9. The points marked as x_1, x_2, x_3 are the elongation distance to break the solder joint for samples without porous Cu as well as samples with P15 and P25 porous Cu, respectively. From this, it is evident that the elongation of solder joint with the addition of porous Cu increased to a larger extent compared to (without degrading the joint strength) the solder joint without porous Cu. The solder joint with higher porosity of P25 porous Cu had a larger elongation than the P15 porous Cu solder joint. The result can be summarized by the following trend: $x_3 > x_2 > x_1$. The increase of Cu atom quantity from porous Cu into interfacial reactions during solidification of molten solder leads to formation of the IMC layer. This interaction between Cu atoms and element from molten solder had possibly restricted the atom dislocation movement and thereby increased the motion resistances of the solder joint, which ultimately generated plastic behavior of the solder joint [13,27–29].

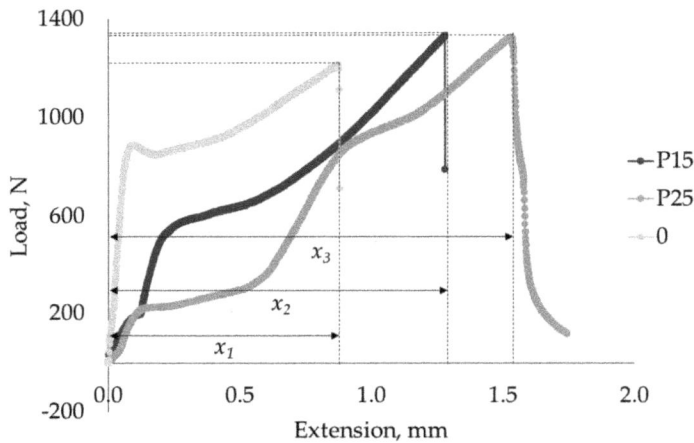

Figure 9. Load-extension curve for solder joints with and without porous Cu soldered at 307 °C with soldering time of 300 s.

3.3.2. Cross-Sectional Analysis

In order to better understand the crack behaviour of solder joint with addition of porous Cu interlayer, the cross section of an as-soldered sample and after tensile test solder joint were observed. Figure 10 shows the cross section micrograph of the solder joint with and without the porous Cu interlayer. The as-soldered condition of solder joint without porous Cu is shown in Figure 10a. It is evident that the crack initiated at the interface of IMC/SAC305 close to the Cu substrate and expanded along the interface as in Figure 10b.

In the case of solder joint with added P15 porous Cu, a large segment of solid cell walls are in contact with each other after being rolled to a very thin layer. As a result, the penetration of molten solder into porous Cu is limited by the blockage from this solid cell wall structure (Figure 10b), and so the crack fracture occurred inside the SAC305, inside the porous Cu and through the interface of porous Cu/IMC (Figure 10c). Despite the smaller pore diameter in P25 porous Cu compared to P15 porous Cu, it was evident that the solid cell walls of P25 porous Cu had less contact with each other as compared to P15 porous Cu. This facilitated better molten solder penetration to reach the inner portion of the porous structure. From Figure 10c, the penetrated molten solder filled the gap inside P25 porous Cu which was evident. It was found that a crack occurred inside the solder, and a few at IMC/SAC305 interface when stress was loaded (Figure 10f). The crack also happened inside porous Cu, which leads to a cleavage fractured, surface as mentioned in Section 3.3.1.

On the other hand, the fracture, along those interfacial boundaries, was accelerated by interfacial stress produced by the differences in physical properties such as thermal expansion. The presence of porous metal in joining assemblies evidently minimize the difference in thermal expansion between the interfaces to prevent the occurrence of joint crack [18]. Therefore, it can be concluded that the use of porous Cu with capability of residual stress absorption can enhance the bond reliability and prevent them from cracking, which was similarly mentioned by Fang et al. in their study through the use of Cu interlayer in joining assembly [17].

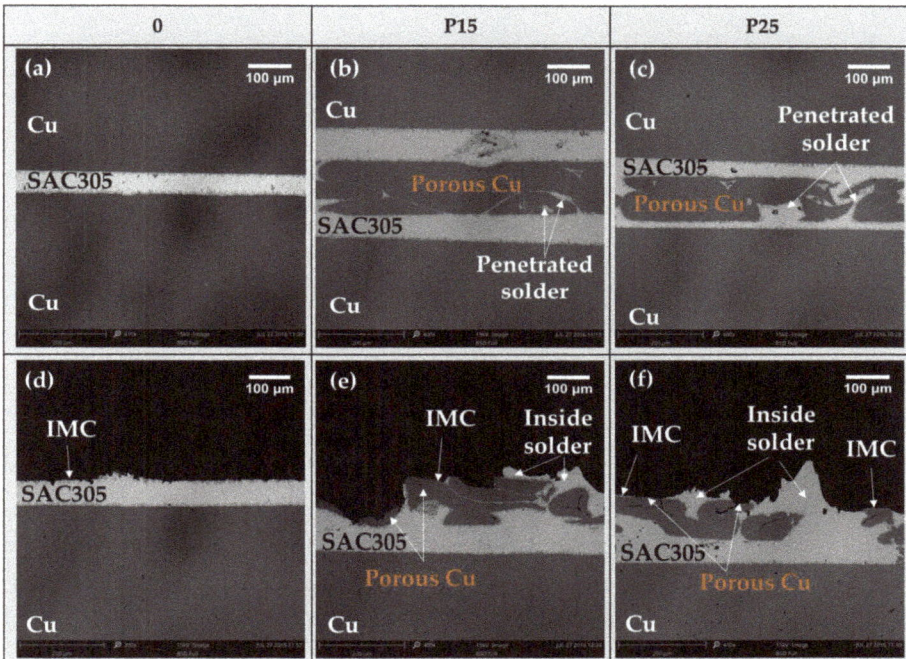

Figure 10. Cross-sectional image of solder joint at soldering temperature of 307 °C with soldering time of 300 s, (**a–c**) as-soldered cross section; and (**d–f**) after crack; for solder joint with 0, P15, and P25 porous Cu.

3.4. Interfacial Structure Analysis

The joint strength of a solder joint is greatly influenced by the metallurgical bonding layer between the solder alloy and the Cu substrate. It is a fact that the interfacial reactions that form IMC have a role to ensure the effective joining between solder/solid interfaces [30]. Figures 11 and 12 show the backscattered electron (BSE) images of IMC layers formed at both sides, namely at the substrate and porous Cu interlayer. The elements of the selected surface are listed in Table 4, together with their respective phases as identified according to the Sn-Cu phase diagram [31]. Basically, the Cu atoms of IMC phase migrated from SAC305 solder alloy and the Cu substrate to form Cu_6Sn_5 phase joining interface [32,33]. As seen in Figure 11a–c, the growth of an elongated scallop-liked layer of Cu_6Sn_5 phase at SAC305/Cu substrate interface was observed for the sample soldered at 307 °C and 300 s with and without the addition of porous Cu. A very thin layer of Cu_3Sn phase was also observed at the boundary of Cu substrate in every sample. This is probably due to the limited diffusion of Sn from Cu_6Sn_5 phase.

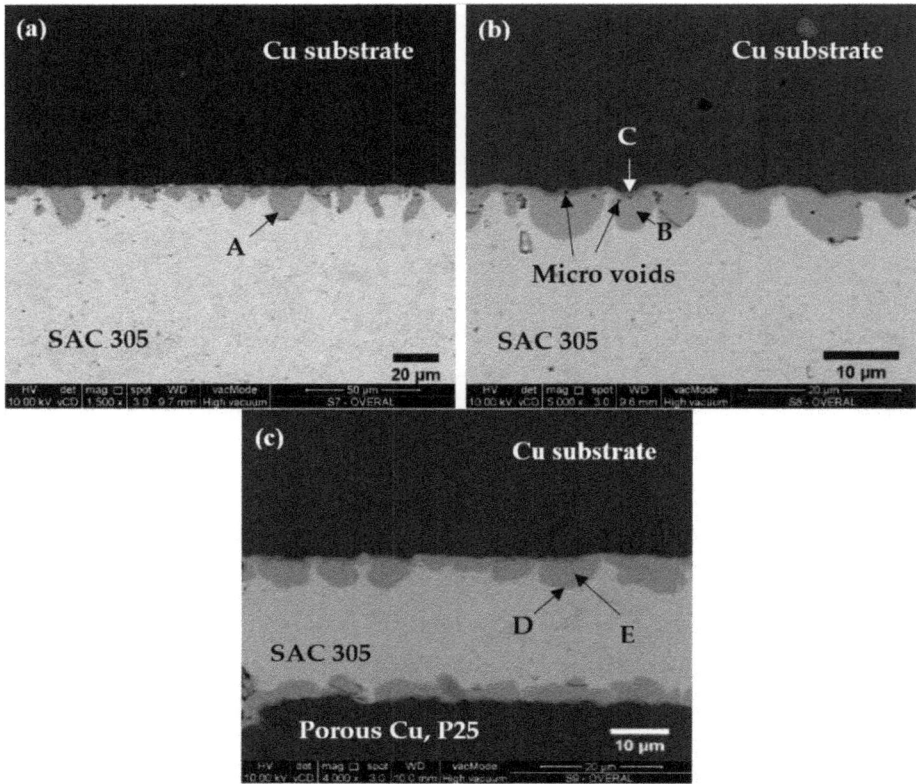

Figure 11. Formation of intermetallic compound (IMC) layer at solder/Cu substrate for solder joint soldered with (**a**) no porous Cu; (**b**) P15 and (**c**) P25 at 307 °C with a holding time of 300 s.

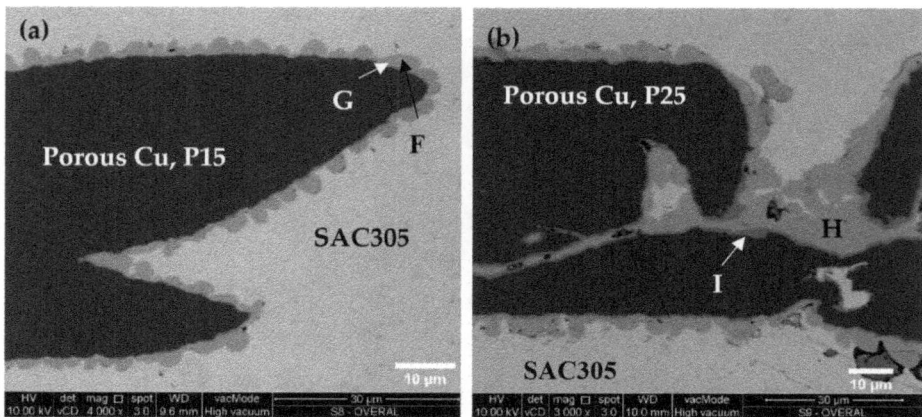

Figure 12. Formation of IMC layer at solder/porous Cu for solder joint soldered with (**a**) P15 and (**b**) P25 at 307 °C with a holding time of 300 s.

Table 4. The element atomic percentage by energy dispersive X-ray (EDX) analysis for Figures 11 and 12.

Point	Element (at. %)		
	Cu	Sn	Phase
A	55.17	44.6	Cu_6Sn_5
B	55.41	44.41	Cu_6Sn_5
C	67.19	32.77	Cu_3Sn
D	56.71	43.16	Cu_6Sn_5
E	72.33	27.67	Cu_3Sn
F	55.42	44.5	Cu_6Sn_5
G	63.4	36.32	Cu_3Sn
H	56.35	43.12	Cu_6Sn_5
I	66.25	33.5	Cu_3Sn

Alternatively, with the addition of the porous Cu interlayer, both typical scallop-liked as well as more uniform and continuous layer of Cu_6Sn_5 phases were observed. Both types of IMC phases were present in the solder joint with porous Cu, with the scallop-shaped phases found to be more dominant for P15 porous Cu solder joint, while the latter phase appeared to be more prominent for the solder joint with P25 porous Cu (Figure 12a,b). The pores of P25 porous Cu was more compressed and compacted due to its smaller solid cell wall (0.06 mm). Smaller pore diameter of P25 porous Cu provided a larger contact area for the dissolution reaction with the molten solder to form IMC Cu_6Sn_5 at porous Cu/SAC305 interface. Therefore, it can be assumed that more diffusion process of Cu-Sn atom occurred at porous Cu/SAC305 interface during penetration of molten solder into P25 porous Cu as compared to P15, leading to the formation of a more uniform and less-developed scallop type IMC structure at the solder joint with P25 porous Cu. This behavior was similarly found in Zou et al.'s investigation, where, with the use of small particles in solder alloy, had increased the surface area to volume ratio of reaction [6].

The structure of porous Cu consists of pores with varying diameter. The pores provide more channels to allow the molten solder to penetrate into the internal porous structure and therefore increase the element diffusion to provide a gripping mechanism. This results in the formation of IMC which forms an adhesion between solder and porous Cu as reported by George et al. [34].

3.5. Measurement of IMC Thickness

The shape of IMC Cu_6Sn_5 phase at the SAC305/Cu substrate is typically hemispherical [35]. Additionally, it is a known fact that solid Cu has high conductive properties. It is mostly utilized as heat exchangers for power electronics. On the contrary, porous Cu has a relatively low amount of metal and therefore possesses low thermal conductivity, which is only adequate for commercial electronics application [22]. The addition of porous Cu interlayer to the solder joint will therefore provide more contact areas for heat transfer from solder alloy during the soldering process, which forms a homogeneous IMC layer at SAC305/porous Cu interface.

Measurements on the IMC thickness at the interfaces would provide additional information that is relevant to the behavior of the solder joints. The average thickness of the IMC layer (Cu_6Sn_5, Cu_3Sn) for every sample is shown in Table 5. The thickness of IMC layer at the interface of SAC305/Cu substrate without porous Cu increased 1 µm with every increase of 20 °C of soldering temperature. For solder joint with porous Cu, the measurement was made at both interfaces of SAC305/Cu substrate and SAC305/porous Cu, and results showed that there was not much variation in the IMC thickness measurement for both the interfaces; except at 307 °C, in which the IMC thickness for P15 porous Cu at SAC305/Cu was about 2 µm thicker than at SAC305/P15 porous Cu interface. Similar observation was also noted for the IMC thickness at the solder joint with P25 porous Cu and SAC305/Cu interface. It can be deduced that the IMC thickness at SAC305/porous Cu reached the limit of deformation around 3–5 µm. Meanwhile, the IMC thickness at SAC305/Cu substrate for solder joint with and without porous Cu showed a similar pattern of increment. This thickness measurement conformed with the readings of typical IMC layer at interface of SAC305/Cu substrate as reported elsewhere [36].

Table 5. Average thickness of intermetallic compound (IMC) layer for every sample.

Soldering Temperature, °C	IMC Thickness, µm				
	No Porous	P15		P25	
		SAC/Cu	SAC/Porous	SAC/Cu	SAC/Porous
267	4.2	3.6	3.3	4.6	3.9
287	5.0	4.8	4.1	5.5	4.4
307	6.4	6.5	4.9	6.8	4.5

4. Conclusions

The effect of porous Cu interlayer addition on the mechanical and microstructural characteristics of SAC305 solder joints formed under different soldering times and temperatures were investigated. The results obtained are summarized as follows:

(1) The joint strength of the Pb-free SAC305 solder joint with the addition of porous Cu interlayer generally increased alongside with increasing soldering time and temperature.

(2) The highest strength for the solder joint with P25 porous Cu addition was recorded at 51 MPa, at soldering time of 180 s and temperature of 307 °C, whereas the 54 MPa highest strength was achieved with P15 porous Cu addition at 300 s with 307 °C.

(3) For solder joints without porous Cu, fracture occured along the interface of solder alloy and Cu substrate. In the case of solder joints soldered with P15 and P25 porous Cu interlayer, fractures occured at three regions, namely at the interface of solder and porous Cu interlayer, inside the porous Cu interlayer as well as inside the solder itself.

(4) Microstructural analysis at the interfacial regions revealed the IMC phase at the SAC305/Cu substrate interface to have a scallop-liked configuration. With addition of porous Cu, both typical scallop-shaped as well as more uniform and continuous layers of IMC phases were observed. The typical scallop-liked configurations were more dominant for the solder joint with P15 porous Cu at the SAC305/P15 porous Cu interface, whereas the latter phase appeared to be more prominent for the solder joint with P25 porous Cu at the SAC305/P25 porous Cu interface.

(5) The IMC layer at the interface of solder alloy and Cu substrate of all specimens was thicker than that at the interface of solder alloy and porous Cu. The IMC layers at these two regions also increased with soldering temperature. The uneven contact area at the interface of porous Cu and solder alloy resulted in the formation of a less pronounced IMC layer.

Acknowledgments: The author greatly appreciated the financial support of this research by the Postgraduate Research Fund, PPP (PG129-2012B) and University Malaya Research Grant, UMRG (RP035A-15AET) from University of Malaya (Kuala Lumpur, Malaysia), Tokai University (Kanagawa, Japan), and contributions from Nagaoka University of Technology (Niigata, Japan) for their support on this study.

Author Contributions: Nashrah Hani Jamadon performed the experiments, analyzed the data and wrote the article. Ai Wen Tan played a key role in reviewing the article. Ariga Tadashi and Miyashita Yukio provided the material for the experiments and gave input on the running of the experiments. Farazila Yusof and Mohd Hamdi supervised the work and discussed the results in detail.

Conflicts of Interest: The authors declare no conflict of interest.

References

1. Suganuma, K. The Current Status of Lead-Free Soldering. *ESPEC Technol. Rep.* **2002**, *13*, 1–8.
2. Chidambaram, V.; Hattel, J.; Hald, J. High-temperature lead-free solder alternatives. *Microelectron. Eng.* **2011**, *88*, 981–989. [CrossRef]

3. Shnawah, D.A.; Said, S.B.M.; Sabri, M.F.M.; Badruddin, I.A.; Che, F.X. High-Reliability Low-Ag-Content Sn-Ag-Cu Solder Joints for Electronics Applications. *J. Electron. Mater.* **2012**, *41*, 2631–2658. [CrossRef]
4. Sabri, M.F.M.; Shnawah, D.A.; Badruddin, I.A.; Said, S.B.M.; Che, F.X.; Ariga, T. Microstructural stability of Sn-1Ag-0.5Cu-*x*Al (*x* = 1, 1.5, and 2 wt. %) solder alloys and the effects of high-temperature aging on their mechanical properties. *Mater. Charact.* **2013**, *78*, 129–143. [CrossRef]
5. Lejuste, C.; Hodaj, F.; Petit, L. Solid state interaction between a Sn-Ag-Cu-In solder alloy and Cu substrate. *Intermetallics* **2013**, *36*, 102–108. [CrossRef]
6. Zou, C.D.; Gao, Y.L.; Yang, B.; Xia, X.Z.; Zhai, Q.J.; Andersson, C.; Liu, J. Nanoparticles of the Lead-free Solder Alloy Sn-3.0Ag-0.5Cu with Large Melting Temperature Depression. *J. Electron. Mater.* **2008**, *38*, 351–355. [CrossRef]
7. Gayle, F.W.; Becka, G.; Badgett, J.; Whitten, G.; Pan, T.; Grusd, A.; Bauer, B.; Lathrop, R.; Slattery, J.; Anderson, I.; et al. High Temperature Lead-Free Solder for Microelectronics. *JOM* **2001**, *53*, 17–21. [CrossRef]
8. Cheng, F.; Gao, F.; Nishikawa, H.; Takemoto, T. Interaction behavior between the additives and Sn in Sn-3.0Ag-0.5Cu-based solder alloys and the relevant joint solderability. *J. Alloy. Compd.* **2009**, *472*, 530–534. [CrossRef]
9. Suganuma, K.; Kim, S.; Kim, K. High-Temperature Lead-Free Solders: Properties and Possibilities. *JOM* **2009**, *61*, 64–71. [CrossRef]
10. Chellvarajoo, S.; Abdullah, M.Z.; Samsudin, Z. Effects of Fe$_2$NiO$_4$ nanoparticles addition into lead free Sn-3.0Ag-0.5Cu solder pastes on microstructure and mechanical properties after reflow soldering process. *Mater. Des.* **2015**, *67*, 197–208. [CrossRef]
11. Chellvarajoo, S.; Abdullah, M.Z. Microstructure and mechanical properties of Pb-free Sn-3.0Ag-0.5Cu solder pastes added with NiO nanoparticles after reflow soldering process. *Mater. Des.* **2016**, *90*, 499–507. [CrossRef]
12. Liu, W.; Lee, N.-C.; Bachorik, P. An innovative composite solder preform for TLP bonding—Microstructure and properties of die attach joints. *Electron. Packag. Technol. Conf. EPTC 2013* **2013**. [CrossRef]
13. Nadia, A.; Haseeb, A.S.M.A. Effects of addition of copper particles of different size to Sn-3.5Ag solder. *J. Mater. Sci. Mater. Electron.* **2011**, *23*, 86–93. [CrossRef]
14. Degischer, H.P.; Kriszt, B. *Handbook of Cellular Metals: Production, Processing, Applications*; Wiley-VCH: Weinheim, Germany, 2002.
15. Qu, Z.; Wang, T.; Tao, W.; Lu, T. Experimental study of air natural convection on metallic foam-sintered plate. *Int. J. Heat Fluid Flow* **2012**, *38*, 126–132. [CrossRef]
16. Liu, Y.; Chen, H.F.; Zhang, H.W.; Li, Y.X. Heat transfer performance of lotus-type porous copper heat sink with liquid GaInSn coolant. *Int. J. Heat Mass Transf.* **2015**, *80*, 605–613. [CrossRef]
17. Fang, F.; Zheng, C.; Lou, H.; Sui, R. Bonding of silicon nitride ceramics using Fe-Ni/Cu/Ni/Cu/Fe-Ni interlayers. *Mater. Lett.* **2001**, *47*, 178–181.
18. Zaharinie, T.; Moshwan, R.; Yusof, F.; Hamdi, M.; Ariga, T. Vacuum brazing of sapphire with Inconel 600 using Cu/Ni porous composite interlayer for gas pressure sensor application. *Mater. Des.* **2014**, *54*, 375–381. [CrossRef]
19. Jamadon, N.H.; Miyashita, Y.; Yusof, F.; Hamdi, M.; Otsuka, Y.; Ariga, T. Formation behaviour of reaction layer in Sn-3.0Ag-0.5Cu solder joint with addition of porous Cu interlayer. *IOP Conf. Ser. Mater. Sci. Eng.* **2014**, *61*, 012020. [CrossRef]
20. Yang, K.-S.; Chung, C.-H.; Lee, M.-T.; Chiang, S.-B.; Wong, C.-C.; Wang, C.-C. An experimental study on the heat dissipation of LED lighting module using metal/carbon foam. *Int. Commun. Heat Mass Transf.* **2013**, *48*, 73–79. [CrossRef]
21. Fakpan, K.; Otsuka, Y.; Mutoh, Y.; Inoue, S.; Nagata, K.; Kodani, K. Creep-Fatigue Crack Growth Behavior of Pb-Containing and Pb-Free Solders at Room and Elevated Temperatures. *J. Electron. Mater.* **2012**, *41*, 2463–2469. [CrossRef]
22. Thewsey, D.J.; Zhao, Y.Y. Thermal conductivity of porous copper manufactured by the lost carbonate sintering process. *Phys. Status Solidi* **2008**, *205*, 1126–1131. [CrossRef]
23. Nawaz, K.; Bock, J.; Jacobi, A.M. Thermal-Hydraulic Performance of Metal Foam Heat Exchangers. Available online: http://docs.lib.purdue.edu/iracc/1283 (accessed on 7 September 2016).
24. Afendi, M.; wan Nordin, W.N.; Daud, R.; Tokuo, T. Strength and Fracture Characteristics of Shear Adhesive Dissimilar Joint. In Proceedings of the International Conference On Applications and Design in Mechnaical Engineering 2012, Penang, Malaysia, 27–28 February 2012.

25. Kim, K.S.; Huh, S.H.; Suganuma, K. Effects of fourth alloying additive on microstructures and tensile properties of Sn-Ag-Cu alloy and joints with Cu. *Microelectron. Reliab.* **2003**, *43*, 259–267. [CrossRef]

26. Shirzadi, A.A.; Zhu, Y.; Bhadeshia, H.K.D.H. Joining ceramics to metals using metallic foam. *Mater. Sci. Eng. A* **2008**, *496*, 501–506.

27. Liu, W.; Wang, C.; Tian, Y.; Chen, Y. Effect of Zn addition in Sn-rich alloys on interfacial reaction with Au foils. *Trans. Nonferrous Met. Soc. China* **2008**, *18*, 617–622. [CrossRef]

28. El-Daly, A.A.; Fawzy, A.; Mohamad, A.Z.; El-Taher, A.M. Microstructural evolution and tensile properties of Sn-5Sb solder alloy containing small amount of Ag and Cu. *J. Alloy. Compd.* **2011**, *509*, 4574–4582. [CrossRef]

29. El-Daly, A.A.; Hammad, A.E. Development of high strength Sn-0.7Cu solders with the addition of small amount of Ag and In. *J. Alloy. Compd.* **2011**, *509*, 8554–8560. [CrossRef]

30. Lee, C.C.; Wang, P.J.; Kim, J.S. Are intermetallics in solder joints really brittle? *Proc. Electron. Compon. Technol. Conf.* **2007**. [CrossRef]

31. Kattner, U.R. Phase diagrams for lead-free solder alloys. *JOM* **2002**, *54*, 45–51. [CrossRef]

32. Fallahi, H.; Nurulakmal, M.S.; Arezodar, A.F.; Abdullah, J. Effect of iron and indium on IMC formation and mechanical properties of lead-free solder. *Mater. Sci. Eng. A* **2012**, *553*, 22–31. [CrossRef]

33. Yang, M.; Cao, Y.; Joo, S.; Chen, H.; Ma, X.; Li, M. Cu_6Sn_5 precipitation during Sn-based solder/Cu joint solidification and its effects on the growth of interfacial intermetallic compounds. *J. Alloy. Compd.* **2014**, *582*, 688–695. [CrossRef]

34. George, E.; Das, D.; Osterman, M.; Pecht, M. Thermal cycling reliability of lead-free solders (SAC305 and Sn3.5Ag) for high-temperature applications. *IEEE Trans. Device Mater. Reliab.* **2011**, *11*, 328–338. [CrossRef]

35. Tan, A.T.; Tan, A.W.; Yusof, F. Influence of nanoparticle addition on the formation and growth of intermetallic compounds (IMCs) in Cu/Sn-Ag-Cu/Cu solder joint during different thermal conditions. *Sci. Technol. Adv. Mater.* **2015**, *16*, 033505.

36. Wu, C.M.L.; Yu, D.Q.; Law, C.M.T.; Wang, L. Properties of lead-free solder alloys with rare earth element additions. *Mater. Sci. Eng. R Rep.* **2004**, *44*, 1–44. [CrossRef]

metals

MDPI

Article

Effect of Welding Parameters on Microstructure and Mechanical Properties of Cast Fe-40Al Alloy

Osman Torun

Bolvadin Vocational School, Afyon Kocatepe University, Afyonkarahisar 03300, Turkey; otorun@aku.edu.tr; Tel.: +90-272-6126353; Fax: +90-272-6116353

Academic Editor: Ana Sofia Ramos
Received: 27 June 2016; Accepted: 20 September 2016; Published: 23 September 2016

Abstract: Friction welding of cast Fe-40Al alloy was carried out at 1000 rmp for various friction times, friction pressures, and forging pressures. The microstructures of the interface of welded samples were analyzed by optical and scanning electron microscopy (SEM). Micrographs demonstrated that excellent welding formed continuously along the interface, except for samples welded for 3 s. Chemical compositions of the interface of the friction welded samples and of the fractured surface of all the specimens were determined using energy dispersive spectroscopy (EDS). After the welding process, shear tests were applied to the welded samples to determine the shear strength of joints. Test results indicated that the maximum shear strength was 469.5 MPa.

Keywords: iron aluminides; friction welding; dynamic recrystallization

1. Introduction

Ordered intermetallic compounds are of interest for high-temperature applications because of their potential for high-temperature stability, high creep resistance, high melting point, and low density. Among the intermetallics, compounds based on the aluminides are of particular interest because of their oxidation resistance due to their ability to form protective oxide films on surfaces [1].

Iron aluminides based on Fe_3Al and FeAl are excellent candidates to be used as structural materials for high-temperature service conditions. The advantages of these materials include their low cost, low density, and high sulfurizing and oxidizing resistance at high temperatures [2–5]. These advantages have enabled several potential uses, including heating elements, furnace fixtures, heat-exchanger piping, and sintered porous gas-metal filters, automobile and other industrial valve components, catalytic converter substrates, and components for molten salt applications [6]. However, these alloys have poor ductility at room temperature, as well as relatively poor high-temperature strength and creep resistance [7].

The development of a joining process is very important for the application of iron aluminides, but the study of welding and/or brazing iron aluminides is very limited in literatures. Joining of iron aluminides plays an important role in practical applications of such alloys. Welding of iron aluminides is difficult due to its inherent low-temperature ductility and poor weldability [8,9]. The low-temperature ductility depends on composition, quenching before testing, grain size, minor alloying additions, dopants on grain boundaries, test environment and strain rate. Grain size refinement is the most attractive way of improving both ductility and strength at the same time. For very fine grain sizes, which are obtained in the presence of dispersoid particles needed to retain the fine grain size, there is some debate about whether such dispersoids may lead to additional improvements in ductility [10–12].

Friction welding is one of the available joining techniques, and it has been used in metals and alloys [13–21]. Many researchers studied the friction welding of similar and dissimilar materials.

Yılbas et al. examined the friction welding of aluminum to steel and copper [20]. They reported that tensile properties were improved for steel-aluminum when the intermetallic thickness extends only 0.2–1 μm; above this value, welds with poor strengths were produced. Tanaka et al. investigated the friction stir welding (FSW) of mild steel to aluminum alloys, and they expressed that the joint strength increased with reduction in the thickness of the intermetallic compound at the weld interface [21]. Watanable et al. studied the friction stir welding of aluminum to steel. They observed formation of a thin intermetallic compound at the upper region of the Fe/Al interface. This phase led to a decrease in the joint strength [22]. Pierpoalo et al. welded aluminum and copper by FSW. They reported that Al_2Cu, $AlCu$, and Al_3Cu_4 phases formed during the friction stir welding process [23].

There have been several published works on the friction welding of Fe_3Al-FeAl type alloys. Sketchley et al. investigated the friction welding of an Fe_3Al-based oxide dispersion-strengthened (ODS) alloy and Inkson et al. studied on friction welding of Fe40Al Grade 3 ODS alloy. They achieved good bonding [24,25]. In addition, author and coworkers reported the successful friction welding of Fe_3Al and dissimilar Fe-28Al alloy and AISI 316 L [26,27]. In this paper, cast Fe-40Al alloys were welded by friction welding under different conditions. Effect of welding parameters on the microstructure and mechanical properties of the cast Fe-40Al alloy were investigated.

2. Experimental Studies

The Fe-40Al alloy was prepared with vacuum arc melting under an argon atmosphere from iron and aluminum with 99.99 wt. % and 99.7 wt. % purity, respectively. The samples were homogenized at 1100 °C for 50 h and cooled in a furnace. Cylindrical cast alloy samples 8 mm in diameter and 50 mm in length were prepared. The friction welding experiments were carried out by a continuous-drive friction welding machine [28] at different forging and friction pressures and friction times under a rotational speed of 1000 rmp (Table 1).

Table 1. Experimental conditions of friction welding.

Friction Speed (rmp)	Forging Pressure (MPa)	Friction Pressure (MPa)	Friction Time (s)	Burn-Off (mm)
1000	140	70	3	1.5
1000	140	70	5	4.2
1000	140	70	7	9.2
1000	140	140	3	4.2
1000	140	140	5	10.9
1000	140	140	7	18.5
1000	200	70	3	4.2
1000	200	70	5	10.9
1000	200	70	7	18.5
1000	200	140	3	5.3
1000	200	140	5	11.7
1000	200	140	7	20.5

The friction welding process is given schematically in Figure 1. After welding, the welded samples were cut perpendicular to the welding interface. The surfaces of the welded samples were ground with 1200 grinding paper and polished with 1 μm diamond paste, then the samples were etched with a mixture consisting of H_2O (30 mL), HNO_3 (30 mL), HCl (20 mL), and HF (20 mL). The microstructures were observed under light microscopy (AKÜ, Afyonkarahisar, Turkey) and scanning electron microscopy (SEM, AKÜ, Afyonkarahisar, Turkey). The chemical compositions of the weld zone and the base alloy were determined by using energy dispersive spectroscopy (EDS, AKÜ, Afyonkarahisar, Turkey). Microhardness values were measured on the welded samples by means of Vickers indenter with a load of 100 g.

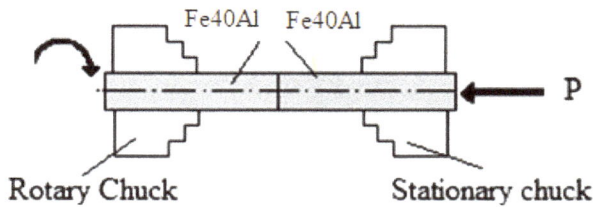

Figure 1. Schematic illustration of the friction welding process.

Shear tests were performed to determine the strength of the weld interface using an electromechanical universal test machine (Shimadzu AG-IS-250, OGÜ, Eskisehir, Turkey) at room temperature. A specially designed specimen holder (Figure 2) was used to measure the shear strength. Three samples were tested for each welding condition. The fractured surfaces of the welds were observed by SEM.

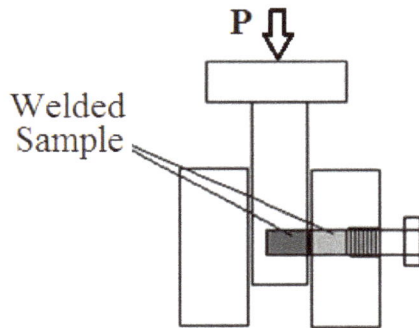

Figure 2. Schematic illustration of the shear test specimen holder.

3. Results and Discussion

3.1. Microstructure

Grain size of the samples was very large because samples were homogenized at 1100 °C for 50 h (Figures 3 and 4). As a result of this situation, ductility of the samples might be decreased. Poor ductility influences welding quality of the friction welded Fe-Al alloy. However, the effect of poor ductility on the welding quality of the friction welding of FeAl alloy has not been reported in the literature up to now. Flash formation was observed in all welded samples because of plastic deformation during welding. Friction and forging pressure and friction time play important roles in flash formation. The burn-off increased with increase in the forging pressure. More plastic deformation occurs because of the higher friction pressure and the longer friction time, which produce higher heat in the weld interface. Thus, the burn-off (axial shortening) increases with the increase in plastic deformation (Table 1). Microstructural observation was carried out for the weld interface using both an optical and a scanning electron microscope with energy dispersive spectroscopy. Figure 3 shows the optical micrographs of the weld interface of welded samples for different experimental conditions.

As seen in Figure 3a, large voids were observed at the welding interface of the welded samples for 3 s under 70 MPa friction pressure. Other welded samples were of sound quality and they did not exhibit any voids or crack formation along the weld interface (Figure 3b–d). The friction time of 3 s is not enough to generate the required heat for friction welding. Therefore, these welded samples exhibited large voids at the welded interface.

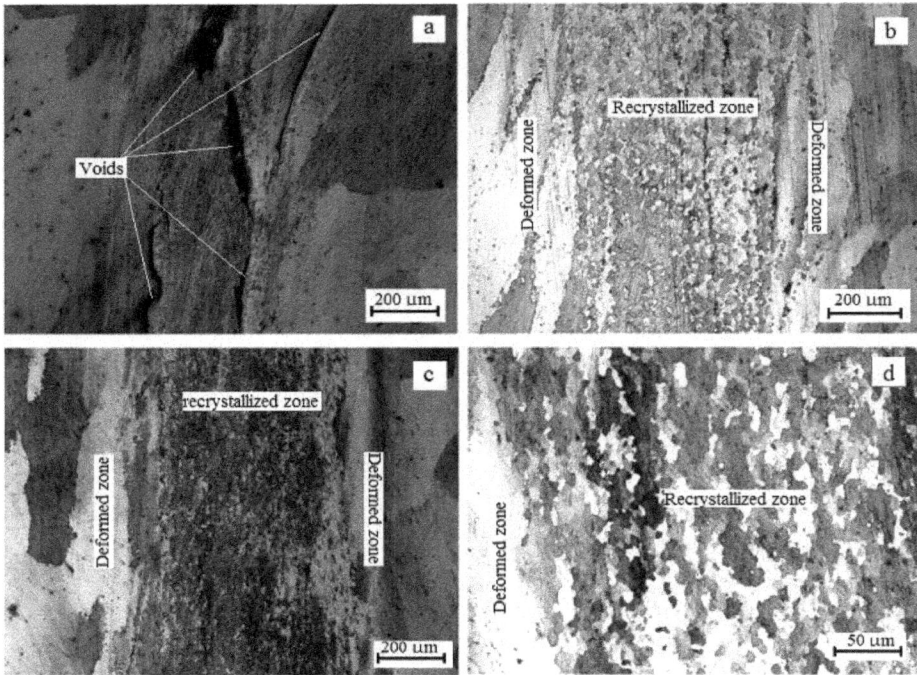

Figure 3. Optical micrographs of the welded samples under 200 MPa forging pressure and 140 MPa friction pressure. (**a**) 3 s; (**b**) 5 s; (**c**) 7 s; (**d**) higher magnification for 5 s.

Two main regions are observable at the interface of all of the welded samples: a dynamically recrystallized zone with very fine grains and a plastically deformed zone (Figure 3b–d). Friction welding involves heavy plastic working at higher temperatures close to the melting point of the base materials. This plastic deformation introduces large number of dislocations in materials. As the density of these dislocations increases, they tend to form subgrain cell structure. These low-angle grains rotate to form high-angle strain-free grains resulting in a zone of very fine equiaxed grains compared to the base materials [29,30]. The width of the recrystallized zone for all of the welded samples was approximately between 400 and 600 μm. The micrographs demonstrate an increase in the width of the recrystallized and the deformed zones with an increase in the friction time and pressure. It is known that some intermetallic phases may form at the welding interface for dissimilar materials joined by the friction welding and the friction stir welding [19–23]. However, similar materials joined by the friction welding may have the same chemical compositions with base alloy at the welding interface [26]. SEM-EDS area analysis was done to determine chemical composition at the weld zone of the welded samples. The analysis revealed the composition of alloy elements at the weld zone (Figure 4a,b). Also, a detailed line-scan analysis of the weld zone is given in Figure 4c,d. Analysis results indicate that compositions of welded zone and the base alloy are similar. There was not any change in the chemical composition of the weld zone of the welded samples due to welding of similar alloy. The analysis results confirmed this finding.

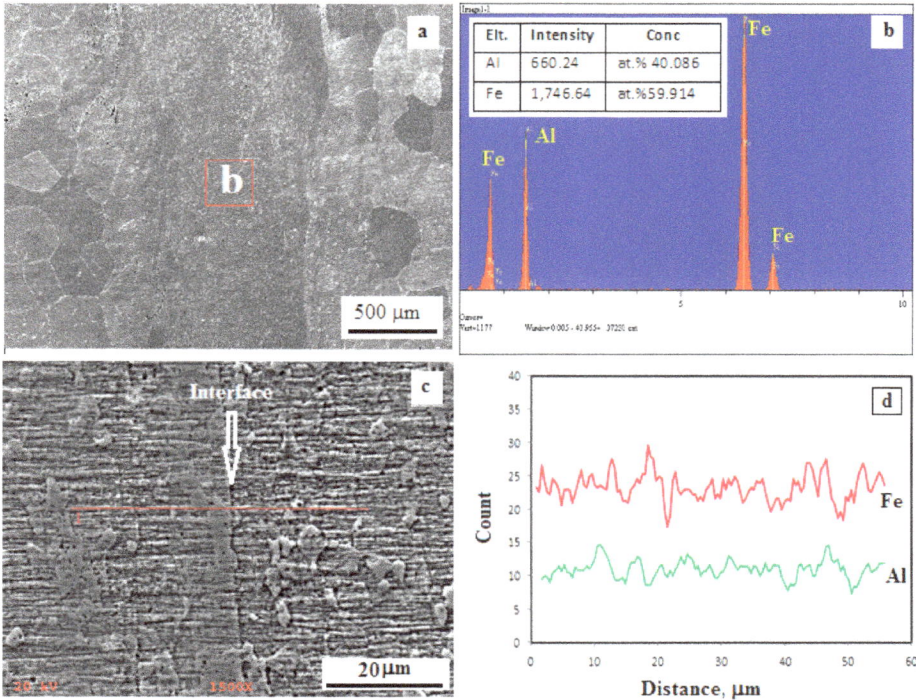

Figure 4. Scanning electron microscopy (SEM)-energy dispersive spectroscopy (EDS) analysis of welded sample for 7 s under 200 MPa forging pressure and 140 MPa friction pressure. (**a**) SEM micrograph; (**b**) EDS analysis; (**c**) SEM micrograph of weld interface; (**d**) line-scan analysis.

3.2. Mechanical Properties

Microhardness values of the welded samples under 140 MPa friction pressure and 200 MPa forging pressure from the weld center to a side are given in Figures 5 and 6. According to the results of the measurements, microhardness profiles for all welding times are found to be similar.

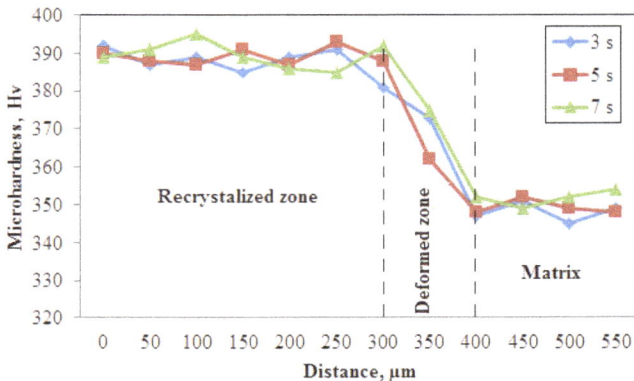

Figure 5. Microhardness profile of welded samples for 7 s under 200 MPa forging pressure and 140 MPa friction pressure from the weld center to a side.

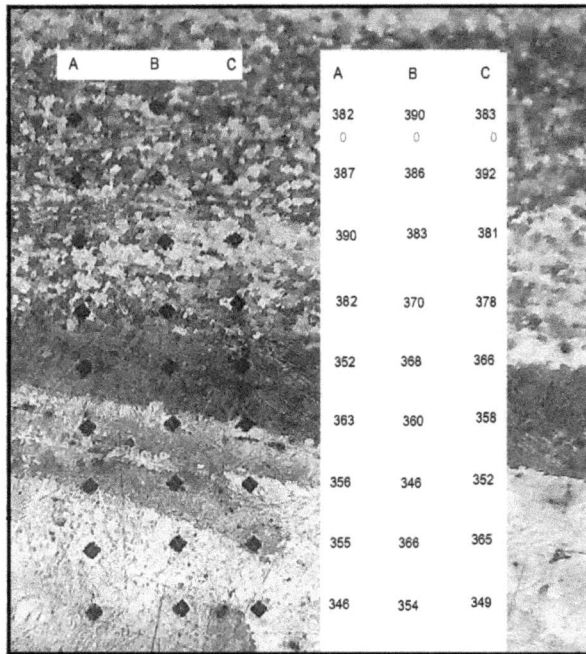

Figure 6. The hardness values of welded samples for 7 s.

Hardness of the recrystallized zone is higher than that of the base alloy due to formation of very fine grains. In addition, hardness of the recrystallized zone can be increased by vacancy hardening due to rapid cooling from a high temperature after friction welding [11,12].

The shear strengths of weld interface of welded samples were determined by using a specially designed testing apparatus. The shear of the welds and the base alloy are shown in Figure 7. Test results demonstrated that the values of the shear strength of the welded samples increased with increasing of the friction time and friction pressure. It can be argued that the shear strength of welds was dependent on the friction time and pressure under the experimental conditions. The shear strength of welded samples for 3 s is much lower due to large voids at the welding interface. The forging pressure affected the shear strength values of the welded samples for 3 and 5 s under friction pressure of 70 MPa. The shear strength values increased with an increase in forging pressure for these times. However, the shear strength of the welds did not show an important change in other welding conditions. Under constant forging pressure conditions, the shear strength of the welds increased with increasing friction pressure. This observation indicates that the increase in the shear strength is related to the magnitude of the accumulated heat input, which depends on the friction time for low and high friction pressures. Test results demonstrate that the shear strength values of the welds are greater than that of the base metal. This situation shows that the structure with very fine grains formed due to dynamic recrystallization during the friction welding has higher strength than the matrix.

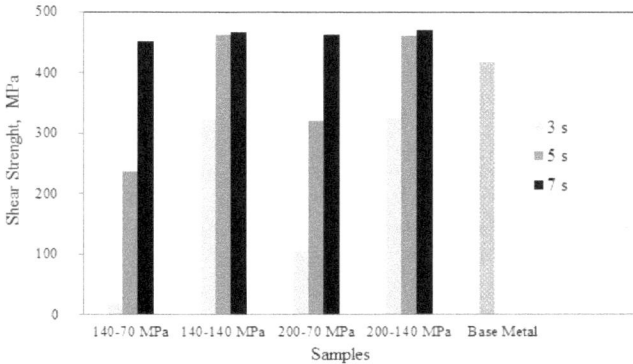

Figure 7. Shear strengths of the welded samples and base alloy.

3.3. Fractographs

The SEM fractographs of the fractured surface of the welded sample for 7 s under 140 MPa friction pressure and 140 MPa forging pressure and the base metal are shown in Figure 8. The welded samples were inserted into the testing apparatus to fracture from the weld interface during the shear test. The shearing process was performed out the weld interface of the welded samples. Fractographs show that the fractured surface of welded sample has very fine grains with respect to the base alloy (Figure 8a). This finding confirms that failure took place at the recrystallized zone during the shear test. EDS analysis taken from the fractured surface revealed that the Fe and Al atoms exist on the fractured surfaces of welded sample for 7 s under 140 MPa friction pressure and 200 MPa forging pressure.

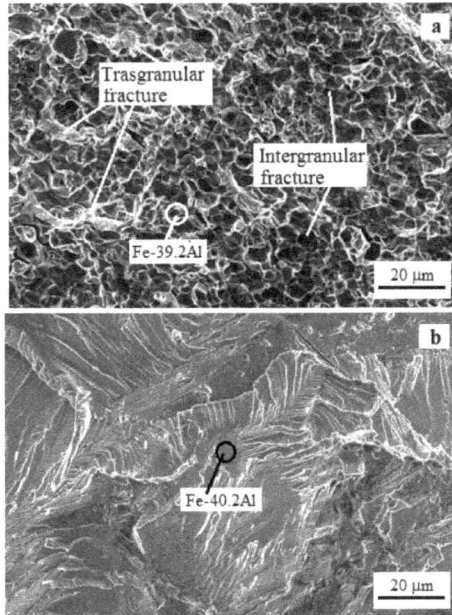

Figure 8. Fractographs of the fractured surface of the welded sample for 7 s under 200 MPa forging pressure and 140 MPa friction pressure and the base alloy. (**a**) The welded sample; (**b**) the base alloy.

4. Conclusions

In this study, cast Fe-40Al alloy was welded under different conditions by friction welding method. Microstructure studies showed the presence of two different regions at the weld interface: the recrystallized zone and the deformed zone. There was variation in the width of the recrystallized zone and the deformed zone with an increase in the friction time and pressure. Analysis results indicate that compositions of welded zone and the base alloy are similar. There was no change in the chemical composition of the weld zone of the welded samples due to welding of similar alloy. The microhardness profiles for all the friction welding times and pressures were similar. Hardness of the recrystallized zone is higher than that of the base alloy due to the formation of very fine grains and vacancy hardening. The shear strengths of the welds increased with increase in the process time and friction pressure. The maximum shear strength was 469.5 MPa. The best welding parameter combination was found to be a friction pressure of 140 MPa, a forging pressure of 200 MPa, and a friction time of 7 s.

Acknowledgments: The author is grateful to Afyon Kocatepe University Scientific Research Committee since this study is supported (Project No.: 10.BOLMYO.01).

Conflicts of Interest: The author declares no conflict of interest.

References

1. Pank, D.R.; Nathal, M.V.; Koss, D.A. *High Temperature Ordered Intermetallics Alloys III*; Liu, C.T., Taub, A.I., Stoloff, N.S., Koch, C.C., Eds.; Materials Research Society: Pittsburgh, PA, USA, 1988; pp. 561–565.
2. Liu, C.T.; Stringer, J.; Mundy, J.N.; Horton, L.L.; Angelini, P. Ordered intermetallic alloys: An assessment. *Intermetallics* **1997**, *5*, 579–596. [CrossRef]
3. Stoloff, N.S.; Liu, C.T.; Deevi, S.C. Emerging applications of intermetallics. *Intermetallics* **2000**, *8*, 1313–1320. [CrossRef]
4. Deevi, S.C.; Sikka, V.K.; Liu, C.T. Processing, properties, and applications of nickel and iron aluminides. *Progress. Mater. Sci.* **1997**, *42*, 177–179. [CrossRef]
5. Deevi, S.C.; Swindeman, R.W. Yielding, hardening and creep behavior of iron aluminides. *Mater. Sci. Eng. A* **1998**, *258*, 203–210. [CrossRef]
6. Stoloff, N.S. Iron aluminides: Present status and future prospects. *Mater. Sci. Eng. A* **1998**, *258*, 1–14. [CrossRef]
7. McKamey, C.G.; DeVan, J.H.; Tortorelli, P.F.; Sikka, V.K. A review of recent developments in Fe_3Al based alloys. *J. Mater. Res.* **1991**, *6*, 1779–1805. [CrossRef]
8. Fasching, A.A.; Ash, D.I.; Edwards, G.R.; David, S.A. Hydrogen cracking behavior in an iron aluminide alloy weldment. *Scr. Metall.* **1995**, *32*, 389–396. [CrossRef]
9. Lee, Y.L.; Shiue, R.K.; Wu, S.K. The microstructural evolution of infrared brazed Fe_3Al by BNi-2 braze alloy. *Intermetallics* **2003**, *11*, 187–195. [CrossRef]
10. Morris, D.G.; Chao, J.; Garcia Oca, C.; Morris-Munnoz, M.A. Obtaining good ductility in an FeAl intermetallic. *Mater. Sci. Eng. A* **2003**, *339*, 232–240. [CrossRef]
11. Morris, D.G.; Morris-Munnoz, M.A. The infuence of microstructure on the ductility of iron aluminides. *Intermetallics* **1999**, *7*, 1121–1129. [CrossRef]
12. Morris, D.G.; Gutierrez-Urrutia, I.; Morris-Munnoz, M.A. Evolution of microstructure of an iron aluminide during severe plastic deformation by heavy rolling. *J. Mater. Sci.* **2008**, *43*, 7438–7444. [CrossRef]
13. Ozdemir, N. Investigation of the mechanical properties of friction-welded joints between AISI 304L and AISI 4340 steel as a function rotational speed. *Mater. Lett.* **2005**, *59*, 2504–2511. [CrossRef]
14. Ateş, H.; Turker, M.; Kurt, A. Effect of friction pressure on the properties of friction welded MA956 iron-based superalloy. *Mater. Des.* **2007**, *28*, 948–953. [CrossRef]
15. Tao, B.H.; Li, Q.; Zhang, Y.H.; Zhang, T.C.; Liu, Y. Effects of post-weld heat treatment on fracture toughness of linear friction welded joint for dissimilar titanium alloys. *Mater. Sci. Eng. A* **2015**, *634*, 141–146. [CrossRef]
16. Rovere, C.A.D.; Ribeiro, C.R.; Silva, R.; Baroni, L.F.S.; Alcântara, N.G.; Kuri, S.E. Microstructural and mechanical characterization of radial friction welded supermartensitic stainless steel joints. *Mater. Sci. Eng. A* **2013**, *586*, 86–92. [CrossRef]

17. Winiczenko, R.; Kaczorowski, M. Friction welding of ductile iron abd stainless steel. *Mater. Process. Technol.* **2013**, *213*, 453–462. [CrossRef]

18. Kırık, I.; Ozdemir, N.; Çalıgülü, U. Effect of particle size and volume fraction of the reinforcement on the microstructure and mechanical properties of friction welded MMC to AA6061. *Kov. Mater.* **2013**, *51*, 221–227.

19. Davis, C.J. Micro friction welding aluminum studs to mild steel plates. *Metal Constr.* **1977**, *9*, 196–197.

20. Yılbaş, B.S.; Şahin, A.Z.; Kahraman, N.; Garni, Z.A.A. Friction welding of St-Al and Al-Cu materials. *J. Mater. Process. Technol.* **1995**, *49*, 431–443. [CrossRef]

21. Tanaka, T.; Morishige, T.; Hirata, T. Comprehensive analysis of joint strength for dissimilar friction stir welds of mild steel to aluminum alloys. *Scr. Mater.* **2009**, *61*, 756–759. [CrossRef]

22. Watanabe, T.; Takayama, H.; Yanagisawa, A. Joining of aluminum alloy to steel by friction stir welding. *J. Mater. Process. Technol.* **2006**, *178*, 342–349. [CrossRef]

23. Pierpaolo, C.; Antonello, A.; Gaetano, S.P.; Valentino, P.; Antonino, S. Microstructural aspects in Al-Cu dissimilar joining by FSW. *Int. J. Adv. Manuf. Technol.* **2015**, *79*, 1109–1116.

24. Sketchley, P.D.; Threadgill, P.L.; Wright, I.G. Rotary friction welding of an Fe_3Al based ODS alloy. *Mater. Sci. Eng. A* **2002**, *329–331*, 756–762. [CrossRef]

25. Inkson, B.J.; Threadgill, P.L. Friction welding of FeAl40 Grade 3 ODS alloy. *Mater. Sci. Eng. A* **1998**, *258*, 313–318. [CrossRef]

26. Torun, O.; Çelikyürek, I.; Baksan, B. Friction welding of cast Fe-28Al alloy. *Intermetallics* **2011**, *19*, 1076–1079. [CrossRef]

27. Çelikyürek, İ.; Torun, O.; Baksan, B. Microstructure and strength of friction-welded Fe–28Al and 316 L stainless steel. *Mater. Sci. Eng. A* **2011**, *528*, 8530–8536. [CrossRef]

28. Karabulut, A.; Tasgetiren, S. Sürekli tahrikli sürtünme kaynak makinesi tasarım ve imalatı. *Mak. Teknol. Elektron. Derg.* **2004**, *3*, 38–46.

29. Misra, M.S.; Ma, Z.Y. Friction stir welding and processing. *Mater. Sci. Eng. R Rep.* **2005**, *50*, 1–78. [CrossRef]

30. Dey, H.C.; Ashfaq, M.; Bhaduri, A.K.; Rao, K.P. Joining of titanium to 304L stainless steel by friction welding. *J. Mater. Process. Technol.* **2009**, *209*, 5862–5870. [CrossRef]

![metals logo] *metals*

MDPI

Article

Tribocorrosion Study of Ordinary and Laser-Melted Ti$_6$Al$_4$V Alloy

Danillo P. Silva [1], Cristina Churiaque [2], Ivan N. Bastos [1] and José María Sánchez-Amaya [2,*]

[1]   Instituto Politécnico, Universidade do Estado do Rio de Janeiro, Rua Bonfim, 25, Nova Friburgo, Rio de Janeiro 8.265-570, Brazil; danillopedrosilva@yahoo.com.br (D.P.S.); inbastos@iprj.uerj.br (I.N.B.)
[2]   Laboratorio de Ensayos, Corrosión y Protección, Departamento de Ciencia de los Materiales, Ingeniería Metalúrgicay Química Inorgánica, Escuela Superior de Ingeniería, Universidad de Cádiz, Avenida de La Universidad de Cádiz, Puerto Real, Cádiz 11519, Spain; cristina.churiaque@uca.es
*   Correspondence: josemaria.sanchez@uca.es; Tel.: +34-956-016762

Academic Editor: Ana Sofia Ramos
Received: 7 August 2016; Accepted: 18 October 2016; Published: 24 October 2016

Abstract: Titanium alloys are used in biomedical implants, as well as in other applications, due to the excellent combination of corrosion resistance and mechanical properties. However, the tribocorrosion resistance of titanium alloy is normally not satisfactory. Therefore, surface modification is a way to improve this specific performance. In the present paper, laser surface-modified samples were tested in corrosion and pin-on-disk tribocorrosion testing in 0.90% NaCl under an average Hertzian pressure of 410 MPa against an alumina sphere. Laser-modified samples of Ti$_6$Al$_4$V were compared with ordinary Ti$_6$Al$_4$V alloy. Electrochemical impedance showed higher modulus for laser-treated samples than for ordinary Ti$_6$Al$_4$V ones. Moreover, atomic force microscopy revealed that laser-treated surfaces presented less wear than ordinary alloy for the initial exposure. For a further exposure to wear, i.e., when the wear depth is beyond the initial laser-affected layer, both materials showed similar corrosion behavior. Microstructure analysis and finite element method simulations revealed that the different behavior between the initial and the extensive rubbing was related to a fine martensite-rich external layer developed on the irradiated surface of the fusion zone.

Keywords: tribocorrosion; Ti6Al4V; laser-treated titanium alloy

1. Introduction

Commercially pure titanium and titanium alloys are largely used in different applications, such as biomedical, aerospace, automotive, petrochemical, nuclear, etc. These vast application fields are ascribed to their high mechanical strength, low specific mass, and good corrosion resistance as a consequence of the formation of very stable, continuous, adherent, and protective oxide films on metal surfaces [1–3]. Moreover, excellent corrosion resistance and high biocompatibility of titanium alloys in physiological medium make them very adequate for biomedical applications such as dental implants and orthopedic prostheses. Titanium's ability to form calcium phosphate in vitro biocompatibility tests in simulated body fluid is considered as a good indicator of in vivo success [4–6]. However, in prostheses, tribochemical reactions can take place at the stem-bone interface [7]. These reactions occur along with micromotion due to body locomotion, leading to important problems to patients, such as the loosening of the implant, osteolysis, release of metal ions, and wear particles.

Despite the diverse satisfactory properties related to titanium based-alloys, its tribological performance is generally poor [8–10]. Tribocorrosion is a highly complex topic, due to an intricate and not well established synergy between mechanical, metallurgical, and electrochemical factors. In this way, in addition to the local electrochemical features of the intermittent passive-active transition in the fretting region, the wear mechanism can be different according to the local interactions between the metal surface and the sliding element. These wear mechanisms are often referred to as abrasive, adhesive, delamination, and oxidative [11], not always being independent nor excluding. For example, the subjacent process, such as the accumulation of dislocations, stress concentration in the metal substrate, as well as scale dependence [12–14], can be relevant in the determination of the mode of wear on damaged regions [14]. Therefore, these issues increase the natural complexity of tribocorrosion phenomena.

Several modifications have been reported in the literature on the improvement of titanium surface properties [5,15–18] by different methods, including laser-assisted methods [10,19,20]. In fact, laser technology can be employed in several industrial processes, comprising welding similar [18] and dissimilar alloys [19], or the modification of surface properties for corrosion resistance improvements [20]. Laser remelting (LR) of Ti_6Al_4V with appropriate fluence provokes microstructural changes leading to an increase in microhardness without jeopardizing the corrosion resistance [21]. Therefore, the laser-modified surface is expected to improve the wear performance. Thus, the tribocorrosion properties of laser remelted titanium alloys deserve further analysis because of the complexity of interfacial phenomena. In the present work, a laser remelting treatment was applied to Ti_6Al_4V and tested under tribocorrosion conditions. Then, a comparative study of ordinary and laser-remelted Ti_6Al_4V alloy was carried out.

2. Materials and Methods

Samples of Ti_6Al_4V alloy (grade 5) were used in tribocorrosion testing. The chemical composition is specified in Table 1. Some Ti_6Al_4V samples were subjected to laser remelting (LR) treatments. Before the laser treatments, the samples were blasted with corundum particle sand cleaned with acetone to homogenize the surface and avoid the possible appearance of defects [22]. A Rofin-Sinar DL028S high power diode laser (Rofin-Sinar, Plymouth, MA, USA) was employed working at the focal distance (69.3 mm from the focusing lens, providing an almost rectangular spot of 2.2×1.7 mm^2) with 2.0 kW laser power and 1.0 m/min scanning rate. To avoid oxidation, samples were placed in a laboratory-made conditioned chamber during laser treatments, in which argon was injected at a flow rate of 10 L/min. Laser remelting treatments consisted of performing 16 linear parallel laser scans of 30 mm long on Ti_6Al_4V samples (35 mm $\times$ 35 mm $\times$ 3 mm). Each scan was separated by 1.5 mm from the previous one to obtain appropriate overlapping. Sixteen seconds was the waiting time between scans to avoid overheating. These treatments allow a remelted area of approximately 30 mm $\times$ 26 mm. Figure 1 exhibits a scheme of the setup used to perform the LR treatments. Laser-treated samples were designed as Ti_6Al_4V-LR, and the base metal denoted as Ti_6Al_4V-BM. In order to perform the tribocorrosion tests, both Ti_6Al_4V-LR and Ti_6Al_4V-BM samples were machined to obtain disks 24 mm diameter. This shape is required to be assembled on the pin-on-disk apparatus [23].

Table 1. Chemical composition of titanium alloy (% mass).

Al	V	Fe	C	Other	Ti
5.67	4.50	0.18	0.10	<0.1	Balance

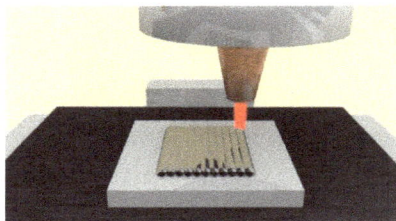

Figure 1. Scheme of the setup used to perform the LR treatments on Ti_6Al_4V samples.

The microstructure and microhardness were evaluated with mounted cross-sections of the samples. Mounted samples were ground, polished (with 0.04 μm colloidal silica suspension, Struers OP-U, Struers, Ballerup, Denmark), etched with Kroll's reagent freshly prepared in laboratory (6.0 mL HNO$_3$, 2.0 mL HF and filled with distilled water to produce 100 mL of solution) for 15 s, rinsed with distilled water, and carefully dried [21]. Microhardness measurements were obtained with a Duramin microhardness tester from Struers (Struers, Ballerup, Denmark), following the standard UNE-EN-ISO 6507-1:2006 [24]. The mechanical load employed was 19.61 N (HV 2), for 19 s.

A finite element method (FEM) has been used to simulate the laser remelting process in Ti$_6$Al$_4$V alloy. The approach estimates the effect on the Ti$_6$Al$_4$V parts of a moving heat source, whose size and shape has been previously optimized for the employed high power diode laser [25]. SYSWELD software (ESI group, Paris, France) was employed to perform the simulations of single linear laser scans.

Before the tribocorrosion tests, titanium samples were slightly abraded with SiC emery paper to #600, subsequently cleaned with distilled water and alcohol, and dried with blast air. In these tests, 4.00 mm diameter alumina spheres (Sóesferas, São Paulo, Brazil), made according to UNE-EN-ISO 3290-2:2014 [26], were used as the counterpart. Each test was performed with a new alumina sphere, applying a normal load of 2.0 N. This load represents a Herztian pressure of 410 MPa. A pin-on-disk tribometer (Homemade, Nova Friburgo, Brazil) was used with a rotation frequency of 1.25 Hz (75 rotations per minute). A load cell (Alfa Instrumentos, São Paulo, Brazil) measured the friction force.

After each tribocorrosion test, the wear track was analyzed by optical microscopy (Olympus, Tokyo, Japan) and its cross-section profile was acquired by profilometry (roughness tester model SJ-210, Mitutoyo, Kawasaki, Japan). Some worn tracks were also examined by a Nanosurf C3000 atomic force microscope (AFM, Nanosurf AG, Liestal, Switzerland) in intermittent mode contact for samples tested in an initial wear of 10 s exposure.

Corrosion behavior was evaluated by three techniques: open circuit potential (OCP, one hour) electrochemical impedance spectroscopy (EIS), and potentiodynamic polarization curves. Electrochemical tests were performed using a Princeton Applied Research Versastat 3 potentiostat (Advanced Measurement Technology Inc., Oak Ridge, TN, USA). Figure 2 depicts the main parts of the electrochemical cell adapted for tribocorrosion measurements. A saturated calomel electrode (SCE) was used as the reference. A platinum wire was used as a counter-electrode in both polarization curves and EIS measurements. All tests were carried out in a three-electrode cell containing 0.90 wt. % NaCl aqueous solution at room temperature. The measured solution pH was 6.5. An electrolyte volume of 150 mL was used in the tribocorrosion cell, and 250 mL for the corrosion tests.

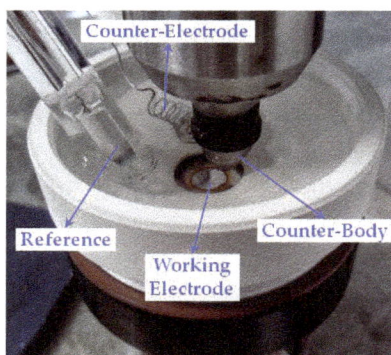

Figure 2. Connections of the electrochemical cell adapted to tribocorrosion testing.

To perform the electrochemical impedance spectroscopy, a frequency range from 10 kHz to 10 mHz with a sine wave perturbation of 8 mV was used under potentiostatic control at the corrosion potential.

The EIS diagrams were fitted by the equivalent electric circuit, as shown in Figure 3, as successfully used for Ti_6Al_4V in previous studies [21,27]. Echem Analyst 6.11® (Gamry Instruments Incorporated, Warminster, PA, USA) was the software used to fit the EIS diagrams. The constant phase element (CPE) was used to fit the impedance of the equivalent circuit. A simplex procedure was used to perform the complex nonlinear regression to obtain the impedance parameters. Re represents the electrolyte resistance; Q and α are the CPE parameter and exponent, respectively; and Rp is the polarization resistance. Mechanical load application was employed for 600 s and, subsequently, the electrochemical measurement was performed. Electrochemical impedance was acquired before, during, and after load application.

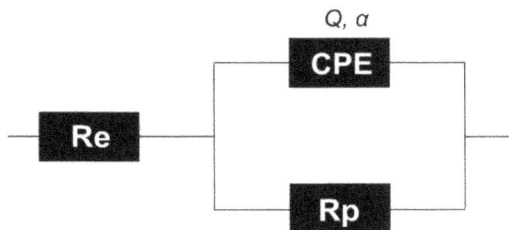

Figure 3. Equivalent electric circuit used to model the EIS diagrams.

Polarization curves were performed at a 0.50 mV/s scan rate and swept from 0.4 V below the open circuit potential (−0.4 V vs. OCP) up to 2 V above the saturated calomel electrode (2.0 V vs. SCE). Mechanically-loaded and non-loaded samples were tested in polarization. Load application was employed for 600 s and, subsequently, the measurement was performed. Friction force was sampled at 10 Hz and the data shown is the averaged signal.

3. Results and Discussion

3.1. Microstructure and Microhardness

Figure 4 shows the microstructure of the base metal (Ti_6Al_4V-BM). It presents a microstructure characterized by two phases: equiaxed α grains with intergranular β particles, in good agreement with previous studies [18,21]. The microstructure of Ti6Al4V samples remelted by laser treatment (Ti_6Al_4V-LR) is quite different than that of Ti_6Al_4V-BM, as depicted in Figure 5. Ti_6Al_4V-LR showed in Figure 5 exhibits a laser-treated region of about 1 mm in depth, from which 700 μm approximately corresponds to the fusion zone (FZ). A detailed examination reveals that FZ microstructure presents acicular martensite (α'), as depicted in Figure 5e,f. This phase is generated as a consequence of the high heating and cooling rates of the laser processing. Its hardness is reported to be between 50 and 100 HV harder than the typical $\alpha + \beta$ microstructure observed in the Ti_6Al_4V base metal [21]. Martensite alpha prime phase (α') generally has high hardness, but relatively low toughness. The HAZ (heat affected zone) has a microstructural mixture of BM ($\alpha + \beta$) and FZ (α'), therefore presenting intermediate hardness values. Mean and standard deviation of Vickers microhardness values measured in Ti_6Al_4V-BM and at the different zones in Ti_6Al_4V-LR are reported in Table 2. The values are in good agreement with those previously reported in the literature, being the FZ of Ti_6Al_4V-LR around 60 HV harder than Ti_6Al_4V-BM. It is worth noting that BM in the laser-remelted sample is slightly harder than the non-treated material, due to the influence of the laser heat treatment, which is not enough to provoke the phase transformation (then, it remains as $\alpha + \beta$) but induces some hardening.

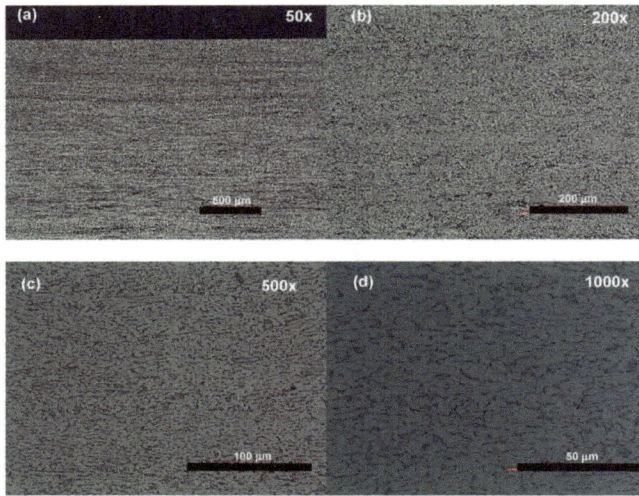

Figure 4. Metallographic images of Ti$_6$Al$_4$V-BM samples at the indicated magnifications: (**a**) 50×, (**b**) 200×, (**c**) 500×, (**d**) 1000×.

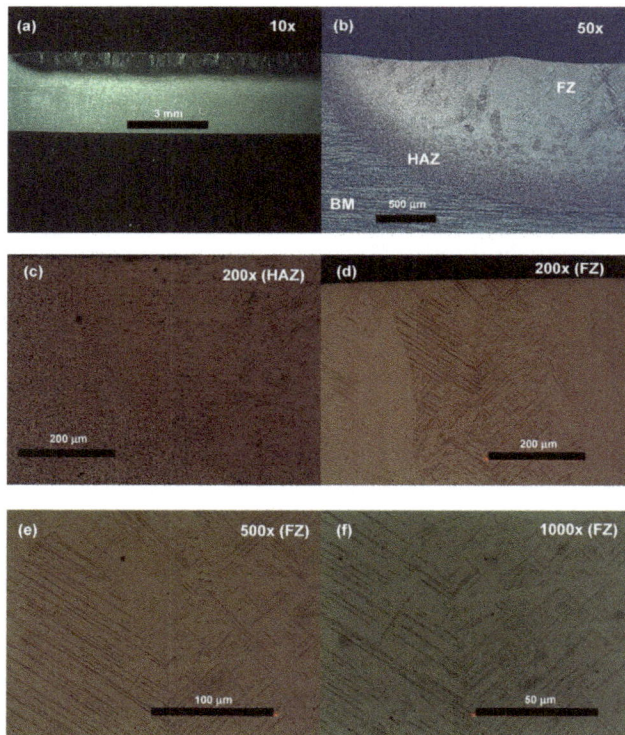

Figure 5. Metallographic images of the indicated zones of the Ti$_6$Al$_4$V-LR samples, at the indicated magnifications: (**a**) 10×, (**b**) 50×, (**c**) 200× (HAZ), (**d**) 200× (FZ), (**e**) 500× (FZ), (**f**) 1000× (FZ).

Table 2. Mean and standard deviation of Vickers microhardness (HV 2) of Ti$_6$Al$_4$V-BM and Ti$_6$Al$_4$V-LR.

FZ (Ti$_6$Al$_4$V-LR)	HAZ (Ti$_6$Al$_4$V-LR)	BM (Ti$_6$Al$_4$V-LR)	Ti$_6$Al$_4$V-BM
410 ± 8	392 ± 11	372 ± 10	353 ± 5

It is interesting focusing on the analysis of the superficial microstructure developed in Ti6Al4V-LR, presented in Figure 6. A thin layer of 8–13 μm is developed from the surface, which seems to be a rich martensitic zone, presenting high microstructural refinement. This acicular martensite is thinner than the one observed at the inner part of the FZ. This thin superficial microstructure is associated with the higher heating and the subsequent higher cooling rate taking place just after the laser treatment, leading to a finer martensite than that formed at the FZ.

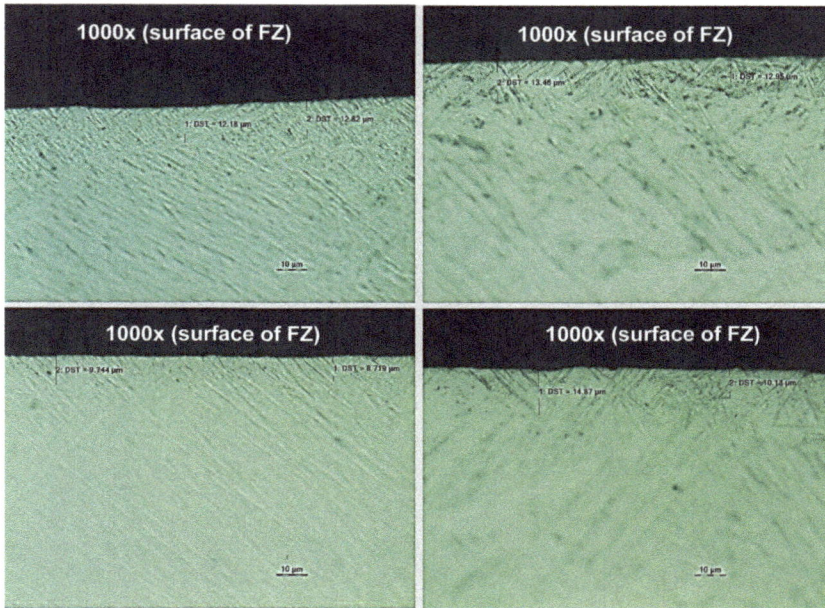

Figure 6. Metallographic images of Ti$_6$Al$_4$V-LR samples at the external surface.

Finite element method simulations were carried out to show that the external surface of the irradiated zone reached a higher temperature and higher cooling rate than the inner part of the fusion zone (Figure 7a). In addition, the simulations also confirm that the external surface of the FZ presents a higher martensite content than the inner part of the FZ (Figure 7b). This feature is shown later to have an important influence on the tribocorrosion behavior. This external martensite rich layer presents an average depth of 10 μm of the fusion zone. Taking into account the presence of this thin layer, the tribocorrosion tests were performed both at initial rubbing and also at extensive rubbing, in order to be sensitive to the layer effects on the results.

(a)

(b)

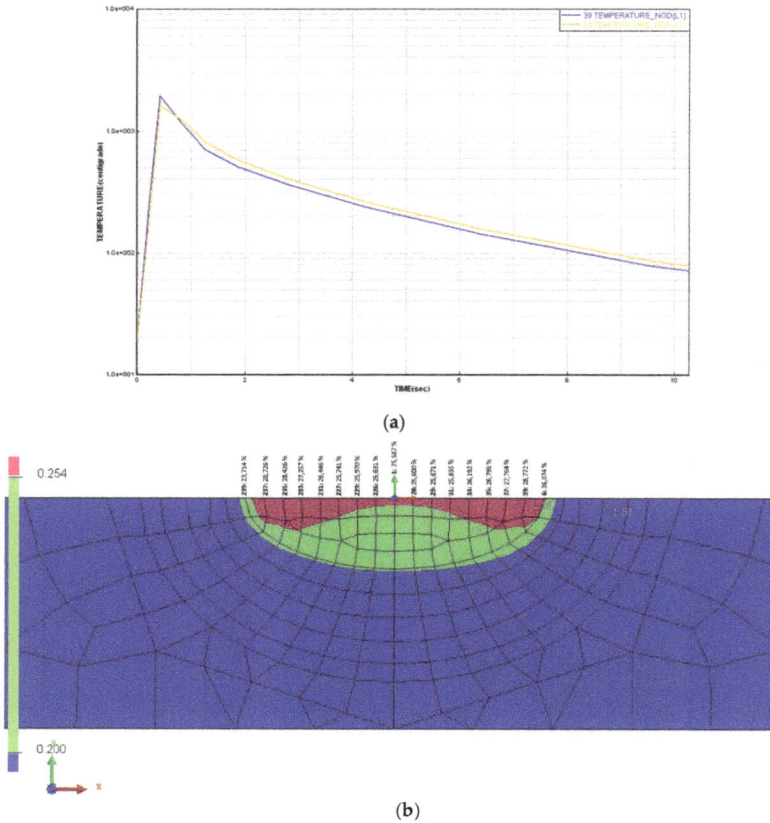

Figure 7. Results obtained from finite element method simulations. (**a**) Temperature evolution of nodes located at the surface (node 39) and the inner part of the FZ (node 14); and (**b**) martensite fraction at different zones of the FZ.

3.2. Tribocorrosion Behavior at Initial Rubbing

Tribocorrosion studies of LR samples have been divided into initial and extensive rubbing, providing the differences aforementioned related to the external martensite-rich thin layer and the inner part of the FZ. The initial rubbing refers to the wear limited to the first instants of rubbing, in which the finer superficial martensite layer could have influence. On the other hand, extensive rubbing studies analyze the effect of severe wear (up to one hour of friction), in which the FZ, or even the base metal is present. This methodology of testing at two rubbing times also allows the study of the layer modification effects on tribocorrosion. As the thin layer can be harder than the HAZ, the damaged (anodic) area might be relatively small for this initial stage of wear.

The first effect of friction evaluated has been noted in the monitoring of the open circuit potential with the initial rubbing, as shown in Figure 8. The open circuit potential after approximately 1800 s exposure is close to -0.45 V vs. SCE, and the steady-state condition can be assumed. Thus, the effect of rubbing on OCP (open circuit potential) can be monitored. When the rubbing (for 2.0 N load) starts, the potential decreases about 0.5 V for the base metal sample (Ti$_6$Al$_4$V-BM) and about 0.4 V for the laser-treated sample (Ti$_6$Al$_4$V-LR). The different potential drops of the samples is thought to be related to their different wear resistance at this initial period of the rubbing. Consequently, the lower potential drop of Ti$_6$Al$_4$V-LR is indicative of its higher wear resistance, surely due to the different superficial microstructure developed in this sample. In fact, the mixed potential presented during sliding is dependent on the anodic (worn)/cathodic (passive) area ratio. Thus, a high ratio gives more negative potential values. Thus, the smaller potential drop of Ti$_6$Al$_4$V-LR suggested a smaller anodic/cathodic area ratio of the laser-treated samples when rubbing is confined in 10 μm of the thin layer. After some minutes of this initial rubbing, the potential behavior of the Ti$_6$Al$_4$V-LR becomes similar to that of Ti$_6$Al$_4$V-BM, most certainly because of the degradation of the superficial rich martensitic layer.

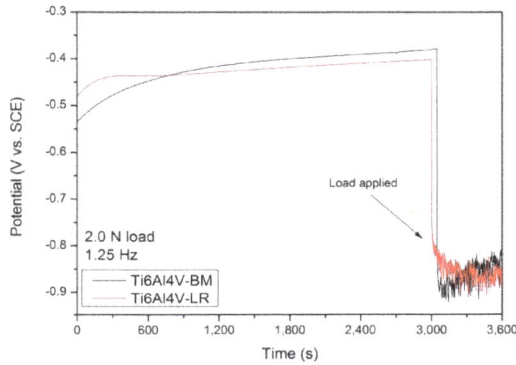

Figure 8. Open circuit potential before and during the mechanical loading.

For comparison, the wear track was also evaluated with AFM. The exposure time was limited to 10 s of friction. In fact, there is a clear difference between Ti$_6$Al$_4$V-BM and Ti$_6$Al$_4$V-LR samples (Figure 9). Initial wear is more intense for Ti$_6$Al$_4$V-BM, in which the wear track is deeper than that of LR. As the surface finish is the same, as well as the alumina counter-body, the shallow track of Ti$_6$Al$_4$V-LR samples is thought to be related to the thin superficial α′ microstructure, providing a tribocorrosion resistance increase at the surface. The representative peak to peak height of Ti$_6$Al$_4$V-LR sample is less than 1 μm, being approximately 2 μm for the Ti$_6$Al$_4$V-BM. The counter-body contact occurs at a width of about 0.1 mm, as depicted in Figure 9b,d. For this initial wear, the laser-treated samples resist more damage than non-treated titanium alloy, which is probably related to the thin superficial martensite microstructure (8–13 μm). These findings suggested that the laser treatment increases the wear resistance of titanium alloy, likely caused by the high hardness [21]. Barril et al. [28] have found that for Ti$_6$Al$_4$V/0.9% NaCl system in fretting corrosion, the initial smeared surface is smaller for lower slip amplitude. In the present paper, in addition to the reported slip difference and even the testing (fretting and pin-on-disk tests), the presence of the thin martensite rich layer seems to reduce the wear.

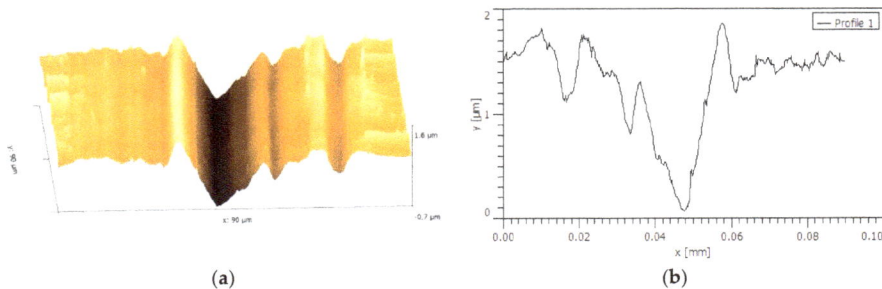

(a)

(b)

Figure 9. *Cont.*

(c)

(d)

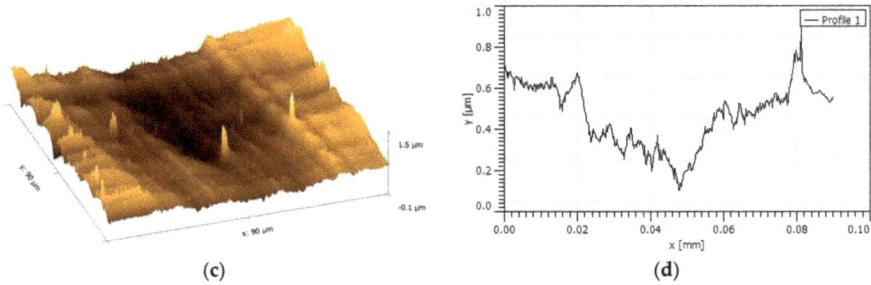

Figure 9. Tracks observed in AFM. Ti_6Al_4V-BM track: (**a**) surface and (**b**) profile; and Ti_6Al_4V-LR: (**c**) surface and (**d**) profile.

3.3. Tribocorrosion Behavior at Extensive Rubbing

The tribocorrosion tests performed under extensive rubbing are analyzed in this section. Thus, the depth profile of the worn surface reaches the inner part of the FZ, going beyond the external martensite-rich thin layer. The tests were performed for 3600 s without load and, subsequently, 600 s of rubbing at open circuit potential. Polarization curves of Ti_6Al_4V-BM and Ti_6Al_4V-LR samples were carried out (Figure 10a) and the effect of applied potential on coefficient of friction was monitored (Figure 10b). In both samples (Ti6Al4V-BM and Ti6Al4V-LR), the corrosion potential decreases from -0.4 V vs. SCE approx. (when no load is applied, Figure 10a) to -0.65 V vs. SCE approx. (when applying the load, Figure 10b). In addition to this decay of potential, the average current density in both samples increases almost two decades when friction is employed. Thus, the average current density increases from 100 nA/cm^2 approx. (in non-rubbing polarization curves, Figure 10a) to 10 $\mu A/cm^2$ approx. (in rubbing polarization curves, Figure 10b). Current scattering under friction is clearly observable. This behavior can be ascribed to the effect of sliding that wears the passive layer of titanium and the subsequent repassivation of the surface.

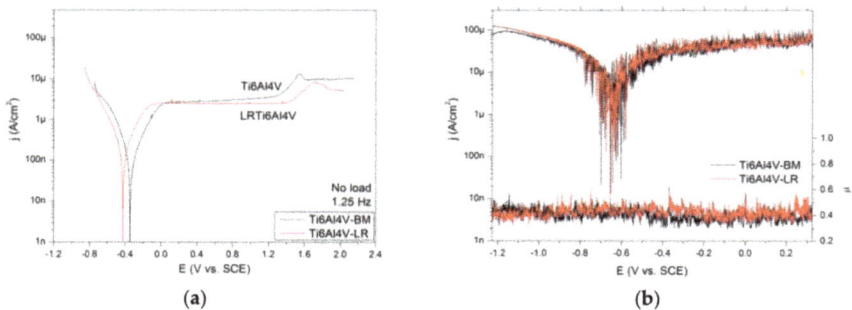

(a)

(b)

Figure 10. (**a**) Polarization curve without load; (**b**) with 2.0 N load: potential versus coefficient of friction, and polarization curve. (Black, Ti_6Al_4V-BM; red, Ti_6Al_4V-LR).

When polarization curves are analyzed under the no rubbing condition, Ti_6Al_4V-BM and Ti_6Al_4V-LR have corrosion potential values around -320 and -420 mV vs. SCE, respectively. Meanwhile, under friction, both samples present corrosion potentials of about -630 mV vs. SCE. The effect of loading in the passivation current densities (measured at 0.0 V vs. SCE) increased, in both samples, from approximately 2 $\mu A/cm^2$ (no rubbing, Figure 10a) to 90 $\mu A/cm^2$ (rubbing, Figure 10b). Moreover, the scattering related to the mean current density is negligible without loading (less than 1 $\mu A/cm^2$), but very high with friction (around 40 $\mu A/cm^2$). These results have been summarized in Table 3.

Table 3. Corrosion potential and passivation current density of Ti$_6$Al$_4$V-BM and Ti$_6$Al$_4$V-LR samples obtained from polarization curves.

Sample	Test Condition	Corrosion Potential (mV vs. SCE)	Passivation Current Density (μA/cm^2)
Ti$_6$Al$_4$V-BM	No rubbing	−320 ± 30	2 ± 1
	Rubbing	−420 ± 20	90 ± 40
Ti$_6$Al$_4$V-LR	No rubbing	−630 ± 30	2 ± 1
	Rubbing	−630 ± 20	90 ± 40

The mean coefficient of friction changes very little with the potential, as shown in Figure 10b. Additionally, there is a significant scattering of the current data. This scattering increases with the potential when anodic processes are more intense. In this case, a possible interaction of dissolution on the wear mechanism would be expected, as related in [23]. However, neither a modification of the coefficient of friction, nor a surface feature modification was observed with the applied potential. Wear grooves were observed in ordinary and laser-treated samples, both related to the abrasion wear mechanism. Then, the wear mechanisms do not seem to change, at least to the potential range of Figure 10b. Moreover, adhesion marks were also observed in the tracks of the samples. As mentioned beforehand, when extensive wear is applied, both samples have similar wear, especially close to the corrosion potential. Moreover, the coefficient of friction is higher than 0.25, regardless of the microstructure or potential. In this case, due to the stress distribution, the plastic flow occurs chiefly on the surface, instead of the subsurface layer [13]. This fact augments the importance of the superficial thin layer produced by laser remelting on tribocorrosion behavior.

In Figure 11, the titanium alloy profiles are shown. It can be seen that wear is very similar in depth and width for both materials. However, titanium track profiles have a serrated shape, unlike the stainless steel-alumina pair [23]. Moreover, in certain sites the tracks were observed to be still deeper, but not detected by profilometry due to the narrow contact area. This suggests that the dissolution and/or wear is not uniform in all points of the track. The tracks are depicted in Figure 12 using optical light microscopy. The alumina counter-body has a diameter of 4.0 mm, but the profiles present a serrated shape with approximately 1 mm.

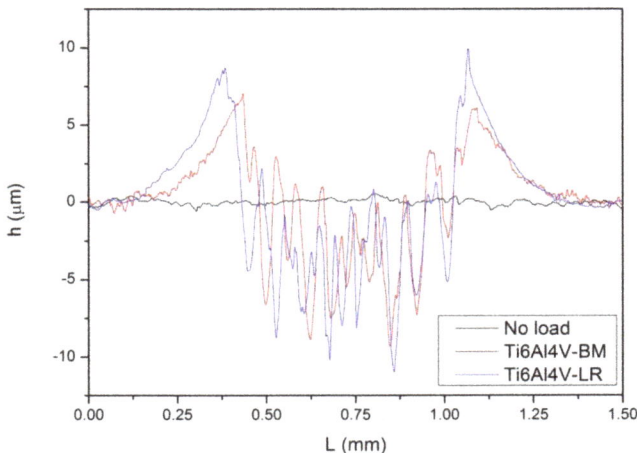

Figure 11. Profile of wear tracks obtained after the polarization curve tests.

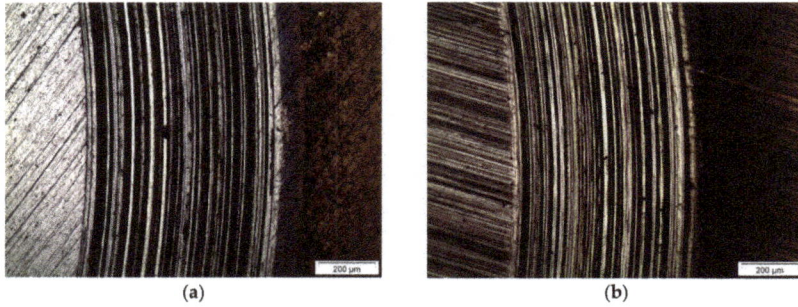

(a)   (b)

Figure 12. Optical images of wear tracks: (**a**) Ti$_6$Al$_4$V-BM; (**b**) Ti$_6$Al$_4$V-LR.

Figure 13 shows the impedance diagrams of titanium alloy samples after an open circuit exposure of 3600 s in Nyquist format. The continuous lines are the fitted response of the equivalent electric circuit with the parameters of Table 4. Both conditions present a high capacitive loop. A capacitive-like loop is normally found in passive alloys. The effect of wear is the reduction of the impedance modulus after the rubbing. However, the impedance of laser samples is slightly superior than that of non-treated surface. The impedance diagrams for base metal are similar to those of Amaya-Vazquez et al. [21], taking into account their lower minimum frequency (1.0 mHz) and the higher open circuit exposure. These factors increased the measured modulus of passive metals. Table 4 shows the adjusted parameters of electrochemical impedance for the circuit of Figure 3.

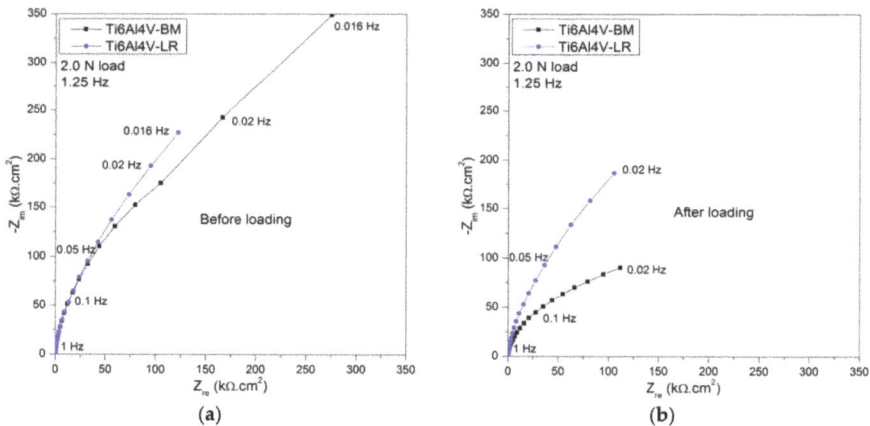

(a)   (b)

Figure 13. Electrochemical impedance spectroscopy represented as Nyquist diagrams: (**a**) before and (**b**) after loading.

Table 4. Mean and standard deviation values of the fitted EIS parameters.

Sample	Test Condition	Re ($\Omega \cdot$cm^2)	α ($\times 10^{-3}$)	Q (μS$\cdot$s$^\alpha$/cm^2)	Rp (k$\Omega \cdot$cm^2)
	Before rubbing	69.7 ± 0.5	945 ± 2	224 ± 10	587.00 ± 18.00
Ti6Al4V-BM	Rubbing	75.7 ± 0.5	850 ± 5	721 ± 18	1.65 ± 0.02
	After rubbing	66.1 ± 0.5	907 ± 2	246 ± 20	175.00 ± 3.00
	Before rubbing	84.7 ± 0.5	935 ± 2	279 ± 20	655.00 ± 18.00
Ti6Al4V-LR	Rubbing	93.1 ± 0.6	841 ± 4	1007 ± 17	3.61 ± 0.04
	After rubbing	79.7 ± 0.5	929 ± 2	270 ± 20	666.00 ± 22.00

Two parameters of the impedance equivalent circuit (Figure 3) were depicted in Figure 14. In general, the closer to unity is the alpha, and the higher the Rp, the better corrosion resistance a given surface has. Before the wear, alpha and polarization resistance are similar in BM and LR samples. During rubbing, both parameters decrease strongly, and Rp decays to a few k$\Omega \cdot$cm^2, corresponding to the bare surface of the alloys. After the rub, the worn surfaces exhibit diagrams with modulus and phase lower than those before loading. The wear provokes an increase of the Q parameter, as can be noted in Table 4. This increase represents a decay of impedance modulus caused by the damage in the passive film. This damage is also observed in Rp and alpha values that reduce the measured impedance. The alpha reduction (Figure 14a) during rubbing can be related to heterogeneities at the electrode surface caused by the friction and the breakdown of the passive film. With the relief of friction, the alpha almost recovers its previous values.

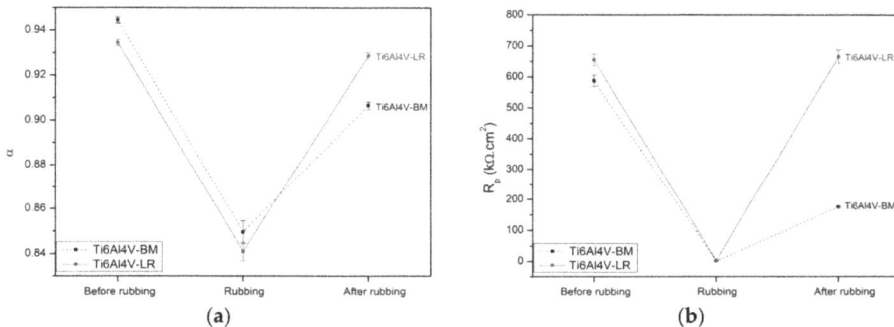

Figure 14. Effect of friction on the electric circuit parameters before, during, and after rubbing: (**a**) alpha and (**b**) polarization resistance.

The CPE circuit has been ascribed to many phenomena, but the roughness can be considered the leading cause of its time dispersion characteristic. We roughly estimated the frequency beyond which the effect of wear roughness are V-shaped grooves on the CPE. We use the findings of [29], specifically Figure 14 of the cited work, to evaluate grosso modo the effect of roughness on the impedance frequency. The roughness factor is $(fr = 1/\cos\theta)$, where θ represents the angle between the groove and the plane surface. In the present case, the factor was approximate to unity, as follows: From a representative peak from the profile of Figure 11, we evaluate tg$\theta \approx 10$ μm/50 μm, $\theta = 11°$, and $fr \approx 1$ at first approximation. Additionally, assuming that the distance between peaks P is 100 μm, the electrolyte conductivity of 0.90% NaCl is 1.45 mS/cm and the interfacial capacitance is 10 μF/cm^2, the roughness affects the impedance above 10 kHz, if the entire electrode has similar finishing. Then, under these assumptions, the effect of wear roughness can be considered as negligible in the measured EIS. The global impedance is evaluated as a result of a parallel connection of impedances from passive and worn surfaces (low impedance). Thus, the impedance diagram mainly expresses the increase of electrochemical activity related to the exposure of bare metal and the passive film, and not the effect of roughness as the profile could suggest. Anyway, the laser-treated samples exhibit better properties than base metal after rubbing, with higher alpha and Rp when compared to the base metal.

Laser peening induces significant residual compressive stress in Ti$_6$Al$_4$V alloy, and a reduction of wear volume of the titanium alloy was also ascribed to the presence of compressive stress [30]. This stress likely occurred in the samples studied in the present work, although it was not measured.

4. Conclusions

The microstructure of ordinary Ti$_6$Al$_4$V (Ti$_6$Al$_4$V-BM) and laser-remelted Ti$_6$Al$_4$V (Ti$_6$Al$_4$V-LR) samples were characterized in 0.90% NaCl at room temperature. Both types of samples were subjected to corrosion (open circuit potential, EIS, polarization curves) and tribocorrosion tests (rubbing on a pin-on-disk tribometer against an alumina sphere). Based on the obtained findings, it is possible to remark the following points:

- Initial rubbing reveals that laser-remelted Ti_6Al_4V samples present better tribocorrosion resistance (shallower track wear) than ordinary non-treated Ti_6Al_4V-BM samples. This initial wear improvement is associated with the development of a fine α' martensitic microstructure on the external surface of the laser samples, with an approximate thickness of 10 μm.
- Although Ti_6Al_4V-BM and Ti_6Al_4V-LR have different microstructures, they showed similar tribocorrosion behavior in extensive rubbing tests.
- Electrochemical impedance diagrams are similar for Ti_6Al_4V-BM and Ti_6Al_4V-LR samples during the rubbing. However, after this phase, and without counter-body contact, a higher EIS modulus is observed for the laser-remelted samples.

Acknowledgments: The authors thank the financial support of Faperj (Brazil, E26/110.644/2012), CNPq (Brazil) and Junta de Andalucía (Spain, project SOLDATIA, Ref. TEP 6180).

Author Contributions: I.N.B. conceived the experiments; C.C. performed the FEM simulations; J.M.S.-A. prepared the LR samples; D.P.S. performed the experiments; J.M.S.-A. and I.N.B. analyzed the data; and D.P.S. wrote the paper.

Conflicts of Interest: The authors declare no conflict of interest.

References

1. Sun, Z.; Annergren, I.; Pan, D.; Mai, T.A. Effect of laser surface remelting on the corrosion behavior of commercially pure titanium sheet. *Mater. Sci. Eng. A* **2003**, *345*, 293–300. [CrossRef]
2. Akman, E.; Demir, A.; Canel, T.; Sınmazçelik, T. Laser welding of Ti6Al4V titanium alloys. *J. Mater. Process. Technol.* **2009**, *209*, 3705–3713. [CrossRef]
3. Prasad, S.; Ehrensberger, M.; Gibson, M.P.; Kim, H.; Monaco, E.A., Jr. Biomaterial properties of titanium in dentistry. *J. Oral Biosci.* **2015**, *57*, 192–199. [CrossRef]
4. Bastos, I.N.; Platt, G.M.; Andrade, M.C.; Soares, G.D. Theoretical study of Tris and Bistris effects on simulated body fluids. *J. Mol. Liq.* **2008**, *139*, 121–130. [CrossRef]
5. Resende, C.X.; Dille, J.; Platt, G.M.; Bastos, I.N.; Soares, G.D. Characterization of coating produced on titanium surface by a designed solution containing calcium and phosphate ions. *Mater. Chem. Phys.* **2008**, *108*, 429–435. [CrossRef]
6. Kokubo, T.; Takadama, H. How useful is SBF in predicting in vivo bone bioactivity? *Biomaterials* **2006**, *27*, 2907–2915. [CrossRef] [PubMed]
7. Runa, M.J.; Mathew, M.T.; Rocha, L.A. Tribocorrosion response of the Ti6Al4V alloys commonly used in femoral stems. *Tribol. Int.* **2013**, *68*, 85–93. [CrossRef]
8. Niinomi, M. Mechanical properties of biomedical titanium alloys. *Mater. Sci. Eng. A* **1998**, *243*, 231–236. [CrossRef]
9. Vieira, A.; Ribeiro, A.; Rocha, L.; Celis, J.P. Influence of pH and corrosion inhibitors on the tribocorrosion of titanium in artificial saliva. *Wear* **2006**, *261*, 994–1001. [CrossRef]
10. Weng, F.; Chen, C.; Yu, H. Research status of laser cladding on titanium and its alloys: A review. *Mater. Des.* **2014**, *58*, 412–425. [CrossRef]
11. Landolt, D. *Corrosion and Surface Chemistry of Metals*, 1st ed.; EPFL Press: Lausanne, Switzerland, 2007.
12. Achanta, S.; Liskiewicz, T.; Drees, D.; Celis, J.-P. Friction mechanisms at the micro-scale. *Tribol. Int.* **2009**, *42*, 1792–1799. [CrossRef]
13. Johnson, K.L. Contact mechanics and the wear of metals. *Wear* **1995**, *190*, 162–170. [CrossRef]
14. Yamamoto, T.; Fushimi, K.; Habazaki, H.; Konno, H. Depassivation-repassivation behavior of a pure iron surface investigated by micro-indentation. *Electrochim. Acta* **2010**, *55*, 1232–1238. [CrossRef]
15. Manhabosco, T.M.; Tamborim, S.M.; dos Santos, C.B.; Müller, I.L. Tribological, electrochemical and tribo-electrochemical characterization of bare and nitrided Ti6Al4V in simulated body fluid solution. *Corros. Sci.* **2011**, *53*, 1786–1793. [CrossRef]

16. Vanzillotta, P.S.; Sader, M.S.; Bastos, I.N.; Soares, G.A. Improvement of in vitro titanium bioactivity by three different surface treatments. *Dent. Mater.* **2006**, *22*, 275–282. [CrossRef] [PubMed]

17. Fazel, M.; Salimijazi, H.R.; Golozar, M.A.; GarsivazJazi, M.R. A comparison of corrosion, tribocorrosion and electrochemical impedance properties of pure Ti and Ti6Al4V alloy treated by micro-arc oxidation process. *Appl. Surf. Sci.* **2015**, *324*, 751–756. [CrossRef]

18. Sánchez-Amaya, J.M.; Vazquez, M.R.A.; Botana, F.J. Chapter 8 Laser Welding of Light Metal Alloys: Aluminium and Titanium Alloys. In *Handbook of Laser Welding Technologies*; Katayama, S., Ed.; Woodhead Publishing Series in Electronic and Optical Materials; Osaka University: Suita, Japan, 2013; No. 41; pp. 215–254.

19. Janasekaran, S.; Tan, A.W.; Yusof, F.; Shukor, M.H.A. Influence of the overlapping factor and welding speed on T-joint welding of Ti6Al4V and Inconel 600 using low-power fiber laser. *Metals* **2016**, *6*, 134. [CrossRef]

20. Kim, J.-M.; Ha, T.-H.; Park, J.-S.; Kim, H.-G. Effect of laser surface treatment on the corrosion behavior of FeCrAl-coated TZM alloy. *Metals* **2016**, *6*, 29. [CrossRef]

21. Amaya-Vazquez, M.R.; Sánchez-Amaya, J.M.; Boukha, Z.; Botana, F.J. Microstructure, microhardness and corrosion resistance of remelted TiG2 and Ti6Al4V by a high power diode laser. *Corros. Sci.* **2012**, *56*, 36–48. [CrossRef]

22. Amaya, J.M.S.; Vazquez, M.R.A.; Rovira, L.Z.; Galvin, M.B.; Botana, F.J. Influence of surface pre-treatments on laser welding of Ti6Al4V alloy. *J. Mater. Eng. Perform.* **2014**, *23*, 1568–1575. [CrossRef]

23. Silva, R.C.C.; Nogueira, R.P.; Bastos, I.N. Tribocorrosion of UNS S32750 in chloride medium: Effect of the load level. *Electrochim. Acta* **2011**, *56*, 8839–8845. [CrossRef]

24. International Organization for Standardization. *ISO 6507-1:2005*; ISO: Geneva, Switzerland, 2005.

25. Churiaque, C.; Vazquez, M.R.A.; Botana, F.J.; Amaya, J.M.A. FEM simulation and experimental validation of LBW under conduction regime of Ti6Al4V alloy. *J. Mater. Eng. Perform.* **2016**, *25*, 3260–3269. [CrossRef]

26. International Organization for Standardization. *ISO 3290-2:2014*; ISO: Geneva, Switzerland, 2014.

27. Zaveri, N.; Mahapatra, M.; Deceuster, A.; Peng, Y.; Li, L.; Zhou, A. Corrosion resistance of pulsed laser-treated Ti-6Al-4V implant in simulated biofluids. *Electrochim. Acta* **2008**, *53*, 5022–5032. [CrossRef]

28. Barril, S.; Mischler, S.; Landolt, D. Influence of fretting regimes on the tribocorrosion behavior of Ti6Al4V in 0.9 wt. % sodium chloride solution. *Wear* **2004**, *256*, 963–972. [CrossRef]

29. Alexander, C.L.; Tribollet, B.; Orazem, M.E. Contribution of surface distributions to constant-phase-element (CPE) behavior: 1. Influence of roughness. *Electrochim. Acta* **2015**, *173*, 416–424. [CrossRef]

30. Kumar, D.; Akhtar, S.N.; Patel, A.K.; Ramkumar, L.; Balani, K. Tribological performance of laser peened Ti-6Al-4V. *Wear* **2015**, *322*, 203–217. [CrossRef]

MDPI AG

St. Alban-Anlage 66

4052 Basel, Switzerland

Tel. +41 61 683 77 34

Fax +41 61 302 89 18

http://www.mdpi.com

Metals Editorial Office

E-mail: metals @mdpi.com

http://www.mdpi.com/journal/metals

www.ingramcontent.com/pod-product-compliance
Lightning Source LLC
Chambersburg PA
CBHW051845210326
41597CB00033B/5778